MW00836868

HYDRAULIC FRACTURING IN UNCONVENTIONAL RESERVOIRS

HYDRAULIC FRACTURING IN UNCONVENTIONAL RESERVOIRS

Theories, Operations, and Economic Analysis

HOSS BELYADI
CONSOL Energy Inc.

EBRAHIM FATHI
West Virginia University

FATEMEH BELYADI
West Virginia University

AMSTERDAM • BOSTON • HEIDELBERG • LONDON
NEW YORK • OXFORD • PARIS • SAN DIEGO
SAN FRANCISCO • SINGAPORE • SYDNEY • TOKYO

Gulf Professional Publishing is an imprint of Elsevier

Gulf Professional Publishing is an imprint of Elsevier
50 Hampshire Street, 5th Floor, Cambridge, MA 02139, United States
The Boulevard, Langford Lane, Kidlington, Oxford, OX5 1GB, United Kingdom

Copyright © 2017 Elsevier Inc. All rights reserved.

No part of this publication may be reproduced or transmitted in any form or by any means, electronic
or mechanical, including photocopying, recording, or any information storage and retrieval system, without
permission in writing from the publisher. Details on how to seek permission, further information about the
Publisher's permissions policies and our arrangements with organizations such as the Copyright Clearance
Center and the Copyright Licensing Agency, can be found at our website: www.elsevier.com/permissions.

This book and the individual contributions contained in it are protected under copyright by the Publisher
(other than as may be noted herein).

Notices
Knowledge and best practice in this field are constantly changing. As new research and experience
broaden our understanding, changes in research methods, professional practices, or medical treatment
may become necessary.

Practitioners and researchers must always rely on their own experience and knowledge in evaluating and
using any information, methods, compounds, or experiments described herein. In using such information
or methods they should be mindful of their own safety and the safety of others, including parties for whom
they have a professional responsibility.

To the fullest extent of the law, neither the Publisher nor the authors, contributors, or editors, assume
any liability for any injury and/or damage to persons or property as a matter of products liability,
negligence or otherwise, or from any use or operation of any methods, products, instructions, or ideas
contained in the material herein.

British Library Cataloguing-in-Publication Data
A catalogue record for this book is available from the British Library

Library of Congress Cataloging-in-Publication Data
A catalog record for this book is available from the Library of Congress

ISBN: 978-0-12-849871-2

For Information on all Gulf Professional Publishing
visit our website at https://www.elsevier.com

Working together
to grow libraries in
developing countries

www.elsevier.com • www.bookaid.org

Publisher: Joe Hayton
Senior Acquisition Editor: Katie Hammon
Senior Editorial Project Manager: Kattie Washington
Production Project Manager: Mohana Natarajan
Designer: Maria Ines Cruz

Typeset by MPS Limited, Chennai, India

CONTENTS

List of Figures		*xiii*
List of Tables		*xix*
List of Equations		*xxi*
Biography		*xxvii*
Preface		*xxix*

1. Introduction to Unconventional Reservoirs   **1**

Introduction   1

Different Types of Natural Gas   4

Natural Gas Transport   5

Unconventional Reservoirs   6

2. Advanced Shale Reservoir Characterization   **13**

Introduction   13

Pore-Size Distribution Measurement of Shale   13

Shale Sorption Measurement Techniques   16

Shale Porosity Measurements   20

Pore Compressibility Measurements of Shale   22

Shale Permeability Measurement Techniques   22

3. Shale Initial Gas-in-Place Calculation   **29**

Introduction   29

Total Gas-in-Place Calculation   29

Density of Adsorbed Gas   33

Recovery Factor   35

4. Multiscale Fluid Flow and Transport in Organic-Rich Shale   **37**

Introduction   37

Multicontinuum Modeling of Shale Reservoirs   38

Interfacial Tension and Capillary Pressure   41

Wettability Effects on Shale Recovery   47

5. Hydraulic Fracturing Fluid Systems   **49**

Introduction   49

Slick Water Fluid System   51

Cross-Linked Gel Fluid System   55

Hybrid Fluid System   56
Foam Fracturing   58
Foam Quality   60
Foam Stability   61
Tortuosity   61
Typical Slick Water Frac Steps   64
Acidization Stage   64
Pad Stage   66
Proppant Stage   68
Flush Stage   69
Frac Fluid Selection Summary   72

6. **Proppant Characteristics and Application Design**   **73**
Introduction   73
Sand   73
Precured Resin-Coated Sand   74
Curable Resin-Coated Sand   74
Intermediate-Strength Ceramic Proppant   76
Lightweight Ceramic Proppant   76
High-Strength Proppant   76
Proppant Size   78
Proppant Characteristics   80
Proppant Particle-Size Distributions   83
Proppant Transport and Distribution in Hydraulic Fracture   83
Fracture Conductivity   87
Dimensionless Fracture Conductivity   88
International Organization for Standardization (ISO) Conductivity Test   89
Non-Darcy Flow   90
Multiphase Flow   90
Reduced Proppant Concentration   91
Gel Damage   91
Cyclic Stress   92
Fines Migration   93
Time Degradation   94
Finite vs Infinite Conductivity   95

7. **Unconventional Reservoir Development Footprints**   **97**
Introduction   97
Casing Selection   98
Conductor Casing   98

Surface Casing   99
Intermediate Casing   100
Production Casing   100
Hydraulic Fracturing and Aquifer Interaction   101
Hydraulic Fracturing and Fault Reactivation   102
Hydraulic Fracturing and Low-Magnitude Earthquakes   105

8. **Hydraulic Fracturing Chemical Selection and Design**   **107**
Introduction   107
Friction Reducer   108
FR Flow Loop Test   108
Pipe Friction Pressure   111
Reynolds Number   112
Relative Roughness of Pipe   112
FR Breaker   114
Biocide   114
Scale Inhibitor   115
Linear Gel   116
Gel Breaker   117
Buffer   118
Cross-Linker   119
Surfactant   120
Iron Control   120

9. **Fracture Pressure Analysis and Perforation Design**   **121**
Introduction   121
Pressure (psi)   121
Hydrostatic Pressure (psi)   122
Hydrostatic Pressure Gradient (psi/ft)   122
Instantaneous Shut-In Pressure (ISIP, psi)   123
Fracture Gradient (FG, psi/ft)   125
Bottom-Hole Treating Pressure (psi)   125
Total Friction Pressure (psi)   126
Pipe Friction Pressure (psi)   127
Perforation Friction Pressure (psi)   127
Open Perforations   128
Perforation Efficiency   129
Perforation Design   130
Numbers of Holes (perfs) and Limited Entry Technique   131
Perforation Diameter and Penetration   132

Perforation Erosion 132
Near-Wellbore Friction Pressure (NWBFP) 133
Fracture Extension Pressure 133
Closure Pressure 133
Net Pressure 137
Surface-Treating Pressure (STP, psi) 139

10. Fracture Treatment Design **143**
Introduction 143
Absolute Volume Factor (AVF, gal/lbs) 143
Dirty (Slurry) Versus Clean Frac Fluid 144
Slurry (Dirty) Density (ppg) 145
Stage Fluid Clean Volume (BBLs) 146
Stage Fluid Slurry (Dirty) Volume (BBLs) 147
Stage Proppant (lbs) 147
Sand per Foot (lb/ft) 148
Water per Foot 148
Sand-to-Water Ratio (SWR, lb/gal) 149
Slick Water Frac Schedule 150
Foam Frac Schedule and Calculations 154
Foam Volume 155
Nitrogen Volume 155
Blender Sand Concentration 156
Slurry Factor (SF) 157
Clean Rate (No Proppant) 157
Clean Rate (with Proppant) 158
Slurry Rate (with Proppant) 159
Nitrogen Rate (With and Without Proppant) 159

11. Horizontal Well Multistage Completion Techniques **169**
Introduction 169
Conventional Plug and Perf 170
Composite Bridge (Frac) Plug 170
Sliding Sleeve 174
Sliding Sleeve Advantages 175
Sliding Sleeve Disadvantages 175
Frac Stage Spacing (Plug-to-Plug Spacing) 176
Shorter Stage Length (SSL) 176
Cluster Spacing 178
Refrac Overview 179

12. **Completions and Flowback Design Evaluation in Relation to Production** **183**

Introduction 183
Landing Zone 184
Stage Spacing 185
Cluster Spacing 185
Number of Perforations, Entry-Hole Diameter (EHD), and Perforation Phasing 186
Sand and Water per Foot 186
Proppant Size and Type 187
Bounded Versus Unbounded (Inner vs. Outer) 188
Up Dip Versus Down Dip 190
Well Spacing 190
Water Quality 192
Flowback Design 192
Flowback Equipment 194
Flowback Equipment Spacing Guidelines 204
Tubing Analysis 204

13. **Rock Mechanical Properties and In Situ Stresses** **207**

Introduction 207
Young's Modulus (psi) 207
Poisson's Ratio (ν) 209
Fracture Toughness (psi/$\sqrt{in}$) 211
Brittleness and Fracability Ratios 212
Vertical, Minimum Horizontal, and Maximum Horizontal Stresses 214
Vertical Stress 214
Minimum Horizontal Stress 216
Biot's Constant (Poroelastic Constant) 217
Maximum Horizontal Stress 218
Various Stress States 220
Fracture Orientation 221
Transverse Fractures 221
Longitudinal Fractures 222

14. **Diagnostic Fracture Injection Test** **225**

Introduction 225
Typical DFIT Procedure 226
DFIT Data Recording and Reporting 226
Before-Closure Analysis 227
Square Root Plot 228
Log—Log Plot (Log (BH ISIP-BHP) Versus Log (Time)) 229
G-function Analysis 235

After-Closure Analysis (ACA)   244

  Horner Plot (One Method of ACA)   245

  Linear Flow-Time Function Versus Bottom-Hole Pressure
  (Another Method of ACA)   247

  Radial Flow-Time Function Versus BHP (Another Method of ACA)   248

15. Numerical Simulation of Hydraulic Fracturing Propagation   251

Introduction   251

Stratigraphic and Geological Structure Modeling   252

Development of Hydraulic Fracturing Simulators   253

Two-Dimensional Hydraulic Fracturing Models   257

Fluid Flow in Hydraulic Fractures   259

Solid Elastic Response   260

Pseudo—Three-Dimensional (Pseudo-3D) Hydraulic Fracturing Models   261

Three-Dimensional Hydraulic Fracturing Models   263

Hydraulic and Natural Fracture Interactions   265

Hydraulic Fracture Stage Merging and Stress Shadow Effects   266

16. Operations and Execution   269

Introduction   269

Water Sources   270

Hydration Unit ("Hydro")   273

Blender   275

Sand Master (Sand Mover, Sand King, or Sand Castle)   276

T-belt   277

Missile   284

Frac Manifold (Isolation Manifold)   285

Frac Van (Control Room)   286

Overpressuring Safety Devices   290

Water Coordination   294

Sand Coordination   295

Chemical Coordination   295

Stage Treatment   296

Tips for Flowback After Screening Out   296

Frac Wellhead   299

17. Decline Curve Analysis   305

Introduction   305

Decline Curve Analysis (DCA)   305

Anatomy of Decline Curve Analysis   306

Primary Types of Decline Curves   309

Arps Decline Curve Equations for Estimating Future Volumes   314

18. Economic Evaluation   325

Introduction   325
Net Cash-Flow Model (NCF)   325
Royalty   326
Working Interest (WI)   327
Net Revenue Interest (NRI)   329
British Thermal Unit (BTU) Content   330
Shrinkage Factor   331
Operating Expense (Opex)   332
Total Opex per Month   335
Severance Tax   338
Ad Valorem Tax   339
Net Opex   340
Revenue   341
NYMEX (New York Mercantile Exchange)   343
Henry Hub and Basis Price   344
Cushing Hub and West Texas Intermediate (WTI)   346
Mont Belvieu and Oil Price Information Services (OPIS)   346
Capital Expenditure Cost (Capex)   347
Opex, Capex, and Pricing Escalations   349
Profit (Net Cash Flow)   350
Before Federal Income Tax (BTAX) Monthly Undiscounted
Net Cash Flow (NCF)   350
Discount Rate   351
Weight of Debt and Equity   353
Cost of Debt   354
Cost of Equity   355
Capital Budgeting   357
Net Present Value (NPV)   357
BTAX and ATAX Monthly Discounted Net Cash Flow (NCF)   362
Internal Rate of Return (IRR)   362
NPV Versus IRR   368
Modified Internal Rate of Return (MIRR)   368
Payback Method   370
Profitability Index (PI)   372
Tax Model (ATAX Calculation)   373

Example Problems   393
References   401
Index   409

LIST OF FIGURES

Figure 1.1	Gas chromatograph.	3
Figure 1.2	Gas resources pyramid.	7
Figure 1.3	Typical shale and coal comparison.	10
Figure 1.4	Gamma ray log.	12
Figure 2.1	Schematic of typical Langmuir isotherm.	17
Figure 2.2	Different adsorption isotherm types.	18
Figure 2.3	Different hysteresis types.	19
Figure 2.4	Darcy's law illustration.	24
Figure 2.5	Core plug pulse-decay permeameter.	24
Figure 2.6	Automated high-pressure, high-temperature (HPHT) pulse-decay permeameter.	25
Figure 2.7	Double-cell Boyle's law porosimeter.	26
Figure 3.1	Shale matrix bulk volume.	30
Figure 3.2	Impact of Langmuir volume on the total gas recovery from horizontal shale gas.	36
Figure 4.1	Schematic of multicontinuum modeling approach.	39
Figure 4.2	Different hydraulic coupling used in multicontinuum approach.	40
Figure 4.3	Dynamic contact angle measurement.	44
Figure 5.1	Complex fracture system illustration.	54
Figure 5.2	Biwing fracture system illustration.	56
Figure 5.3	20 lb linear gel system.	58
Figure 5.4	Foam quality vs lb of proppant per gallon of foam.	60
Figure 5.5	Tortuosity.	63
Figure 5.6	Frac design basis.	63
Figure 5.7	Densometer.	71
Figure 6.1	Curable resin-coated proppant at standard conditions.	75
Figure 6.2	Curable resin-coated proppant under reservoir conditions.	75
Figure 6.3	Ceramic proppant.	77
Figure 6.4	100 mesh sand size.	78
Figure 6.5	Visual estimation of roundness and sphericity.	81
Figure 6.6	Test sieve shaker.	83

Figure 6.7   Effect of proppant size on dimensionless productivity   86
                  index for different reservoir permeability.

Figure 6.8   Effect of different proppant size and volume   86
                  combination on well dimensionless productivity
                  index.

Figure 6.9   Fracture width.   87
Figure 6.10   Matrix and hydraulic fracture interactions.   88
Figure 6.11   Fracture conductivity testing.   89
Figure 6.12   Relative permeability curve.   90
Figure 6.13   Proppant placement in hydraulic fracturing.   92
Figure 6.14   Proppant crushing embedment.   93
Figure 6.15   Dimensionless frac conductivity vs effective drainage   95
                  radius.

Figure 7.1   Various casing string illustrations.   101
Figure 7.2   Fracture growth limitation.   103
Figure 7.3   Geometry of single hydraulic fracture and fault plain.   104
Figure 7.4   Numerical simulation of change in slip tendency   105
                  around pressurized multistage hydraulic fractures
                  using finite element technique.

Figure 8.1   Schematic of flow loop apparatus.   109
Figure 8.2   Flow loop test results.   110
Figure 8.3   Linear base gel (polymer chains).   117
Figure 8.4   Cross-linked gel.   119
Figure 9.1   ISIP illustration.   123
Figure 9.2   ISIP selection.   124
Figure 9.3   Fracture extension pressure.   135
Figure 9.4   Closure pressure determination from injection   136
                  fall-off test.

Figure 9.5   Net pressure interpretation.   138
Figure 11.1   Composite bridge plug.   172
Figure 11.2   Perforation guns.   173
Figure 11.3   Inside view of perforation guns.   173
Figure 11.4   Frac ball inside of a composite bridge plug for frac   174
                  stage isolation.

Figure 11.5   Plug and cluster spacing example.   174
Figure 12.1   Perforation phasing.   186
Figure 12.2   Bounded versus unbounded example.   189

Figure 12.3 Up dip versus down dip.   190
Figure 12.4 Choke manifold.   195
Figure 12.5 Sand trap.   197
Figure 12.6 Horizontal separators.   200
Figure 12.7 Four-phase horizontal separator from inside.   200
Figure 12.8 Liquid level controller (LLC).   200
Figure 12.9 Back pressure regulator (BPR).   201
Figure 12.10 Mechanical pop-off.   201
Figure 12.11 Oil tanks (upright tanks).   203
Figure 12.12 Flare.   203
Figure 12.13 Nodal analysis.   205
Figure 13.1 Young's modulus example.   209
Figure 13.2 Poisson's ratio illustration.   212
Figure 13.3 Wells drilled perpendicular to max horizontal stress.   222
Figure 13.4 Longitudinal versus transverse fractures.   223
Figure 13.5 Well location and transverse fractures.   223
Figure 14.1 Typical fracture injection test.   227
Figure 14.2 BHP versus square root of time.   229
Figure 14.3 Square root of time example.   230
Figure 14.4 Log–log plot.   232
Figure 14.5 Log–log plot example.   233
Figure 14.6 Log–log plot example 2.   234
Figure 14.7 Pressure-dependent leak-off.   236
Figure 14.8 G-function plot with PDL signature example.   238
Figure 14.9 Height recession behavior.   239
Figure 14.10 Height recession leak-off.   240
Figure 14.11 G-function plot with height recession signature   241
                 example.
Figure 14.12 Tip extension leak-off.   242
Figure 14.13 Horner analysis.   246
Figure 14.14 Horner analysis example.   247
Figure 14.15 Linear flow-time function plot (ACA).   248
Figure 14.16 Radial flow-time function plot.   249
Figure 15.1 Khristianovic–Geertsma de Klerk (KGD) fracture   258
                 geometry, schematic diagram.
Figure 15.2 Perkins and Kern (PKN) fracture geometry,   258
                 schematic diagram.

Figure 15.3   Radial fracture geometry.   259
Figure 15.4   Schematic of pseudo-3D fracture model.   262
Figure 15.5   Different hydraulic fracturing regimes.   263
Figure 16.1   In-ground pit.   272
Figure 16.2   Above-ground storage tank (AST).   272
Figure 16.3   Tank batteries.   273
Figure 16.4   Centrifugal pump from inside.   276
Figure 16.5   Sand masters.   277
Figure 16.6   Sand masters and T-belt.   278
Figure 16.7   Hopper and blender screws.   278
Figure 16.8   Sand screws with proppant in the hopper.   279
Figure 16.9   Blender tub and blender screws.   283
Figure 16.10 Chemical totes.   283
Figure 16.11 High-pressured iron and low-pressured hose on the   285
                  missile.
Figure 16.12 Three-leg frac manifold.   286
Figure 16.13 An overview of frac site.   287
Figure 16.14 Pressure chart.   288
Figure 16.15 Net bottom-hole pressure (NBHP) chart.   289
Figure 16.16 Chemical chart.   289
Figure 16.17 Frac equipment setup.   291
Figure 16.18 Pressure spike and pump trips.   291
Figure 16.19 Mechanical pop-off.   292
Figure 16.20 Pressure transducer.   293
Figure 16.21 Pressure transducer on a pump.   293
Figure 16.22 Tubing head with production tubing hung inside the   300
                  tubing hanger.
Figure 16.23 Hydraulic valve.   301
Figure 16.24 Flow cross, 2″ and 4″ sides.   302
Figure 16.25 Pneumatic vs. hydraulic ESD.   302
Figure 16.26 Manual valve.   303
Figure 16.27 Four-way entry frac head (goat head).   303
Figure 16.28 Typical frac wellhead.   304
Figure 17.1   Nominal versus effective decline.   307
Figure 17.2   Gas well production decline with various b.   308
Figure 17.3   Exponential decline.   310

Figure 17.4 Hyperbolic decline.									310
Figure 17.5 Hyperbolic versus modified hyperbolic decline.		312
Figure 17.6 Secant versus tangent decline rates.					315
Figure 18.1 Net cash-flow (NCF) model.							326
Figure 18.2 Crossover point illustration.							366

LIST OF TABLES

Table 1.1	Typical Natural Gas Components	2
Table 1.2	General Uses for Natural Gas Components	2
Table 1.3	BTU of Each Natural Gas Component	3
Table 1.4	Weighted Average BTU Factor Example	4
Table 1.5	Different Types of Kerogen	9
Table 1.6	Vitrinite Reflectance Values and Reservoir Relationship	9
Table 1.7	Typical TOC of North American Shale Plays	10
Table 3.1	Cumulative Gas Productions, Initial Gas in Place, and Gas Recovery Factor Obtained for Different Langmuir Volume Conditions	36
Table 5.1	Specific Gravity (SG) of HCl Acid	64
Table 6.1	Proppant Comparisons	77
Table 6.2	Standard Sieve Openings	82
Table 8.1	Relative Roughness	113
Table 9.1	Limited-Entry Design Example	131
Table 9.2	Bottom-Hole Pressure (BHP) Versus Sqrt(Time) Example	136
Table 10.1	Slick Water Schedule Example	152
Table 10.2	Completed Slick Water Schedule Answer	153
Table 10.3	Foam Design Schedule Example	167
Table 13.1	Brittleness and Fracability Ratios Example	214
Table 13.2	True Vertical Depth (TVD), Poisson's Ratio, and Pore Pressure Gradient	219
Table 17.1	Reservoir Drive Mechanism Versus b Values	308
Table 17.2	Cumulative and Monthly Production Volumes Example	318
Table 17.3	Monthly Production Rate Example	322
Table 17.4	Hyperbolic Example Summary	323
Table 18.1	Net Revenue Interest (NRI) Example	330
Table 18.2	Gas, CND, and NGL Production Volumes	337
Table 18.3	Net Opex Example	341
Table 18.4	Net Revenue Example	343
Table 18.5	NYMEX and Basis Forecast Example	345
Table 18.6	NPV Example	359

Table 18.7   Net Present Value Summary   360

Table 18.8   Present Value Example Summary   361

Table 18.9   IRR Example   364

Table 18.10 NPV at Various Discount Rates Example   365

Table 18.11 IRR Example   365

Table 18.12 Net Present Value (NPV) Profile   366

Table 18.13 MIRR Example   369

Table 18.14 Payback Period Example   370

Table 18.15 Discounted Payback Period Example Problem   371

Table 18.16 Discounted Payback Period Example Answer   372

Table 18.17 Profitability Index Example Problem   373

Table 18.18 Profitability Index Example Answer   373

Table 18.19 ACR2 7–Year   374

Table 18.20 Taxable Income Example Problem   376

Table 18.21 Taxable Income Example Answer   376

Table 18.22 ATAX Monthly Undiscounted NCF Example   377
                Problem

Table 18.23 ATAX Monthly Undiscounted NCF Example Answer   378

Table 18.24 ATAX and BTAX NPV Profile Example   391

LIST OF EQUATIONS

Equation 2.1	Washburn	14
Equation 2.2	NMR exponential function	15
Equation 2.3	Langmuir isotherm (gas content)	17
Equation 2.4	Linearized form of Langmuir equation	18
Equation 2.5	Freundlich equation	18
Equation 2.6	Combined Langmuir equation	18
Equation 2.7	Pore compressibility	22
Equation 2.8	Darcy's law	23
Equation 2.9	Permeability	23
Equation 3.1	Total OGIP	30
Equation 3.2	Free OGIP	30
Equation 3.3	Adsorbed layer correction	31
Equation 3.4	Van der Waals equation of state	33
Equation 3.5	Van der Waals equation of state correction factor for gas molecules volumes	34
Equation 3.6	Adsorbed gas density	34
Equation 3.7	Recovery factor	35
Equation 4.1	Effective stress	37
Equation 4.2	Material balance in organic matters	39
Equation 4.3	Material balance in inorganic matters	40
Equation 4.4	Material balance in fracture system	40
Equation 4.5	Gas sorption kinetics	41
Equation 4.6	Interfacial tension	42
Equation 4.7	Young–Dupré equation	43
Equation 4.8	Equilibrium contact angle	44
Equation 4.9	Capillary pressure	45
Equation 4.10	Amott wettability index	47
Equation 5.1	Linear gel conversion to gpt	57
Equation 5.2	Foam quality	60
Equation 5.3	Original acid volume	65
Equation 5.4	Nolte method pad volume	66
Equation 5.5	Shell method pad volume	66
Equation 5.6	Kane method pad volume	66
Equation 5.7	Flush volume	70
Equation 5.8	Casing capacity	70

Equation 6.1   Proppant settling velocity $Re \leq 2.0$ 84
Equation 6.2   Proppant settling velocity ($2 < Re < 500$)   84
Equation 6.3   Proppant settling velocity ($Re \geq 500$)   84
Equation 6.4   Corrected proppant settling velocity   85
Equation 6.5   Frac fluid viscosity   85
Equation 6.6   Fracture conductivity   87
Equation 6.7   Dimensionless fracture conductivity   88
Equation 6.8   Proppant stress   94
Equation 7.1   Slip tendency   104
Equation 8.1   Pipe friction pressure   111
Equation 8.2   Newtonian fluid Reynolds number   112
Equation 8.3   Relative roughness   112
Equation 8.4   Fanning friction factor for laminar flow   112
Equation 8.5   Darcy friction factor for turbulent flow   113
Equation 8.6   Fanning friction factor for turbulent flow   113
Equation 9.1   Pressure   121
Equation 9.2   Hydrostatic pressure   122
Equation 9.3   Hydrostatic pressure gradient   122
Equation 9.4   Instantaneous shut-in pressure   124
Equation 9.5   Frac gradient   125
Equation 9.6   Bottom-hole treating pressure equations   126
Equation 9.7   Total friction pressure   126
Equation 9.8   Perforation friction pressure   128
Equation 9.9   Number of open perfs (holes)   129
Equation 9.10 Perforation efficiency   129
Equation 9.11 Q/N in limited entry design   130
Equation 9.12 Near-wellbore friction pressure   133
Equation 9.13 Fracture extension pressure   133
Equation 9.14 Time to reach closure   135
Equation 9.15 Net pressure, equation 1   137
Equation 9.16 Net pressure   137
Equation 9.17 Surface-treating pressure   139
Equation 9.18 Hydraulic horsepower   140
Equation 9.19 Rearranged BHTP   140
Equation 10.1 Absolute volume factor   143
Equation 10.2 Clean rate   144
Equation 10.3 Slurry density   145
Equation 10.4 Slurry volume   147

Equation 10.5 Stage proppant   148
Equation 10.6 Sand per foot   148
Equation 10.7 Water per foot   149
Equation 10.8 Sand-to-water ratio (SWR)   149
Equation 10.9 Standard cubic feet of nitrogen per barrel of liquid 154
Equation 10.10 Foam volume   155
Equation 10.11 Nitrogen volume   155
Equation 10.12 Blender sand concentration   156
Equation 10.13 Slurry factor (SF)   157
Equation 10.14 Clean rate (no proppant)   157
Equation 10.15 Clean rate (with proppant)   158
Equation 10.16 Slurry rate with proppant   159
Equation 10.17 Nitrogen rate (no proppant)   159
Equation 10.18 Nitrogen rate (with proppant)   160
Equation 12.1 Critical drawdown pressure   193
Equation 12.2 Liquid capacity   202
Equation 13.1 Static Young's modulus from core analysis   208
Equation 13.2 Formation modulus   208
Equation 13.3 Dynamic Young's modulus from log analysis   208
Equation 13.4 Static Young's modulus conversion   209
Equation 13.5 Poisson's ratio, core analysis   210
Equation 13.6 Poisson's ratio, log analysis   210
Equation 13.7 R_v calculation   210
Equation 13.8 Brittleness ratio   212
Equation 13.9 Incompressibility constant   213
Equation 13.10 Rigidity constant   213
Equation 13.11 Fracability ratio   213
Equation 13.12 Average formation density   215
Equation 13.13 Vertical stress   215
Equation 13.14 Minimum horizontal stress   216
Equation 13.15 Biot's constant   217
Equation 13.16 Biot's constant estimation   217
Equation 13.17 Breakdown pressure for penetrating fluid   218
Equation 13.18 Breakdown pressure for nonpenetrating fluid   218
Equation 14.1 G-function time   235
Equation 14.2 Anisotropy   237
Equation 14.3 Effective permeability using G-function plot   243

Equation 14.4   Fluid efficiency using G-function time   243

Equation 14.5   Horner time   245

Equation 14.6   Reservoir transmissibility   245

Equation 14.7   Linear flow-time function   247

Equation 14.8   Radial flow-time function   248

Equation 14.9   Transmissibility using radial flow-time function   249

Equation 15.1   Stress   254

Equation 15.2   Mass conservation in fracture   260

Equation 15.3   Carter's leak-off model   260

Equation 15.4   Equilibrium condition   261

Equation 15.5   Constitutive law   261

Equation 15.6   Displacement   261

Equation 15.7 K_m definition   262

Equation 15.8 C_m defenition   263

Equation 16.1   Bernoulli's principle for incompressible flow   273

Equation 16.2   Round per minute (rpm) calculation   280

Equation 16.3   Amount of chemical needed per stage   295

Equation 17.1   Decline curve differential equation   306

Equation 17.2   Effective decline   307

Equation 17.3   Power law exponential decline model   312

Equation 17.4   Stretched exponential decline   312

Equation 17.5   Cumulative time relationship   313

Equation 17.6   Duong decline model   313

Equation 17.7   Determination of parameters a and m in Duong   313
                     decline model

Equation 17.8   Monthly nominal exponential   314

Equation 17.9   Exponential decline rate   314

Equation 17.10   Monthly nominal tangent hyperbolic   315

Equation 17.11   Monthly nominal secant hyperbolic   315

Equation 17.12   Hyperbolic decline rate   316

Equation 17.13   Monthly hyperbolic cumulative volume   316

Equation 17.14   Monthly nominal decline   319

Equation 17.15   Annual effective decline   319

Equation 18.1   Net revenue interest (NRI)   329

Equation 18.2   Adjusted gas pricing   330

Equation 18.3   Total shrinkage factor   331

Equation 18.4   Total Opex per month   336

Equation 18.5   Severance tax per month   338

Equation 18.6   Ad valorem tax   340

Equation 18.7   Net Opex   341

Equation 18.8   Monthly shrunk net gas production   342

Equation 18.9   Monthly shrunk net natural gas liquid (NGL) production   342

Equation 18.10   Monthly shrunk net CND production   342

Equation 18.11   Net revenue   342

Equation 18.12   Net Capex   348

Equation 18.13   Nominal interest rate   349

Equation 18.14   Profit (excluding investment)   350

Equation 18.15   BTAX monthly undiscounted NCF   350

Equation 18.16   Weighted average cost of capital (WACC)   353

Equation 18.17   Capital asset pricing model (CAPM)   355

Equation 18.18   Net present value (NPV)   358

Equation 18.19   BTAX or ATAX monthly discounted NCF   362

Equation 18.20   Internal rate of return (IRR)   363

Equation 18.21   Modified internal rate of return (MIRR)   368

Equation 18.22   Payback period   370

Equation 18.23   Profitability index   372

Equation 18.24   Monthly depreciation calculation for tax model   374

Equation 18.25   Taxable income @ investment date   375

Equation 18.26   Taxable income after investment   375

Equation 18.27   Corporation tax   376

Equation 18.28   ATAX monthly undiscounted NCF   377

BIOGRAPHY

Hoss Belyadi began his career as a completions engineer designing and pumping Marcellus, Utica, and Upper Devonian Shale plays, and is currently a senior reservoir engineer at CONSOL Energy specializing in production and completions optimization, completions and reservoir modeling, and project evaluation. Mr. Belyadi is also an adjunct faculty member at West Virginia University teaching hydraulic frac stimulation design. Mr. Belyadi has been a member of Society of Petroleum Engineers (SPE) since 2006 and has authored and coauthored several SPE papers on completions optimization and production enhancement. Hoss earned his BS and MS, both in petroleum and natural gas engineering, from West Virginia University.

Ebrahim Fathi is currently an assistant professor of petroleum and natural gas engineering at West Virginia University. In 2014, Dr. Fathi received the Rossiter W. Raymond Memorial Award from the American Institute of Mining, Metallurgical, and Petroleum Engineers (AIME), and Outstanding Technical Editor Award from SPE in 2013. Ebrahim has authored several peer reviewed journal publications and conference papers. He has been heavily involved in research on various aspects of unconventional reservoir developments, including multistage hydraulic fracturing of horizontal wells and prediction of fault reactivation. His research has also included laboratory measurement and characterization of fluid transport and storage in organic-rich shales, innovative research developing a new generation of flow simulators for unconventional reservoirs, and the area of computational fluid dynamics. He earned a BS in exploration of mining engineering and an MS in exploration of petroleum engineering, both from Tehran University, as well as a PhD and postdoc in petroleum engineering from the University of Oklahoma.

Fatemeh Belyadi is currently an assistant professor of petroleum and natural gas engineering at West Virginia University. Dr. Belyadi was formerly with Exterran Energy Solution Company and Kish Oil and Gas Company (KOGC), where she specialized in process equipment, gas plant, refinery reactor design, and reservoir management. SPE publishes a large portion of her professional activities, and she regularly attends and presents papers at SPE conferences. Her research interests include multiphase fluid flow, wellbore integrity, designing and managing drilling fluids, and enhanced oil recovery especially from stripper wells. Dr. Belyadi received her BS, MS, and PhD in petroleum and natural gas engineering from West Virginia University.

PREFACE

Unconventional oil and gas reservoirs are playing an important role in providing clean energy, environmental sustainability, and increased security for all nations. Application of horizontal well drilling and multistage hydraulic fracturing treatments allow enormous amounts of hydrocarbon to be released from different shale oil/gas reservoirs. Due to production from these tremendous resources in North America and the potential proliferation of production and development technologies, the United States plays a crucial role in changing the global energy landscape in many ways, leading to a growing interest in unconventional oil/gas resources all over the world. However, very limited knowledge on shale reservoir characteristics and difficulties associated with hydraulic fracturing and production strategies due to the ultratight and multiscale nature of shale structure give rise to limited production from these substantial resources. Thus, there is a critical need to develop new technologies that can improve ultimate recovery and minimize the environmental impact and footprint associated with these activities, in addition to meeting the needs of industry, governments, and academia.

Having experience in both industry and academia, we integrated the most recent literature in the area of shale reservoir characterization and hydraulic fracturing with our personal experience in teaching and performing hydraulic fracturing jobs in unconventional reservoirs. One of the primary reasons this book was written is to place the complex nature of hydraulic fracturing in unconventional reservoirs into a practical approach that can be applied as a workflow for designing fracture treatments in various shale basins across the world.

The book is focused on theories, best practices, operation and execution, and economic analysis of hydraulic fracturing in unconventional reservoirs. However, it covers broad topics including the introduction to unconventional reservoirs, advanced shale reservoir characterization, and shale gas-in-place calculation. The book expands basic theories of hydraulic fracturing and advanced topics in shale reservoir stimulation. The discussions encompass different fluid systems, proppant design, unconventional development footprints, chemical selection and design, fracture treatment and perforation design, fracture pressure analysis, horizontal multistage completions techniques, completions and flowback design evaluation in relation to production, rock mechanical properties and in situ stress, diagnostic fracture injection test, numerical simulation of hydraulic fracturing propagation, operation and execution, decline curve analysis, and finally detailed economic analysis in unconventional shale reservoirs.

Hydraulic Fracturing in Unconventional Reservoirs: Theories, Operations, and Economic Analysis serves as a reference tool and guide for field engineers in the oil and gas industry and geoscientists interested in unconventional reservoir development. Its practical writing style, field examples, and practices make it a valuable teaching material for undergraduate and postgraduate students at universities and research institutes.

CHAPTER ONE

Introduction to Unconventional Reservoirs

INTRODUCTION

Oil and natural gas are extremely important. Our society is dependent upon fossil fuels. They alone afford many of our greatest everyday comforts and conveniences. From the packaging used for our foods, to the way we heat our homes, to all of our various transportation needs, without fossil fuels our way of life would come to a screeching halt. In light of current technological advancements oil and natural gas will be the major player in the energy industry for years to come. Other sources of energy such as wind, solar, electricity, biofuel, and so forth will eventually contribute along with fossil fuels to meet the growing global energy demand. When compared to different fossil fuels natural gas is the cleanest because it emits much smaller quantities of CO_2 when burnt. Natural gas is a hydrocarbon mixture that primarily consists of methane (CH_4). It also includes varying amounts of heavier hydrocarbons and some nonhydrocarbons (as presented in Table 1.1). General usages of natural gas components are also presented in Table 1.2.

Natural gas can be found in pockets as structural or stratigraphic gas reservoirs or in oil deposits as a gas cap. Gas hydrates and coalbed methane are considered as a major source of natural gas. Natural gas is measured by MSCF, which is 1000 standard cubic feet (SCF) of gas. Combustion of 1 cubic foot of natural gas produces an equal amount of 1000 British thermal units (BTUs), the traditional unit for energy. One BTU by definition is the amount of energy needed to cool or heat one pound of water by one degree Fahrenheit. Each hydrocarbon has a different BTU and the heavier the hydrocarbon the higher the BTU becomes. Table 1.3 shows the BTU/SCF and BTU factor for each natural gas component. As can be seen below, methane has a BTU of 1012. If the price of gas is assumed to be \$4/MMBTU, 1 MSCF of pure methane would be valued at \$4.048/MSCF. To measure the actual

Hydraulic Fracturing in Unconventional Reservoirs
DOI: http://dx.doi.org/10.1016/B978-0-12-849871-2.00001-0
© 2017 Elsevier Inc.
All rights reserved.

Table 1.1 Typical Natural Gas Components

Natural Gas Components	Chemical Formula	Short Formula	
Methane	CH_4	C_1	Light ends
Ethane	C_2H_6	C_2	
Propane	C_3H_8	C_3	Heavier
i-Butane	C_4H_{10}	i-C_4	hydrocarbons
n-Butane	C_4H_{10}	n-C_4	
i-Pentane	C_5H_{12}	i-C_5	
n-Pentane	C_5H_{12}	n-C_5	
Hexane$^+$	C_6H_{14}	C_6^+	
Nitrogen	N_2	N_2	Inert/no heat
Carbon dioxide	CO_2	CO_2	content
Oxygen	O_2	O_2	

Table 1.2 General Uses for Natural Gas Components

General Uses for Natural Gas Components	
Methane	Cooking, heating, fuel, hydrogen gas production for oil refining, and ammonia production
Ethane	Ethylene for plastics, petrochemical feedstock
Propane	Residential and commercial heating, cooking fuel, petrochemical feedstock
i-Butane	Refinery feedstock, blend in gasoline, petrochemical feedstock
n-Butane	Petrochemical feedstock, gasoline blend stock
i-Pentane	"Natural gasoline" blended into gasoline, jet fuel, naphtha cracking
n-Pentane	"Natural gasoline" blended into gasoline, jet fuel, naphtha cracking
Hexane$^+$	"Natural gasoline" blended into gasoline, jet fuel, naphtha cracking
Nitrogen	Air is 78% N_2
Carbon dioxide	Air is 0.04% CO_2
Oxygen	Air is 21% O_2

BTU of natural gas, a gas sample is taken from a producing well. This sample is then taken to the lab, and by using a device called a gas chromatograph the natural gas composition (mol %) can be measured by component. After measuring the gas composition of the natural gas sample, the approximate weighted average BTU of the gas can be calculated. It is important to note that natural gas is sold by volume

Table 1.3 BTU of Each Natural Gas Component

Natural Gas Components	BTU/SCF	MMBTU per MSCF (BTU Factor)
Methane	1012	1.012
Ethane	1774	1.774
Propane	2522	2.522
i-Butane	3259	3.259
n-Butane	3270	3.27
i-Pentane	4010	4.01
n-Pentane	4018	4.018
Hexane$^+$	4767	4.767
Nitrogen	—	—
Carbon dioxide	—	—
Oxygen	—	—

Figure 1.1 Gas chromatograph.

and heat content. Therefore, the heat content (weighted average BTU) of natural gas must be measured and calculated for sales purposes. Fig. 1.1 shows the gas chromatograph instrument.

●●●

Example

A gas sample was taken from a producing well site and transferred to the lab. Using a gas chromatograph, the composition of the natural gas sample was measured. The result is reported in Table 1.4 as mol % for each component. Calculate the approximate BTU of the gas sample, discarding compressibility factor because the compressibility factor will slightly change the BTU.

Table 1.4 Weighted Average BTU Factor Example

	Known		Chromatograph	Simple Product
Natural Gas Components	BTU/SCF	MMBTU per MSCF (BTU Factor)	Mol%	Product BTU Factor and Mol%
Methane	1012	1.012 ×	88.2187	= 0.8928
Ethane	1774	1.774 ×	9.3453	= 0.1658
Propane	2522	2.522 ×	1.4754	= 0.0372
i-Butane	3259	3.259 ×	0.1768	= 0.0058
n-Butane	3270	3.27 ×	0.2125	= 0.0069
i-Pentane	4010	4.01 ×	0.0586	= 0.0023
n-Pentane	4018	4.018 ×	0.0236	= 0.0009
Hexane$^+$	4767	4.767 ×	0.0313	= 0.0015
Nitrogen	—	—	0.3323	—
Carbon dioxide	—	—	0.0932	—
Oxygen	—	—	0.0323	—
Total (weighted average BTU factor)				**1.113**

To calculate the weighted average BTU of gas, take the mol% (measured from the gas chromatograph) and multiply it by the BTU factor of each component. The summation of product of mol% and BTU factor will yield the weighted average BTU factor. The BTU of the gas sample is 1113 (not corrected for compressibility), but the BTU factor is 1.113. If the price of gas is \$4/MMBTU the value of the gas based on the heat content is actually $4 \times 1.113 = \$4.452/\text{MSCF}$.

DIFFERENT TYPES OF NATURAL GAS

Natural gas can be found in different forms such as natural gas liquid (NGL), compressed natural gas (CNG), liquefied natural gas (LNG), and liquefied petroleum gas (LPG). NGLs refer to the components of natural gas that are liquid at surface facilities or gas processing plants. For the purpose of this book, NGLs consist of ethane, propane, butane, pentane, and hexane$^+$, but do not include methane. Iso-pentane, n-pentane, and hexane$^+$ are also called "natural gasoline." CNG is the compression of natural gas to less than 1% of the volume occupied in standard atmospheric pressure. CNG is stored and transported in cylindrical and spherical high-pressure containers. LPG consists of only propane and butane and has been liquefied at low temperatures and moderate pressures. LPG has many uses including heating, cooking, refrigeration, motor fuel, and so on. A simple example of LPG is a propane tank used for grilling. In addition to the aforementioned types of natural gas, terms like associated or nonassociated gas are also used in the oil and gas industry. Associated gas refers to the gas associated with oil deposits either as free gas or dissolved in solution. Nonassociated gas is not in contact with significant quantities of liquid petroleum. Nonassociated gas is sometimes referred to as dry gas.

NATURAL GAS TRANSPORT

Natural gas can be transported using three different methods. The first method is via pipeline, which is currently used across the United States. The second method is by liquefying natural gas, and the third method is by converting natural gas to hydrates and transporting the hydrates. In the case of LNG, natural gas is cooled to $-260°F$ at atmospheric pressure to condense. The main purpose of LNG is ease of storage and transportation. LNG occupies approximately 1/600th of the volume of gaseous natural gas. LNG is transported through ocean tankers. Another advantage of liquefying natural gas is the removal of oxygen, sulfur, carbon dioxide (CO_2), hydrogen sulfide (H_2S), and water from natural gas.

One of the main disadvantages of converting natural gas to LNG is the cost. However, technological advancements can decrease the cost and make the process economically feasible. In some places, the construction of pipeline facilities could be more expensive because of the lack of infrastructure. A disadvantage or risk of LNG is when cooled natural gas comes in contact with water it can result in rapid phase transition explosion. In this type of explosion a massive amount of energy is exchanged between water at a normal temperature and LNG at $-260°F$. This transfer of energy causes rapid phase transition, which is also known as cold explosion. When the gas reservoir is far from pipelines, the third method of gas transport, which is converting gas to gas hydrates, can be used. The economy plays a major role in choosing the gas–transport technique. In some cases, as studied by Gudmundsson and Hveding (1995), it is economically more viable to convert gas to gas hydrates, and then transport natural gas as frozen hydrate. One major concern in gas hydrate transport is the hydrate stability. Mid-refrigeration at $-20°F$ prevents gas dehydration. This is due to the generation of an ice shell around the hydrate that prevents early gas dehydration. There are several centers around the world working on pilot and laboratory-scale experimental studies of gas hydrate transport, including British Gas, Ltd., and the Japanese National Marine Research Institute.

UNCONVENTIONAL RESERVOIRS

As time passes, more technological advancements will result in more commercial production of oil and natural gas. For example, shale was a known resource decades before it could be exploited in an economically feasible process to produce significant amounts of natural gas. The development of drilling horizontal wells and using multistage hydraulic fracturing have made the exploitation of previously untapped resources not only possible, but profitable reserves for small and big operators. These new extraction methods have led to the shale reservoirs playing a major role in the oil and gas industry. These burgeoning technologies will enable us to extend the life of Earth's finite natural gas resources. Therefore, in 50 years, if the question of "how much oil and gas is left on this earth" is proposed, the answer would be another 50 years. Technology continuously advances and improves as such, that they will cause oil and gas to be recovered more efficiently and economically. For example, the development of unconventional shale reservoirs has added a tremendous amount of reserves and value to the oil and gas industry.

Unconventional oil and gas reservoirs are playing an important role in providing clean energy, environmental sustainability, and increased security. The US Energy Information Administration (EIA) predicted that shale gas production would increase 23% in 2010 and 49% by 2035. The US Geological Survey in 2008 estimates the mean undiscovered volume of hydrocarbon in only the Bakken formation in the United States portion of the Williston Basin of Montana and North Dakota to be 3.65 billion barrels of oil, 148 million barrels of NGL, and 1.85 trillion cubic feet of associated/dissolved natural gas. The United States will play a critical role in changing the global energy landscape because of production from these resources. The potential for transferring the production and development technologies has led to growing interest in unconventional oil/gas resources all over the world as reflected in the World Shale Map published by the Society of Petroleum Engineers (SPE) in the Journal of Petroleum Technology (JPT, March 2014).

Due to the tight and multiscale nature of shale structures, knowledge of shale characteristics is limited and there are difficulties associated with stimulation and production strategies causing diminished production from these substantial resources (between 5% and 10% with current technology from shale oil resources) (Hoffman, 2012). A conventional enhanced oil recovery technique, such as water flooding, is also a suboptimal method for stimulation because of the ultralow permeability. The current industry standard practice is to decrease the well spacing and increase the number of stages in hydraulic fracturing treatments to increase production. This approach raises serious environmental concerns for governmental entities. There is a critical need to develop new technologies that improve recovery and minimize the environmental impact associated with these activities. In the absence of such technology, our prediction and optimization of field-scale production in this new generation of clean energy will likely remain limited.

Unconventional gas resources are different than conventional resources in that they are technically difficult to produce because of low permeability or poorly understood production mechanisms. There are also challenges associated with the risk analysis and economics of these resources. Fig. 1.2 shows the resource pyramid where gas resources are divided in three categories of "good," "average," and "poor" based on their formation permeability. The majority of the "good" resources have already been produced and we are now looking into "average" and "poor" resources. As the oil and gas industry moves to produce from "average"

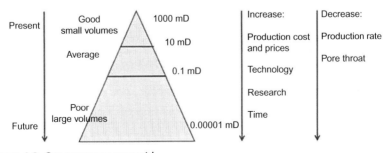

Figure 1.2 Gas resources pyramid.

and "poor" resources, more advanced technology, time, and research must be devoted to produce from these resources.

Unconventional gas reservoirs fall into the "poor" resource category and are comprised of mainly tight gas sands, coalbed methane, shale gas, and gas hydrates. Gas sands, coalbed methane, and shale gas are currently being produced. Natural gas hydrates, with perhaps the largest volume of gas in place, pose the greatest future challenges with respect to technology, economics, and environment. Tight gas sand, shale gas, and coalbed methane can be distinguished based on their total organic contents (TOCs). TOC is represented by the weight percent of organic matter. Shale gas reservoirs require a value of at least 2% to be economically feasible for investment. Shale reservoirs with a TOC of more than 12% are considered to be excellent.

Tight gas sands have a minimum amount of TOC—less than 0.5%. Most of the gas presented in tight gas sands is free gas. Shale gas reservoirs have a TOC of between 0.5% and 40% and coalbed methane reservoirs are mainly made of organic matter (more than 40%). Among these unconventional gas resources, coalbed methane, and shale gas reservoirs are very similar. They are both sedimentary rock with organic materials having low to ultralow permeability and a multiscale pore structure. Coal is a mixture of various minerals and organic materials exhibiting an intricate pore network. Coalification is defined as the process of gradual change in the physical and chemical properties of coal as pressure and temperature increase during geological time. Coalification, also known as metamorphism, delineates different ranks of coal. As coal reaches a higher rank, it contains more carbon content and volatile components, and less moisture.

Shale is the most common sedimentary rock and is composed of fine-grained and clay-sized particles. The more quartz in the matrix of a shale sample, as compared to clay minerals, leads to a more brittle or fracable shale formation. Shale sediments with potential for natural oil and gas production are generally rich in a type of organic matter known as kerogen (Kang et al., 2010). The color of shale ranges from gray to black depending on the organic content. Oftentimes, as shale gets darker, more organic material will be present. Shale can be presented as a source rock or cap rock in unconventional and conventional reservoirs. Source rock is what generates oil and gas; it is known as black shale when it has a high TOC. Often organic-rich black shale has a high TOC and gas content, and low water saturation. During digenesis, most of the organic content of shale and coal is transformed to large molecules known as kerogen. Increasing the temperature and reducing the microbial activities transform kerogen to bitumen, which has smaller and more mobile molecules. Kerogen is made of maceral, which is equivalent to the minerals in inorganic material. Of the four different Kerogen types, type I is simultaneously the most valuable and vulnerable because it has the highest capacity to produce liquid. Type II is also a good source for hydrocarbon liquid production. However, kerogen type III produces mainly gas, except when it is mixed with type II. Kerogen type IV is highly oxidized and has no hydrocarbon generation potential. Waples (1985) categorized different kerogen types based on their original organic matter and maceral (as illustrated in Table 1.5). In addition to kerogen type and TOC, the thermal maturity (TM) of shale is also a key parameter in shale reservoir evaluation. TM is a measure of the heat-induced process of converting organic matter to oil or natural gas. TM measures the degree to which a formation has been exposed to the high heat needed to break down organic matter into hydrocarbon. This parameter is quantified based on vitrinite reflectance (% Ro), which measures the maturity of the organic matter. Vitrinite reflectance varies from 0.7% to 2.5 + %. A vitrinite reflectance of greater than 1.4% indicates the hydrocarbon is dry. A TM closer to 3% indicates overmaturation resulting in gas evaporation. Table 1.6 summarizes vitrinite reflectance and its

Table 1.5 Different Types of Kerogen

Kerogen Type	Maceral	Origin of Maceral
I	Alginite	Freshwater algae
II	Exinite; cutinite	Pollen, spores; land-plant cuticle; land-plant resins
	Resinite; liptinite	All land-plant lipids, marine algae
III	Vitrinite	Woody and cellulosic material from land plants
IV	Inertinite	Charcoal; highly oxidized or reworked material of any origin

Table 1.6 Vitrinite Reflectance Values and Reservoir Relationship

Reservoir	Vitrinite Reflectance Values (%)
Immature	<0.60
Oil window	0.60−1.10
Condensate/wet Gas	1.10−1.40
Dry gas window	>1.40

significance in various reservoir fluid windows. The range of vitrinite reflectance for different reservoir fluid windows (oil, gas, and condensate) may vary depending on the kerogen type.

Both shale and coal have multiscale pore structures important for gas transport and production that consist of primary pores (inorganic materials with free and adsorbed gas) and secondary pores (in inorganic materials). Fig. 1.3 shows schematics and sample pictures of coalbed methane and shale from the Black Warrior Basin and Marcellus. Fig. 1.3 illustrates that the coalbed methane matrix consists of mainly organic materials, whereas the shale matrix organic materials are represented as islands inside of the inorganic matrix. Table 1.7 shows the typical TOC of North American shale gas plays.

It is important to examine the different natural fracture systems present in coalbed methane and shale reservoirs. Coalbed methane has a uniform fracture network making it easy to model using dual porosity and dual permeability models, conventionally called "cubic sugar" models. In contrast, shale matrixes possess a nonuniform fracture system that requires

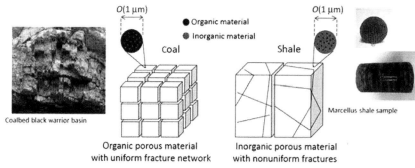

Figure 1.3 Typical shale and coal comparison. *Modified from Kang, S.M., Fathi, E., Ambrose, R.J., Akkutlu, I.Y., Sigal, R.F. 2010. CO$_2$ applications. Carbon dioxide storage capacity of organic-rich shales. SPE J. 16 (4), 842−855.*

Table 1.7 Typical TOC of North American Shale Plays

Shale or Play	Average TOC (Weight %)
Barnett	4
Marcellus	1−10
Haynesville	0−8
Horn river	3
Woodford	5

sophisticated numerical models such as quad-porosity and double permeability models. Natural fractures are very important in economically producing coalbed and shale formations. The connection of hydraulic fractures (created during a frac job) with natural fractures in the reservoir creates the necessary channels for optimum production. Therefore, a moderate presence of natural fractures is necessary to economically produce from shale reservoirs.

In addition to the amount and quality of shale organic content, water saturation must also be less than 45% for production to be economically feasible. Water saturation of Marcellus Shale is typically less than 25% while Bakken Shale in North Dakota has a varying water saturation of 25−60%. The clay content of shale is another important parameter to investigate for shale reservoir evaluation. Clays are soft and loose materials formed as a result of weathering and erosion over time. The clay minerals most often found in shale gas reservoirs are illite, chlorite, montmorillonite, kaolinite, and smectite. Some clay swells when in direct contact with water, and this can cause a reduction in the efficiency of hydraulic fracturing. A moderate clay content (of less than 40%) is needed for a marketable production in shale reservoirs. Rock mechanical properties such as brittleness, Young's modulus, and Poisson's ratio also play an important role when designing a fracturing job. A high Young's modulus and a low Poisson's ratio is the goal in hydraulically fracturing a zone. Rock brittleness is often used as an indication of a formation fracability. Formation density must be determined to decide where to land the horizontal well. For this purpose, a density log is commonly used to determine the density of the formation. The lower the density of the formation, the better suited the zone is for landing the well. In addition, lower density is typically indicative of higher organic content.

A gamma ray log is one of the most common logs used in drilling operations. It can detect the presence of shale inside the tubing or casing, and it can be run in salt−mud or nonconductive mud such as oil or synthetic-based mud. A gamma ray log measures the natural radiations in the formation. Sandstone and limestone have a lower gamma ray, and shale has a higher gamma ray. In a gamma ray log light emissions are counted and ultimately displayed as counts per second (CPS) versus depth on a graph. The unit for a gamma ray is converted from CPS to gamma-ray, American Petroleum Industry unit (GAPI) and is shown as GAPI on the log. When uranium is the driver in Marcellus Shale, a higher gamma

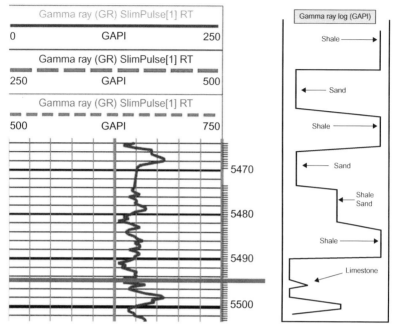

Figure 1.4 Gamma ray log.

ray is often associated with a higher TOC and organic content in the rock. When uranium is not the driver, density logs can be used to determine the zones with higher organic contents. Fig. 1.4 shows a gamma ray log and interpretation.

Reservoir pressure, also known as pore pressure, is another important parameter in commercial production from shale gas reservoirs. Reservoir pressure needs to be above normal, which is defined as any reservoir with a pressure gradient greater than 0.465 psi/ft. Areas that have above normal reservoir pressure gradients are considered optimal for production enhancements. The highest ultimate recoveries will be from abnormal reservoir pressures. Reservoir pressure can be calculated using build-up tests or more often calculated using diagnostic fracture injection tests (DFITs).

This book will focus on many critical considerations regarding shale development, namely shale reservoir characterization, modeling, hydraulic fracturing, enhanced shale oil and gas recovery, and economic analysis.

CHAPTER TWO

Advanced Shale Reservoir Characterization

INTRODUCTION

Unconventional shale reservoir characterization is important for accurate estimation of original oil- and gas-in-place (OOIP and OGIP) and production rates. Production from unconventional reservoirs is a function of reservoir matrix porosity, permeability, hydrocarbon saturation, pore pressure, contact area, and conductivity provided by hydraulic fracturing and effective enhanced oil recovery techniques (Rylander et al., 2013). Characterization often includes laboratory measurements of pore volume, permeability, molecular diffusivities, saturations, and sorption capacity of selected shale samples. Conventional methods of sampling and measuring these properties have limited success due to the tight and multiscale nature of the core samples. Therefore, new experimental techniques are needed to analyze shale samples.

PORE-SIZE DISTRIBUTION MEASUREMENT OF SHALE

As shale oil and gas resources gain popularity it is critical to search for more information regarding their rock and fluid characteristics. One such piece of critical information is the porosity of shale rocks. Knowing the total and effective porosity of shale resources is crucial to determine OOIP and OGIP and gas storage capacity. In addition to shale matrix porosity, understanding pore shapes and connectivity can provide information about how fast oil and gas can be produced and how oil and gas flow will be impacted as reservoir pressure changes. Therefore, to retrieve the most accurate storage capacity of a reservoir, the pore-size distribution must be analyzed and interpreted.

© 2017 Elsevier Inc.
All rights reserved.
13

Pore sizes are classified in four main categories by the International Union of Pure and Applied Chemistry (IUPAC) and are defined as macropores, mesopores, micropores, and ultramicropores. These have diameters of greater than 50 nm, between 2 and 50 nm, between 0.7 and 2 nm, and less than 0.7 nm, respectively. One of the main characteristics of organic-rich shale is the matrix micropore structure that controls the oil and gas storage and transport in these tight formations. Using focused ion beam scanning electron microscopy (FIB/SEM) Ambrose et al. (2010) showed that a significant portion of the pores associated with gas storage are found within shale organic material known as kerogen. Kerogen has a pore-size distribution between 2 and 50 nm, with the average kerogen pore-size typically below 10 nm (Akkutlu and Fathi, 2012; Adesida et al., 2011). The range of pore sizes shows that the organic-rich shale can also be considered as organic nanoporous material.

There are different pore-size distribution measurement techniques, each capable of capturing different ranges of pore sizes. To capture the whole range of pore-size distribution, a combination of different pore-size measurement techniques are required. The earliest work on pore-size distribution measurements goes back to 1945 by Drake and Ritter. They injected mercury into the porous material and used the intrusion pressure and volume of mercury displaced to obtain the pore-size distribution. A high-pressure mercury injection in shale samples is a common technique to find the pore-size distribution. In this technique, the pressure profile is collected during mercury injection, and will be used in Washburn Eq. (2.1) (Washburn, 1921) to obtain the pore diameter.

$$D = \frac{-4\sigma\cos\theta}{P}$$

Equation 2.1 Washburn.

D is the pore diameter, σ is surface tension, and θ is contact angle. In the case when mercury is used for the experiment, a contact angle of 130 degrees and surface tension of 485 dyne/cm are commonly used.

Nuclear magnetic resonance (NMR) is used in the industry to estimate pore-size distribution and rock matrix grain sorting. In this technique, a sample saturated with brine is exposed to NMR where collecting the single fluid relaxation time reflects the pore-size distribution and grain sorting of the sample matrix. The assumption is that water molecules inside pores excited by an NMR pulse will diffuse, hiding in the pore walls much like Knudsen diffusion. Given enough time, these fluid—rock molecular collisions lead to relaxation of the NMR signal, which can be modeled with exponential function such as Eq. (2.2).

$$N(t) = \omega_0 e^{-1/T}$$

Equation 2.2 NMR exponential function.

In this equation ω_0 is the total relaxation time and T is a function of total bulk relaxation, surface relaxation, and molecular diffusion gradient effect. For simplicity, T is considered a function of surface relaxation that is related to fluid—rock molecular collision. Fluid—rock molecular collision is a function of pore radius, pressure, temperature, and fluid type. In the case of a sample saturated with water, a linear relationship between relaxation time and pore diameter can be developed and used for pore-diameter estimation. Micropores are detected in an NMR signal by the shortest T value while mesopores have a middling length, and macropores the longest T value.

Recent advancements in imaging techniques and the availability of three dimensional images of organic-rich shales in different scales have made it possible to investigate the fundamental physics governing fluid flow, storage, and phase coexistence in organic nanopores. These advanced technologies offer new opportunities to unlock this abundant source of oil and natural gas. FIB/SEM is used to image the microstructure of shale samples (Ambrose et al., 2010). FIB/SEM is also used to provide detailed information on microstructure, rock, and fluid characteristics of organic-rich shale samples. The FIB system is used to remove very thin slices of material from shale rock samples, while the SEM provides high-resolution images of the rock's structure, distinguishing voids, and minerals. Curtis et al. (2010) used the FIB/SEM technique, and measured pore-size distribution of different shale samples. He concluded that small pores were dominant based on their number; however, large pores still provided the major pore volume in the samples investigated. Scanning transmission electron microscopy imaging (STEM) is also used to image and measure pore-size distribution of shale samples. STEM has similar resolution as FIB/SEM.

It is also possible to use adsorption—desorption data to characterize the pore structure of different materials. Ida Homfray and Z. Physik 1910 were the pioneers using sorption behavior of different gases to characterize the charcoal pore structure. Currently, low-temperature nitrogen adsorption techniques are widely used to determine the pore-size distribution of shale samples, estimate an effective pore size, and determine sorption behavior of shale samples.

SHALE SORPTION MEASUREMENT TECHNIQUES

Sorption is a physical or chemical process in which gas molecules become attached or detached from the solid surface of a material. There are physical and chemical sorption processes. Physical sorption is caused by electrostatic and van der Waals forces, while chemical sorption (high-heat sorption) is the result of a strong chemical bond (Ruthven, 1984). As free gas pressure increases, the amount of the sorbed gas will increase. This is referred to as the adsorption process. Desorption is the process that occurs when free gas pressure drops and adsorbed gas molecules start desorbing from a solid surface. Sorption isotherms are often used to determine maximum adsorption capacity and the amount of adsorbed gas at different pore pressures. Here we are concerned with sorption behavior of clay minerals and organic materials such as coal and shale.

Among several models describing equilibrium sorption behavior, the Henry's law isotherm is the simplest. It considers the linear relationship between adsorbed and free gas. That is, $C_\mu = KC$ where C_μ is the adsorbed gas concentration, K is the Henry's constant, and C is the free gas concentration. Even though the relationship between adsorbed and free gas concentrations is not linear, Henry's law has been used extensively because of its simplicity. There are other isotherm models presented, including Gibbs, potential theory, and Langmuir. The Gibbs model defines the sorption process by the equation of state in terms of two-dimensional films. Several authors including Sunders et al. (1985) and Stevenson et al. (1991) have used this model for the gas sorption measurement in the coal. The potential theory model defines sorbed volume as the thermodynamic sorption potential. The Gibbs and potential theory models were largely implemented for coal gas sorption measurements (Yee et al., 2011). The Langmuir model is defined as the equilibrium between condensation and evaporation. The Langmuir model consists of three different types of isotherms including Langmuir, Freundlich, and the combination of both (Langmuir and Freundlich) isotherms (Yang, 1987).

Irvin Langmuir (1916) developed the theory of Langmuir isotherm, which is the most common model used in the oil and gas industry

describing the sorption relationship. The main assumptions for deriving the Langmuir equation are as follows:
- In each adsorption site, one gas molecule is adsorbed.
- There is no interaction between adsorbed gas molecules at the neighboring site.
- The energy at the adsorption site is equal (homogenous adsorbent).

The Langmuir isotherm has been extensively considered as $C_\mu = abC/(1 + aC)$. In this case, a is the Langmuir equilibrium constant, and b represents complete monolayer coverage of the open surface by the gas molecules. The Langmuir equilibrium equation is a special form of the multilayer Brunauer, Emmett, and Teller adsorption equation, $C_\mu = abC/(1 - /(1 + b(C - 1)))$. The Langmuir equation is rearranged as Eq. (2.3).

$$V = V_L \frac{P}{P + P_L}$$

Equation 2.3 Langmuir isotherm (gas content).

V is the adsorbed gas volume (gas content) in scf/ton at pore pressure P (psi). V_L is the maximum monolayer adsorption capacity of the sample in scf/ton. P_L is the Langmuir pressure (psi), which is the pore pressure at which half of the adsorbed sites are taken (Fig. 2.1). The Langmuir model could be presented in a linear form by taking the reciprocal of the terms on both sides of the equation above (Mavor et al., 1990; Santos and Akkutlu, 2012; Fathi and Akkutlu, 2014).

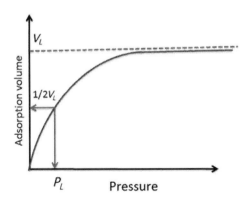

Figure 2.1 Schematic of typical Langmuir isotherm.

$$\frac{1}{V} = \frac{1}{V_L} + \left(\frac{P_L}{V_L}\right)\frac{1}{P}$$

Equation 2.4 Linearized form of Langmuir equation.

The Freundlich isotherm is given by Eq. (2.5) (Yee et al., 2011):

$$V = KP^n$$

Equation 2.5 Freundlich equation.

The combined Langmuir/Freundlich isotherm is presented as follows (Yee et al., 2011):

$$V = V_L\frac{KP^n}{1 + KP^n}$$

Equation 2.6 Combined Langmuir equation.

The relationship between adsorbed gas volume and free gas pressure is nonlinear at equilibrium conditions, homogeneous conditions, and isotropic media. Experimental studies on the sorption behavior of different materials show six different adsorption isotherm types as illustrated in Fig. 2.2 (Sing, 1985).

In Fig. 2.2, the adsorption amount is plotted versus relative pressure, which is the ratio of absolute pressure to the saturation pressure. Saturation pressures have been found empirically for many gases and can be found by increasing the pressure of a gas until it condenses. As absolute pressure approaches saturation pressure the adsorption is maximized. The desorption isotherm can also be obtained by reaching the maximum

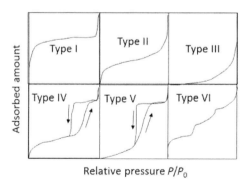

Figure 2.2 Different adsorption isotherm types.

adsorption, and then systematically reducing the pressure and plotting the quantity of molecules desorbed versus the pressure. The shape of sorption isotherms can also be used to characterize the pore structure of the material. Type I isotherms are typically representative of microporous materials with monolayer adsorption as discussed in Langmuir-type adsorption. The adsorption isotherm of natural gas in organic-rich shale typically follows the type I adsorption isotherm. Types II and IV adsorption isotherms are very similar except type IV experiences hysteresis, or a deviated curve on the desorption isotherm that could be related to condensation and type II has a larger saturation pressure. These are often indicative of nonporous or macroporous materials. Type II adsorption isotherms can be seen when monolayer and multilayer adsorption exist on solid surfaces. Types III and V are also very similar in shape. Type V shows hysteresis on the desorption curve unlike type III. Type III isotherms are usually representative of large pores while type V is representative of mesopores. Type VI adsorption isotherm corresponds to multilayer adsorption on a completely uniform surface without pores. IUPAC has introduced four different types of hysteresis as shown in Fig. 2.3. Type I represents uniform distribution of pores with no interconnecting channels. Type II shows interconnecting channels, and types III and IV mainly represent slit-like pores. Types II and IV are different in the sense that the former does not show adsorption reaching the plateau while Type IV shows limited adsorption even at a very high pressure (Sing, 1985).

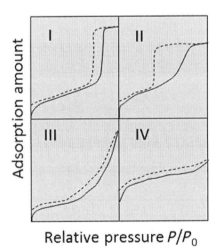

Figure 2.3 Different hysteresis types.

There are different experimental techniques available to measure the adsorption capacity of the sample including volumetric, gravimetric, chromatographic, pulse, and dynamic adsorption methods. The last two techniques are an extension to the traditional volumetric technique. Among all of these different techniques, the volumetric technique is the most commonly used in the oil and gas industry to measure the sorption capacity of shale. The low-temperature nitrogen adsorption technique is one of the volumetric techniques used to measure the sorption capacity of the sample. This technique as discussed earlier can also be used to determine the pore-size distribution of shale and to characterize the effective porosity of the shale sample. Volumetric sorption measurement techniques usually consist of a double-cell gas expansion porosimeter in a constant temperature unit. The experiment is performed at multiple stages and involves the following steps:

I. accurate measurement of sample and reference cell pressures at initial condition (in the case where the adsorption measurement reference cell has higher pressure)

II. bringing the sample in the sample cell to equilibrium pressure with reference cell and measuring new equilibrium pressure

III. charging the reference cell to new initial condition and repeating the experiment in elevated pressures to recover the whole isotherm curve

Sorbed gas quantity then will be calculated using material balance and a compressibility equation of state. Crushed samples are usually used for adsorption measurements. However, it is not possible to perform the experiment under reservoir stress conditions, using crushed samples. Kang et al. (2010) used a new five-stage adsorption measurement technique where they performed the measurements using core plugs under actual reservoir conditions.

SHALE POROSITY MEASUREMENTS

Porosity is defined as total pore volume over bulk volume, and effective porosity is effective pore volume divided by bulk volume. Effective pore volume is defined as interconnected pore volumes. Effective pore volumes of the samples can be obtained using the difference between bulk and grain densities. One needs to first obtain the bulk density, and then measure the grain density of the sample. Sample bulk volume can be obtained by immersing the sample in a mercury bath and measuring the mercury displacement. Bulk density of the sample can then be obtained by measuring the dry sample weight. To obtain the grain density the sample will be crushed and low-pressure gas pycnometry is used to measure the grain density. For this purpose gas, typically helium, is introduced to the gas pycnometer and a change in pressure with and without the sample is used to obtain the grain density. This calculation is based on Boyle's law and the real gas compressibility equation of state. If shale samples are used, this method might overestimate the effective pore volume since helium has a much smaller molecular size when compared to methane molecules. Helium can access pores that are inaccessible to methane molecules. Since we are interested in finding the porosity of the shale sample to methane (methane is the dominant gas component) an experiment needs to be performed using methane gas that requires additional safety considerations. In addition to molecular size, methane has a much higher adsorption capacity that leads to a reduction in the pore-size diameter and therefore the effective pore volume. Organic-rich shale samples act as a molecular sieve for the gas measurement.

The high-pressure mercury injection is also conventionally used to measure the sample's effective pore volume. In this case, mercury is injected into penetrometer and pressure increases until mercury invades and fills all connected pore volumes. The sample effective pore volume is then equal to the volume of displaced mercury. If a shale sample is used then the mercury intrusion will start at high pressure, typically 10,000 psi, due to very small pore-size distribution. To invade all of the interconnected pores pressure has to rise to more than 60,000 psi. At this pressure the instrument cannot detect the contribution of micropores and some of the mesopores on pore volume.

Several other techniques are available for the measurement of total and effective pore volumes based on different principles such as thermogravimetry, NMR spectrometry, SEM, and low-temperature adsorption. When the sample under investigation is shale all of these techniques have their own limitations. These measurements have limitations due to the fact that they are not performed under reservoir conditions (effective in situ stress and temperature) (Akkutlu and Fathi, 2012).

It is believed that pore volume associated with organic matter is linked to the thermal maturity of the shale. Therefore, thermal maturity can impact both storage (porosity) and transport (permeability) potential of the organic-rich shales (Curtis et al., 2013).

PORE COMPRESSIBILITY MEASUREMENTS OF SHALE

Pore compressibility is defined as the change in pore volume of a sample with respect to pressure at a constant temperature and denoted by C_p:

$$C_p = \frac{-1}{V_o} \times \frac{dV}{dP}$$

Equation 2.7 Pore compressibility.

In this equation the relative change of sample pore volume with respect to some referenced volume (usually at standard conditions) is measured. The relationship is inverse since increasing the pressure at a constant temperature will result in a reduction in volume. Pore compressibility can also be used as an indication of rock mechanical properties such as bulk modulus. Therefore, accurate measurement of heterogeneous rock pore compressibility is not an easy task. The problem becomes more complicated considering the change in pore pressure and overburden pressure during oil and gas production. This will result in dynamic pore compressibility. There have been several studies on the relationship between pore compressibility and mineralogy of different consolidated and unconsolidated formations such as Newman (1973), Anderson (1988), Zimmerman (1991), and Cronquist (2001). However, except for special cases, no distinct and universal relation is found. Therefore, most of the relationships developed are used for qualitative and comparison studies. If shale samples are used, finding the correlation is harder due to the quasibrittle/ductile characteristics of shale samples. For this purpose, a special experimental setup for shale samples was designed by Kang et al. (2010). Later, Santos and Akkutlu (2012) used modified pulse-decay permeameter and measured the pore compressibility of shale samples using the two–stage gas expansion technique. The details of the experimental setup are shown in Fig. 2.5.

SHALE PERMEABILITY MEASUREMENT TECHNIQUES

Shale reservoirs are known to have ultralow matrix permeability. Permeability is defined as the ability of rock to transmit fluid and is measured based on Darcy's unit (1 darcy is equivalent to 9.869233×10^{-13} m^2). The shale body can be divided into the shale matrix and fractures. Effective permeability of shale, which is the combination of matrix and natural fracture permeability, can be measured using well test analysis, diagnostic fracture injection tests, advanced steady state, or pressure pulse-decay permeability measurement techniques. Darcy's law describes the fluid flow through porous media, which is a proportional relationship between the discharge rate through a porous medium, geometry of the media, length and cross section, viscosity of the fluid, and pressure gradient over the length of media. The negative sign in the equation is necessary since fluid always flows from high pressure to low pressure. Eq. (2.8) shows that Darcy's law can be rearranged to find the absolute permeability K using Eq. (2.9).

$$Q = \frac{-kA \times (\Delta P)}{\mu \times L}$$

Equation 2.8 Darcy's law.

Where Q is the flow rate (m^3/s), K is absolute permeability (m^2), ΔP is the pressure gradient across the core sample (Pascal), μ is the fluid viscosity (Pa.s), and L is the sample length (m).

Darcy's equation can be rearranged to solve for permeability:

$$K = \frac{-Q \times \mu \times L}{A \times (\Delta P)}$$

Equation 2.9 Permeability.

Fig. 2.4 shows the schematic of Darcy's experiment, in which the sample with a cross section of A and a length of L is exposed to a pressure gradient of $\Delta P(P_1 > P_2)$ between point 1 and 2. An incompressible fluid such as water will be injected with the constant flow rate of Q until a steady state condition is reached. At the steady state condition, K (sample absolute permeability) can be obtained using Eq. (2.9). If the sample is shale, conventional steady state methods of permeability measurements are not practical because of very low flow rates and the extremely long time needed to reach the steady state condition. Therefore, unsteady state methods based on pressure pulse-decay measurement have been extensively used to estimate permeability of the shale samples (Brace, 1968; Ning, 1992; Finsterle and Persoff, 1997). The unsteady state methods are

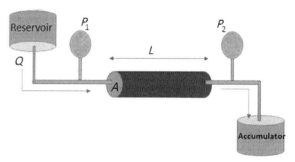

Figure 2.4 Darcy's law illustration.

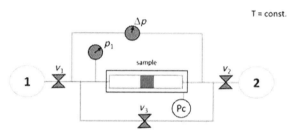

Figure 2.5 Core plug pulse-decay permeameter.

faster and can be used to measure permeabilities as low as 10E-9 md
(Ning, 1992). It is crucial to perform the experiment under reservoir
conditions since pore pressure, temperature, and confining stress conditions could lead to changes in shale rock characterization.

New pulse–decay permeameters perform the experiment under high
pressure and high temperature. They are designed and assembled to
precisely measure shale matrix, fracture, and effective permeability using a
pulse-decay technique. This technique is used at pressure and temperatures up to 10,000 psia and 340°F under different effective stress conditions. In the pulse-decay permeability measurement technique, the shale
core plug (after preparation) will be placed in a core holder and brought
to equilibrium pressure conditions. Different pulses will then be applied
to the system and pressure decay upstream and pressure buildup downstream will be recorded with high accuracy as illustrated in Fig. 2.5. In
Fig. 2.5 temperature is kept constant to reservoir temperature, and confining pressure is applied to resemble reservoir overburden stresses.
Different history-matching algorithms can then be used to match the
pressure profiles and extract the permeability values. Depending on the
magnitude of the pressure pulse, the fracture, matrix, or effective

Figure 2.6 Automated high-pressure, high-temperature (HPHT) pulse-decay permeameter.

permeability can be obtained. Small pulses likely carry the impact of fractures and are used to measure fracture permeability. On the other hand, large pulses are affected by both fracture and matrix and can be used to extract effective matrix permeability. Fig. 2.6 shows the schematic of the setup for performing the experiment with different pulse magnitudes.

Generally the slope of pressure versus time in a semilog plot is used to estimate the matrix permeability. Yamada (1980) developed analytical solutions for transient behavior of pressure during pulse decay. However, his solution was only valid under very specific and simplified conditions. The most common method used by the industry is the technique that was introduced by Jones (1997). He modified the conventional pulse–decay setup by using equal upstream and downstream volumes and added two large dead volumes to reduce the time required to reach equilibrium pressure. In Jones' technique, the adsorption capacity of shale and the possibilities of solid or surface transport were neglected. Akkutlu and Fathi (2012); Fathi and Akkutlu, (2013) introduced new sets of governing equations to simulate gas transport and adsorption in shale gas reservoirs and used that in a nonlinear history–matching algorithm to obtain unique shale rock properties.

To use the pressure decline curves obtained from pressure pulse-decay techniques, the sample pore volume and porosity at different pressures are needed. A double-cell Boyle's law porosimeter can be used to provide a precise estimation of these quantities as discussed earlier. Interpretations of the data obtained from transient techniques introduce a large margin of uncertainty due to the nonuniqueness of the results and reproducibility (except if more advanced techniques are used). To avoid the complications in interpretation of the pulse-decay techniques and to perform the experiment in a much shorter time, most of the commercial laboratories are using a crushed sample permeability measurement known as the Gas Research Institute (GRI) technique presented by Luffel (1993). In this case, a double-cell porosimeter is used to provide the crushed sample permeability. It is believed that by crushing the sample, the effect of natural fractures will be removed and the permeability measured by this technique can be a good representation of the shale matrix permeability. Fig. 2.7 illustrates the GRI technique used for crushed sample permeability measurements. The GRI permeability measurement technique on crushed rock is highly affected by particle size, average pressure of the experiment, and gas type (Tinni et al., 2012; Fathi et al., 2012). This results in a discrepancy in permeability values of up to three orders of magnitude measured by different commercial labs using the same samples (Miller, 2010; Passey et al., 2010). In addition, recently injecting mercury in different sizes of crushed shale samples and imaging using micro computed tomography, Tinni et al. (2012) showed that crushing the shale samples does not remove the microcracks from the matrix. Therefore, the permeability measured by the GRI technique is *not* shale matrix

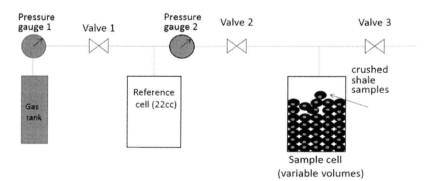

Figure 2.7 Double-cell Boyle's law porosimeter.

permeability, but rather a combination of matrix and fracture permeability without the impact of confining stresses.

Recently, Zamirian et al. (2014) designed a new pseudo-steady state permeability measurement experiment to overcome difficulties presented with conventional steady state permeability measurements, such as the extremely long time required to reach a steady state, and the inability to measure extremely low flow rates. The laboratory system is referred to as Precision Petrophysical Analysis Laboratory (PPAL). The experimental setup is very similar to the pulse-decay permeameter. In this setup, a pressure gradient will be applied between the upstream and downstream after the initial equilibrium pressure has been reached. As pressure builds up downstream, an ultraprecise pressure differential gauge measures the pressure difference between the sample and downstream. As pressure in downstream builds up by 0.5 psi, a bypass valve opens, discharging the gas to keep the downstream pressure constant. Rate of pressure buildup in downstream versus time is then used in Darcy's law to calculate matrix effective permeability. To reach a steady state condition, usually more than 50 cycles of 0.5 psi pressure buildup in the downstream is required. This technique enables us to measure flow rates as small as 10^{-6} cm^3/s.

CHAPTER THREE

Shale Initial Gas-in-Place Calculation

INTRODUCTION

Initial hydrocarbon-in-place calculation is crucial in determining the economic feasibility of shale oil and gas reservoirs and reserve estimation. In ultralow permeability reservoirs, transient flow regime can last for a long period of time. Therefore, having a good understanding of original oil and gas in place helps for determining the long-term production forecast and it will decrease the uncertainty when performing the reserve estimation. There are different techniques developed for original oil- and gas-in-place calculations in unconventional reservoirs. These are either numerical or analytical techniques based on volumetric or material balance calculations. The volumetric method is the most common technique, requiring detailed information regarding reservoir rock and fluid properties such as porosity, compressibility, saturations, and formation volume factor. This information is mostly extracted from well logs or obtained using experimental techniques as discussed earlier in chapter two. In the volumetric approach, the shale matrix will be divided into grain volume (e.g., clay minerals, nonclay inorganic minerals, and organic materials), volume occupied by water, oil and free gas, volume occupied by clay-bound water, some dead ends, and isolated pores (Hartman et al., 2011). Fig. 3.1 shows the schematic of bulk volume of shale sample.

TOTAL GAS-IN-PLACE CALCULATION

In the case of shale gas reservoirs, the gas storage can be divided into free, adsorbed, absorbed, and dissolved gas. Free gas is stored in natural and induced fractures, in inorganic macropores, and organic meso- and micropores, while adsorbed gas is stored mainly at the solid surface of organic materials and some by clay minerals. The amount of

© 2017 Elsevier Inc.
All rights reserved.

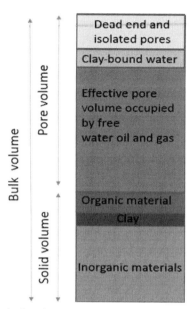

Figure 3.1 Shale matrix bulk volume.

absorbed gas as discussed earlier is assumed to be negligible. However, there are new studies investigating the impact of gas absorption in original gas-in-place (OGIP) calculation (Ambrose et al., 2012). Gas dissolved in water and hydrocarbon cannot be distinguished from adsorbed gas with current experimental techniques and is usually considered as part of adsorbed gas in gas-in-place calculations. Therefore, OGIP embraces free G_{free} and adsorbed gas G_{ads} scf/ton.

$$OGIP = G_{free} + G_{ads}$$

Equation 3.1 Total OGIP.

The volumetric approach is used to measure organic pore volumes, and therefore, the free gas amount. However, due to adsorption in organic nanopores part of the pore volume will be occupied by adsorbed gas and is not available for free gas storage. Ambrose et al. (2010) proposed a model where free gas pore volume is corrected to include the adsorption layer thickness as follows:

$$G_{free} = \frac{32.0368}{B_g} \left[\frac{\varphi(1 - S_w - S_o)}{\rho_b} - \Psi \right]$$

Equation 3.2 Free OGIP.

In this equation, B_g represents the gas formation volume factor, φ is the effective rock porosity, S_w is the water saturation, and S_o is oil saturation, ρ_b is the bulk density of shale in g/cc, and Ψ is the correction factor for adsorbed layer thickness defined as follows:

$$\Psi = \frac{1.318 \times 10^{-6} M \rho_b}{\rho_s} \left[V_L \frac{P}{P + P_L} \right]$$

Equation 3.3 Adsorbed layer correction.

M is the molecular weight of the single-component gas or apparent molecular weight of the gas mixture, ρ_s is adsorbed gas density, P is the pore pressure, P_L and V_L are Langmuir pressure and volume, respectively. Adsorbed gas density is a parameter that requires more detailed studies to obtain. Different analytical and numerical techniques are suggested to obtain adsorbed gas density including the application of the Van der Waals equation of state or molecular dynamic technique. G_{ads} as discussed earlier assumes monolayer adsorption. The Langmuir equation is as follows:

$$G_{ads} = V = V_L \frac{P}{P + P_L}$$

Ambrose et al. (2010) also extended the calculation to a multicomponent single-phase case. In general, ignoring the adsorbed layer effect in OGIP calculations might result in more than 30% overestimation.

●●●───

Example

Calculate **adsorbed gas in place (MSCF)** with the following properties for a **half-foot** section of the reservoir. The following data was obtained from core and log analyses:

A = 640 acres = 640 acres is also referred to one section
h = 0.5'
Bulk density = 2.6 g/cc (obtained from log)
V_L = 60 SCF/ton (obtained from core analysis)
P_L = 800 psia (obtained from core analysis)
P = 4400 psia (obtained from diagnostic fracture injection tests or DFIT)

$$G_{ads} = V = V_L \frac{P}{P + P_L} = \frac{60 \times 4400}{800 + 4400} = 50.77 \text{ SCF/ton}$$

The amount of adsorbed gas for a half-foot of reservoir section given can then be obtained as follows:

$$\text{Adsorbed gas} = A \times h \times \rho_b \times G_{ads}$$

$$= 640 \,(\text{acres}*43,560 \, \text{ft}^2/\text{acres}) \times 0.5 \,(\text{ft}) \times 2.6 \left(g/cc* \frac{1}{3.531e^{-5} \, \text{ft}^3/\text{cc}} \right)$$

$$\times 50.77 \left(\frac{\text{SCF}}{\text{ton} \frac{907,185 \, g}{\text{ton}}} \right) = 1359 \times A \times h \times \rho_b \times G_{ads} = 57,405 \, \text{MSCF}$$

This indicates that a half-foot section of the reservoir contains 57.4 MMSCF of adsorbed gas in place in addition to the free gas in place.

Example

Calculate free gas and adsorbed gas in place (BCF) given the following information (assume 100% methane gas):

$A = 640$ acres, $h = 100'$, bulk density = 2.35 g/cc, adsorbed gas density = 0.37 g/cc, $V_L = 60$ SCF/ton, $P_L = 700$ psia, $P_R = 4800$ psia, $S_w = 20\%$, $S_O = 0\%$, porosity = 10%, $B_{gi} = 0.0038$

$$\Psi = \frac{1.318 \times 10^{-6} M \rho_b}{\rho_s} \left[V_L \frac{P}{P + P_L} \right] = \frac{1.318 \times 10^{-6} \times 16.04 \times 2.35}{0.37} \left[\frac{60 \times 4800}{700 + 4800} \right] = 0.00703$$

$$G_{free} = \frac{32.0368}{B_g} \left[\frac{\varphi(1 - S_w - S_o)}{\rho_b} - \Psi \right] = \frac{32.0368}{0.0038} \left[\frac{0.1 \times (1 - 0.2)}{2.35} - 0.00703 \right]$$

$$= 227.7357 \, \text{scf/ton}$$

$$\text{Free gas in place} = 43,560 \times A \times h \times \rho b \times G_{free}$$

$$= 43,560 \times 640 \times 100 \times 2.35 \times \left(g/cc* \frac{1}{3.531e^{-5} \, \text{ft}^3/\text{cc}} \right)$$

$$\times 227.7357 \left(\frac{\text{SCF}}{\text{ton} \frac{907,185 \, g}{\text{ton}}} \right)$$

$$= \mathbf{4.66 \times 10^{10} \, SCF = 46.6 \, BCF}$$

$$G_{ads} = V = \frac{V_L \times P_R}{P_L + P_R} = \frac{60 \times 4800}{700 + 4800} = 52.36 \text{ SCF/ton}$$

Adsorbed gas $= 1359 \times A \times h \times \rho_b \times G_{ads} = 1359 \times 640 \times 100 \times 2.35 \times 52.36 = 10.7$ BCF

Total GIP $= 46.6 + 10.7 = 57.3$ BCF

This example indicates that 57.3 BCF of the total gas is present in one section (640 acres) of the reservoir. This does not mean that the entire amount of gas can be recovered. In unconventional shale reservoirs, the recovery factor (RF) can range anywhere from 10% to 80% depending on the reservoir properties and the completions design. For example, if recovery factor is assumed to be 25% for this particular reservoir, only 25% of 57.3 BCF can be recovered per section. Therefore, 14.325 BCF of gas can be recovered from this reservoir. Finally, 14.325 BCF is also called estimated ultimate recovery (EUR) per 640 acres.

DENSITY OF ADSORBED GAS

As discussed earlier, adsorbed gas density is required to calculate the OGIP in shale gas reservoirs. However, this is not an easy quantity to measure in the laboratory. Dubinin in 1960 proposed to use the Van der Waals equation of state as a means to calculate the adsorbed gas density. The Van der Waals equation of state relates the density of gases to pressure, temperature, and volume and is one of the earliest attempts to modify the ideal gas equation of state.

$$\left(P + \frac{a}{V^2}\right)(V - b) = RT$$

Equation 3.4 Van der Waals equation of state.

In the Van der Waals equation of state, Eq. (3.4), the interaction forces and volume of gas molecules that are neglected in the ideal gas equation of state are considered using the correction factors a and b, respectively. Dubinin (1960) suggested that in the cases where adsorption is of importance, the b constant in the Van der Waals equation of state is equal to the volume taken by the adsorbed phase ν divided by the actual adsorbed gas amount μ as follows:

$$\frac{\nu}{\mu} = b$$

Therefore, the adsorbed gas density can be written as

$$\rho_s = \frac{M\mu}{v} = \frac{M}{b}$$

where M is the molecular weight of the gas and b is the Van der Waals coefficient. The coefficient b can be obtained using first and second derivative of pressure in the Van der Waals equation of state with respect to volume at critical temperature as follows:

$$b = \frac{RT_c}{8P_c}$$

Equation 3.5 Van der Waals equation of state correction factor for gas molecules volumes.

In this equation, R is the universal gas constant and T_c and P_c are critical temperature and pressure of the gas. Therefore, the adsorbed gas density using Eq. (3.5) can be obtained as follows:

$$\rho_s = \frac{8P_c M}{RT_c}$$

Equation 3.6 Adsorbed gas density.

The critical properties of the pure components are constant values that can be obtained from the physical properties table (see engineering data book; GPSA, 1987). In the case of gas mixtures, the apparent molecular weight of gas mixture and the pseudocritical pressure and temperature can be used in Eq. (3.6). Pseudocritical properties of gas mixture are defined as the weighted average of the critical properties of pure components in the mixture. Ideal adsorption solution (IAS) theory in this case can also be used to calculate the gas mixture adsorption if the individual adsorption properties of the pure components are known (Myers and Prausnitz, 1965). Recently, molecular dynamic simulations have been extensively used to investigate the gas mixture adsorption and adsorbed phase density and thickness when multicomponent gas mixtures are considered (see Kim et al., 2003 and Rahmani Didar and Akkutlu, 2013, for more detailed discussions).

Overall, there is a substantial understanding of the application of different equations of state to investigate the phase transitions in bulk fluids where the system size is not of importance. However, as the volume of the system reduces to the meso- and microscales, the phase equilibriums become size dependent, where the wall confinement effects significantly change the thermodynamic properties of the fluids. Experimental and numerical investigations on equilibrium and non-equilibrium thermodynamical properties of fluids in nanoporous materials show dramatic deviations from their bulk values obtained using pressure—volume—temperature (PVT) measurements. Results of recent studies show that as pore size decreases to the nanoscale, critical temperature, freezing, and melting points decrease. It has also been observed that water viscosity is significantly reduced with critical pressure and interfacial tensions.

RECOVERY FACTOR

The recovery factor equation is as follows:

$$RF = \frac{EUR}{IGIP}$$

Equation 3.7 Recovery factor.

RF = Recovery factor, %
EUR = Estimated ultimate recovery, BCF
IGIP = Initial gas in place, BCF

In old OGIP calculation techniques, the pore volume occupied by adsorbed gas was neglected, which could lead to up to 30% overestimation of the OGIP depending on the amount of total organic contents (TOC) and nanoorganic pore-size distribution (Ambrose et al., 2012). Belyadi (2014) studied the impact of adsorbed gas on OGIP, total gas production, and recovery factor using information from the Marcellus Shale in West Virginia and Pennsylvania. Using compositional reservoir simulation, she showed that an increase in adsorbed gas amount increases the initial gas in place, and therefore, total gas production. However, total gas recovery decreased by increasing the amount of adsorbed gas during a specific time period of production. An increase in the adsorbed gas amount leads to a longer transient regime, while a decrease in adsorbed gas amount leads to a faster boundary dominated regime. Table 3.1 shows the details of the calculations and Fig. 3.2 compares total gas recovery from a single horizontal well with 13 hydraulic fracture stages assuming

Table 3.1 Cumulative Gas Productions, Initial Gas in Place, and Gas Recovery Factor Obtained for Different Langmuir Volume Conditions

Langmuir Volume (m) Mscf/Ton	Total Gas Production (G_p) (BCF)	Initial Gas in Place (IGIP) (BCF)	Total Gas Recovery Factor Fraction
0	4.21	5.77	0.73
0.05	4.65	7.53	0.62
0.089	4.98	8.89	0.56
0.1	5.06	9.28	0.55

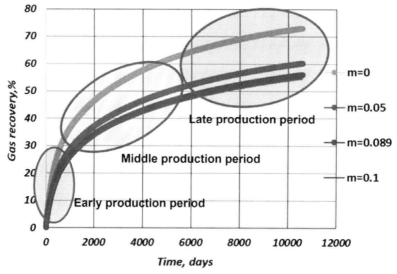

Figure 3.2 Impact of Langmuir volume on the total gas recovery from horizontal shale gas.

different Langmuir volumes (*m*). It can be observed that during the early production period, Langmuir volume has a minor impact on the total gas recovery because during this period the production is controlled by the hydraulic fractures and mainly free gas is produced. At middle and late time production periods, higher Langmuir volume leads to lower recovery factor. This is due to fact that even though more gas in place and gas production can be obtained by having more adsorption, most of the adsorbed gas is not available for production due to ultralow matrix permeability resulting in lower recovery factor. In the next chapter, we will discuss the effect of adsorption and organic nanopore size distribution on fluid flow and transport in shale reservoirs.

CHAPTER FOUR

Multiscale Fluid Flow and Transport in Organic-Rich Shale

INTRODUCTION

Gas transport and storage in organic-rich shale reservoirs are not very well understood. Previously dual-porosity single-permeability models were used to model fluid flow and transport in shale reservoirs. This approach follows conventional reservoir simulators developed for naturally fractured carbonate or coalbed methane reservoirs. To include the sorption rate and mass exchange between matrix and fractures, bidisperse models were developed with diffusion rates introduced as a controlling factor for sorption rates (Gan et al., 1972; Yang, 1997; Shi and Durucan, 2003). In these models, instantaneous adsorption/desorption of gas to and from the organic materials are assumed. The resistance to flow is considered to be governed by transfer function, which is a function of pressure gradient between two media, matrix transport, and a shape factor. This basically follows the approach introduced earlier by Warren and Root (1963) to model mass exchange between matrix and fractures. In this approach, two main assumptions have been made. First, the presence of a uniformly distributed fracture network with a known fracture matrix interface that clearly confines the matrix block is assumed. Second, the application of a matching parameter called the shape factor, which controls the mass transfer between matrix and fracture, is used. However, in the organic-rich shale reservoirs, multiscale pore structure presents nonuniformly distributed natural fractures with different dimensions through the reservoir matrix. The magnitude of the contribution of these multidimensional natural fractures in the oil and gas storage and transport is a function of reservoir effective stress, which is defined as follows:

$$\sigma_e = \sigma_n - \alpha P$$

Equation 4.1 Effective stress.

Hydraulic Fracturing in Unconventional Reservoirs
DOI: http://dx.doi.org/10.1016/B978-0-12-849871-2.00004-6
© 2017 Elsevier Inc.
All rights reserved.

In Eq. (4.1), σ_e is the effective stress; σ_n is the normal stress applied to the rock, or overburden pressure; and α is the Biot's poroelastic coefficient. Several authors applied discrete fracture network modeling technique to investigate fluid flow, transport, and storage in these formations. However, their approach does not only require detailed information on the explicit distribution of fractures, but also requires an accurate approximation of pressure distribution in the matrix. In some cases, a general parabolic equation to describe the pressure distribution in the matrix is used.

MULTICONTINUUM MODELING OF SHALE RESERVOIRS

Recently multicontinuum models have been used to simulate fluid flow and transport in shale gas reservoirs that can bypass some major difficulties associated with the application of discrete fracture network modeling. In this approach, first the number of components in the reservoir will be identified. Next, the hydraulic properties such as nature of fluid flow and transport in each component will be recognized. Finally, different coupling scenarios describing the mass exchange between components are investigated. Unlike discrete models, this approach does not need to explicitly define the spatial distribution of each component. At any location in the space, all of the components of the multicontinuum model are present and their contribution to the flow will be identified though coupling and mass exchange terms between each component. Fig. 4.1 shows how the conceptual shale matrix model is separated to three continua (inorganic, organic, and fractures) and then uniformly combined to generate the multicontinuum structure.

There are three different types of coupling introduced in the literature to generate the multicontinuum structure, including series, parallel, and selective couplings. Series coupling is defined where the coupling is in the order of hydraulic conductivity. Kang et al. (2010) showed that in samples under investigation in the Barnett Shale, the coupling between organic—inorganic and fractures mostly occurred in a series fashion. After initial gas production from fractures, the organic materials supply gas to inorganic materials, which exchange mass with the system of

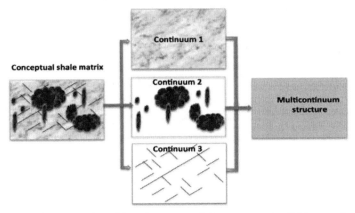

Figure 4.1 Schematic of multicontinuum modeling approach: *light gray*, inorganic matrix; *brown islands* (black in print versions), organic materials (kerogen); *Continuum 3*, discrete natural fracture system.

natural fractures. This behavior can be seen by two distinct changes in the slope of the pressure decay test. In the case of parallel coupling, both organic and inorganic materials are in hydraulic communication and both will supply gas to the system of natural fractures. Selective coupling occurs when two continuums are not hydraulically connected but exchange mass with the third continuum. Fig. 4.2 shows schematics of different possible hydraulic couplings in shale reservoirs.

Recent experimental and numerical studies have shown that the shale matrix can be divided in organic, inorganic, and fractures (Kang et al., 2010; Akkutlu and Fathi, 2012). The transport in organic materials can be represented by free and solid (surface) diffusions, while the transport in inorganic materials is mostly governed by free gas diffusion and convection (Darcy flow). Eq. (4.2) shows the mass balance in organic micropores, and Eq. (4.3) illustrates material balance in inorganic macropores of shale matrix.

$$\frac{\partial(\varepsilon_{kp}\phi C_k)}{\partial t} + \frac{\partial[\varepsilon_{ks}(1-\phi-\phi_f)C_\mu]}{\partial t}$$

$$= \frac{\partial}{\partial x}\left(\varepsilon_{kp}\phi D_k\frac{\partial C_k}{\partial x}\right) + \frac{\partial}{\partial x}\left(\varepsilon_{ks}(1-\phi-\phi_f)D_s\frac{\partial C_\mu}{\partial x}\right)$$

Equation 4.2 Material balance in organic matters.

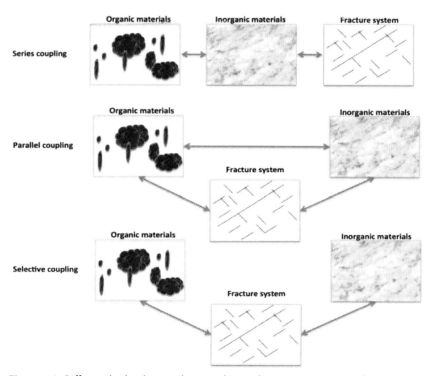

Figure 4.2 Different hydraulic coupling used in multicontinuum approach.

$$\frac{\partial[(1 - \varepsilon_{kp})\phi C]}{\partial t} = \frac{\partial}{\partial x}\left[(1 - \varepsilon_{kp})\phi D \frac{\partial C}{\partial x}\right] + \frac{\partial}{\partial x}\left((1 - \varepsilon_{kp})\phi C \frac{k_m}{\mu}\frac{\partial p}{\partial x}\right) - W_{km}$$

Equation 4.3 Material balance in inorganic matters.

Free gas mass balance in fracture networks can also be shown as:

$$\frac{\partial(\phi_f C_f)}{\partial t} = \frac{\partial}{\partial X}\left(\phi_f K_L \frac{\partial C_f}{\partial X}\right) + \frac{\partial}{\partial X}\left(C_f \frac{k_f}{\mu}\frac{\partial p_f}{\partial X}\right) - W_{mf}$$

Equation 4.4 Material balance in fracture system.

where W_{km} and W_{mf} are mass exchange terms between different continua and defined as follows:

$$W_{kmi} = \Omega_m \Psi_{ki}(C_i - C_{ki})$$

$$W_{mfi} = \Omega_f \Psi_{mi}(C_{fi} - \overline{C}_i)$$

In the above equations the subscripts k, m, and f refer to quantities related to the organic (kerogen), inorganic matrix, and fracture respectively. The variables x and t are the space and time coordinates. $C(x, t)$ and $C_\mu(x, t)$ represent the amounts of free and adsorbed gas in terms of moles per pore volume and moles per organic solid volume respectively. P is the pore pressure and ϕ and ϕ_f are the total matrix and fracture porosity respectively. ε_{ks} is the total organic content in terms of organic grain volume per total grain volume and ε_{kp} is organic pore volume per total matrix pore volume. D represents total free gas diffusion such as bulk plus Knudsen diffusion. D_s is solid or surface diffusion coefficient. k the absolute permeability, K_L represents the macrodispersion coefficient in the fracture network, and μ the dynamic gas viscosity. Ω is the shape factor and Ψ is the transport function in source media, and $\overline{C}$ is referred to as average free gas concentration (Akkutlu and Fathi, 2012).

To describe the gas sorption behavior in organic-rich shale reservoirs, Fathi and Akkutlu (2009) suggested nonlinear adsorption kinetics as shown in Eq. (4.5). This approach was also suggested earlier by Srinivasan et al. (1995) and Schlebaum et al. (1999) to study the carbon molecular sieve and organic contaminant fraction in soil. They argued that the nonlinear sorption kinetic could significantly impact the diffusion processes. The general nonlinear sorption kinetics model in shale can be presented as follows:

$$\frac{\partial C_\mu}{\partial t} = k_{desorp}\left[K(C_{\mu s} - C_\mu)C_k - C_\mu\right]$$

Equation 4.5 Gas sorption kinetics.

In this equation, $C_{\mu s}$ represents the maximum monolayer gas adsorption, K is the ratio of adsorption to desorption rate that is known as equilibrium coefficient, and k_{desorp} is the gas desorption rate. In a limiting case where the system reaches equilibrium conditions, i.e., $\partial C_\mu/\partial t = 0$, Eq. (4.5) will simplify to single-component monolayer Langmuir isotherm as described earlier.

INTERFACIAL TENSION AND CAPILLARY PRESSURE

Interfacial tension (IFT) is the enhancement in intermolecular attraction forces of one fluid facing another and has dimension of force per unit length. IFT is responsible for many fluid behaviors such as interfacial behaviors of vapor—liquid and liquid—liquid. Laplace's law indicates that there is a linear relationship between the pressure difference between two phases and the radius of interface curvature. Laplace's law has been used to study both interfaces between a liquid and its own vapor (surface tension) and between different fluids (IFT) as presented in Eq. (4.6).

$$\Delta P = \frac{\sigma}{r}$$

Equation 4.6 Interfacial tension.

where σ is the surface tension, r is the radius of curvature, and ΔP is the pressure difference between the inside and outside of the interface. This linear relationship is used to obtain the surface tension by simulating series of bubbles with various sizes, measuring their radius, and inside and outside densities (liquid and gas). Different experimental techniques have been used in the oil and gas industry to measure the IFT including the capillary rise and du Noüy ring technique. In the du Noüy ring method, the IFT can be obtained as follows:

$$\sigma = \delta \times \frac{g_c}{2\pi d}$$

where σ is the IFT in dynes per centimeter, g_c is the gravitational constant (980 cm/s^2), d is the ring diameter in cm, and δ is grams-force measured with analytic balance.

In the capillary rise method, the height of the liquid rise in a capillary can be obtained as follows:

$$h = \frac{2\sigma \times \cos \theta}{r\rho g_c}$$

In this equation, r is the radius of capillary in cm, ρ is the density of the denser fluid in g/cc, and $\cos \theta$ is the cosine of the angle between the surface inside the capillary and the capillary wall.

The IFT measurements are also a function of temperature of the experiment. As the temperature increases, the IFT drops. As discussed earlier, phase behavior and phase coexistence properties of fluids under confinement are different than those in the bulk system, especially in organic-rich shales. Recent studies using molecular dynamic simulations revealed that IFTs under pore-wall confinements decrease manyfold and the results are highly sensitive to the temperature (Singh et al., 2009).

Bui and Akkutlu (2015) also showed that the surface tension of methane is smaller under confinement using molecular dynamic simulation and is a function of liquid saturation and pore width.

Example

IFT measurement of synthetic oil is done using du Noüy ring where the ring diameter was 1.55 cm and the force measured by analytic balance was 0.38 grams-force. Calculate the IFT:

$$\sigma = \delta \times \frac{g_c}{2\pi d} = 0.38 \times \frac{980}{2 \times 3.1416 \times 1.55} = 38$$

The experiment is repeated this time with a capillary rise experiment. The height of the rise in the capillary with a radius of 0.2 cm was 0.36 cm and the density of the synthetic oil is measured to be 0.99 g/cc. Calculate the IFT using capillary rise technique. Assume $\cos \theta = 1.0$.

$$\sigma = \frac{hr\rho g_c}{2 \cos \theta} = \frac{0.36 \times 0.2 \times 0.99 \times 980}{2 \times 1} = 35$$

Most often, we are interested in phase coexistence in porous media where two immiscible fluids are in contact. In this case, depending on the chemical properties of each fluid and formation solid surface, one fluid tends to have a higher affinity to wet the formation solid surface. The fluid that adheres to the solid surface is called the wetting phase and the other fluid is called the nonwetting phase. The Young–Dupré equation describes the relationship between imbalance forces of fluid–fluid and fluid–solid interactions as follows:

$$A_T = \sigma_{nw-s} - \sigma_{w-s} = \sigma_{nw-w} \cos \theta_{eq}$$

Equation 4.7 Young–Dupré equation.

where $\sigma_{nw-s}, \sigma_{w-s}, \sigma_{nw-w}$ are IFTs between nonwetting phase and solid, wetting phase and solid, and nonwetting phase and wetting phase, respectively, and θ_{eq} is the equilibrium contact angle between solid surface line and liquid. Fig. 4.3 shows the schematic of three-phase gas (air), liquid (water), and solid interactions and quantities of the Young–Dupré equation. The contact angle between liquid and solid is not constant and will change as a function of liquid volume. If we were to inject a small amount of liquid into the droplet using a needle, the contact line between liquid and solid can stay constant; however, the contact angle will increase

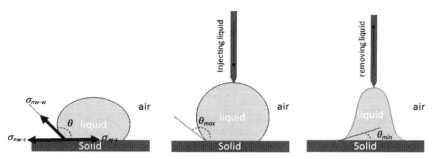

Figure 4.3 Dynamic contact angle measurement.

(maximum angle θ_{max}). On the other hand, removing the liquid from the droplet with constant contact line between the solid and liquid will result in a decrease in the contact angle (minimum angle θ_{min}) as shown in Fig. 4.3. The equilibrium contact angle can be calculated using the Tadmor (2004) equation as follows:

$$\theta_{eq} = arccos\left(\frac{r_{max}\ cos\ \theta_{max} + r_{min}\ cos\ \theta_{min}}{r_{max} + r_{min}}\right)$$

Equation 4.8 Equilibrium contact angle.

where the r_{max} and r_{min} are defined as follows:

$$r_{max} = \left(\frac{sin^3\theta_{max}}{2-3\ cos\ \theta_{max}+cos^3\theta_{max}}\right)^{1/3} \quad and \quad r_{min} = \left(\frac{sin^3\theta_{min}}{2-3\ cos\ \theta_{min}+cos^3\theta_{min}}\right)^{1/3}$$

The capillary pressure is defined as the difference between the pressure in the nonwetting phase and the wetting phase, and can be obtained as follows:

$$P_c = P_{nw} - P_w$$

Considering the presence of water and oil in the porous media, if the contact angle between solid and water phase falls between 0 and 70 degrees, we call the formation water wet. If the contact angle is between 70 and 110 degrees, the formation is called neutrally wet, and if the contact angle is greater than 110 degrees, the formation is called oil wet. Capillary pressure in the formation is mainly a function of formation wettability, saturation of different phases, and pore geometry.

The Young–Laplace equation describes this relationship at equilibrium condition where there is no flow as follows:

$$P_c = \frac{2\sigma \cos \theta}{r}$$

Equation 4.9 Capillary pressure.

where r is the capillary radius. There are different techniques available to measure the capillary pressure including porous diaphragm method, mercury injection method, centrifuge method, and dynamic method.

The mercury injection method is the most common and rapid technique to measure capillary pressure. In this technique, mercury as a nonwetting fluid is forced to the sample, and the pressure required to get excess mercury volume into the core sample is recorded. The mercury saturation is calculated from a known injection and pore volumes of a sample. This technique has a major disadvantage because the sample cannot be used for further analysis after it is exposed to the mercury. Special considerations are also needed to convert the capillary data obtained from the mercury/air system to the reservoir fluid system. The mercury injection technique has been conventionally used to measure the capillary pressure of shale samples. In this case, mercury is injected to the crushed shale samples with increasing pressure of up to 60,000 psi. Three different volumes will be invaded by mercury including closure or conformance volume, i.e., the volume that the mercury needs to fill to overcome the sample surface roughness, pore volume of the sample, and the volume caused by relative change in the sample volume due to compression exposed by mercury. Crushing the shale samples introduces an artificial interparticle volume that will be occupied at low pressure by mercury. This volume is considered as the closure volume and needs to be corrected. Actual intrusion of mercury in shale pore volume occurs after injection pressure exceeds the capillary pressure required for mercury to enter large pores. This will continue until all possible pores of the sample are invaded by mercury. Mercury can enter pores as small as 3 nm, however, there are shale samples with a significant amount of pores of less than 1 nm that cannot be invaded by mercury even at 60,000 psi injection pressure. To accurately measure the capillary pressure, it is crucial to be able to distinguish between the end of closure and start of intrusion, i.e., the pressure at which mercury starts invading the larger pores in the sample. This can be identified by the rapid change in the slope of mercury injection pressure vs. mercury saturation. In extremely tight shale samples with the majority of pores in the order of nanometer, identifying this point is extremely difficult and leads to a significant error in capillary pressure measurements. As mercury is injected to the sample and before it reaches the minimum pressure required to invade the larger pores, the mercury pressure applies external stress on both pore and bulk volumes of the sample. Depending on the difference between pore and grain compressibility and pore-throat volume, this external stress on the sample increases the intraparticle volumes due to sample compression and can impair the actual intrusion pressure. Detailed study on pore compressibility of shale as a function of pressure is required to advance the understanding of capillary measurement using mercury injection in ultratight samples (Bailey, 2009).

As previously discussed, the mercury injection technique uses crushed samples. Therefore, it is not performed at in situ reservoir conditions. To be able to perform the capillary pressure measurement at the reservoir condition, the advanced high-pressure/high-temperature porous diaphragm method is proposed in which a core plug is used in a resistivity core holder. The core sample is then exposed to confining and pore pressure, and reservoir temperature is achieved through the application of a heating jacket. In this technique a low-permeability porous plate saturated with core sample fluid will be set at the down-stream of the core holder and a precise pump is used to inject different fluids to perform imbibition or drainage testing. The average fluid saturation of the core sample is determined volumetrically from the displaced fluid received from downstream and fluid volume injected upstream. The equilibrium condition is tested through resistivity measurements performed in axial and radial directions of the core sample. Equilibrium in electrical resistance is assumed when the variation is less than 0.5% in 1 hour. The experiment will be repeated at different differential pressures between upstream and downstream and at elevated temperatures. Even though the porous diaphragm method can be applied at reservoir condition and provides more accurate results, calibration, and preparation of the test; also, performing the experiment is very time consuming.

Recently, there have been several studies using nonequilibrium molecular dynamics on the flow of hydrocarbons in organic nanocapillaries to understand the physics of capillary pressure and IFT in organic-rich shale reservoirs. The main drive behind these studies was the difficulty associated with direct measurements of these properties in shale samples and huge uncertainty associated with these techniques (Feng and Akkutlu, 2015).

WETTABILITY EFFECTS ON SHALE RECOVERY

As discussed earlier, wettability is defined as the relative adhesion of fluid to the solid surface. Wettability is conventionally measured using three different techniques including contact angle, Amott wettability index, and US Bureau of Mines (USBM) wettability index measurements. In the Amott wettability index test, the sample is imbibed with water to its residual oil saturation first and then immersed in oil for 20 hours. The amount of water displaced by spontaneous drainage of water is then measured as a volume of water (V_{wsp}). Next, water is drained to its residual water saturation and the maximum amount of water recovered is recorded as the total water volume (V_{wt}). The sample is then immersed in brine for 20 hours and oil volume displaced by natural water imbibition is measured as V_{osp}. The remaining oil in the sample is then forced out by injecting brine to the sample to its residual oil saturation to measure the total oil volume V_{ot}. The Amott wettability index can then be obtained using Eq. (4.10).

$$I_w = \frac{V_{osp}}{V_{ot}} - \frac{V_{wsp}}{V_{wt}}$$

Equation 4.10 Amott wettability index.

In this equation, I_w is the Amott wettability index and ranges between -1 and 1, in which -1 stands for oil-wet, 0 stands for neutral-wet, and $+1$ indicates the water-wet formation. The wettability characteristic of the formation highly impacts the hydrocarbon recovery and multiphase flow in porous media and is a function of the solid surface chemistry and microscale roughness of the surface. Due to the wide range of applications of wettability characteristics of the material, there have been several studies on either alteration or restoration of the solid surface wettability. Wettability of the solid surface can be changed by altering the solid surface chemistry or by changing the microscale surface roughness. The chemistry of the solid surface can be altered using different techniques including oxidation of the solid surface, deposition of nonwetted material at the solid surface, or application of the electric fields. However, they are poorly developed techniques to change the microscale roughness of the solid surfaces (Aria and Gharib, 2011). In the oil and gas industry, to change the wettability of the formation, different techniques such as treating the solid surface with a coating agent (e.g., organosilanes), using naphthenic acid or asphaltenes, and adding surfactants to the injected fluids are used. In the case of restoration of the wettability in the core samples, toluene followed by ethanol to extract the toluene is also used. In some cases, drying the core sample or aging in crude oil for 100 hours at 65°C is also recommended. In organic-rich shale reservoirs, these techniques are not practical. This is due to the complex pore structure, extremely low permeability, and heterogeneity in mineral compositions of organic-rich shales. Recently, nondestructive techniques such as nuclear magnetic resonance (NMR) are used to study the wettability characteristics of the shale reservoirs and to monitor sequential imbibition and drainage processes in shale samples. Odusina et al. (2011) used a total of 50 samples from different US shale basins and measured their wettability using the NMR technique. In this approach, they first performed NMR study on the samples as received. Next, they immersed the samples in brine at room-temperature conditions and ran NMR after 48 hours of spontaneous imbibition. The samples were then immersed for 48 hours in dodecane and NMR analysis was done on the drained samples. They found that shale samples in general show mixed wettability, with organic material contributing to oil-wetting characteristics. To have a better understanding of the wettability characteristics in organic-rich shale reservoirs, a combination of different direct and nondestructive approaches are required.

CHAPTER FIVE

Hydraulic Fracturing Fluid Systems

INTRODUCTION

Hydraulic fracturing or fracing has become one of the most important parts of completing a well. Hydraulic fracturing is essentially the act of pumping sand, water, and specific chemicals at a very high rate and pressure to break the rock and release the hydrocarbon. Hydraulic fracturing stimulation is used to increase the permeability and reduce the skin damage caused by drilling. Unconventional shale reservoirs are known for having an extremely low permeability. In an attempt to increase the production volumes of unconventional shale reservoirs, hydraulic fracturing is performed on every well. Without hydraulic fracturing, reservoirs with low permeability will never produce at an economically feasible rate.

The first use of hydraulic fracturing was in 1947; however modern hydraulic fracturing referred to as "slick water multistage horizontal stimulation" or "slick water frac" was first performed in the Barnett Shale, located in Texas, in 1998 using more water and higher pump rate than previously attempted techniques. The introduction of slick water horizontal frac made the production of low–permeability reservoirs promising. This is when the industry started looking at various shale plays across the United States and the world. The industry is moving from conventional resources with high permeability, which are hard to find, but easy to produce, to resources such as shale that are much easier to find, but more difficult to produce. Conventional resources are hard to find, but once the appropriate reservoirs are found no hydraulic fracturing is typically necessary to increase the permeability. The permeability of conventional resources is usually high enough that the hydrocarbon trapped in the reservoir will automatically flow into the wellbore right after perforation. In contrast, unconventional resources would not be economically

© 2017 Elsevier Inc.
All rights reserved.

feasible to produce without hydraulic fracturing. There are many different applications for hydraulic fracturing and they are as follows:

1. Increase the flow rate from low-permeability reservoirs such as shale formation in general.
2. Increase the surface area or the amount of formation contact with the wellbore.
3. Reduce the number of infill wells with horizontal hydraulic fracturing stimulation.
4. Connect hydraulic fractures with existing natural fractures.
5. Increase the flow rates from wells that have been damaged (near wellbore skin damage) because of drilling.
6. Decrease the pressure drop around the well, which will cause reduction in sand production.

The first application listed above (the most important application) is the essence of hydraulic fracturing since the main reason behind hydraulic fracturing is to increase the permeability of the reservoir. Not only does the flow rate increase in a naturally fractured formation with low permeability, but also hydraulic fracturing will connect the natural fractures and faults (if present) in the formation. When the formation is hydraulically fractured, the amount of the formation in contact with the wellbore will increase, and as a result the flow rate will also increase. Porous and permeable reservoir rocks filled with lots of hydrocarbon are any company's dream to obtain. However, poorly cemented sandstone formations with high permeability can cause lots of issues when it comes to production. In this type of formation, sand grains flow into the wellbore with the produced hydrocarbon and cause various issues. These issues can lead to severe pipe erosion/damage, flow line blockage, and finally reduction in production. Various completion techniques such as gravel packing, frac packing, and expandable sand screens can be used to fight this problem. Gravel packing is essentially the placement of a steel screen and packing the surrounding annulus with specific and designed size gravel. The designed size gravel prevents the passage of formation sand into the wellbore. Hydraulic fracturing can be used in conjunction with the conventional gravel-packing technique in a process called *frac packing*. In the frac-packing process, hydraulic fracturing occurs after the placement of a gravel pack to create a good conduit for the flow of hydrocarbon at some distance from the wellbore. Therefore, hydraulic fracturing can also have a positive impact in conventional high-permeability sandstone formations with sand production issues. There are various types of hydraulic fracture fluid systems in the industry and every formation requires a specific system. The most commonly used frac fluid systems are described in the following sections.

SLICK WATER FLUID SYSTEM

This type of fluid system is well known in the industry and is currently being used in the Marcellus Shale, Barnett Shale, Utica/Point Pleasant, and many other low-permeability reservoirs. In this technique, water, sand, and specific chemicals are pumped downhole to create a complex fracture system within the reservoir. The main goal in low-permeability reservoirs using water frac is to create a complex fracture system and maximum surface area. When there is not enough surface area created in low-permeability rocks, the well productivity is not maximized. This is why this system uses a huge amount of water to create the maximum possible surface area. Additionally, *rate* is the drive needed to create the complex fracture system within the formation. Using more rate yields better surface area and as a result better production. Some operators limit their rate to prevent fracture height growth and paying less for hydraulic horsepower if and only if partial cost of the job or the contract depends on the horsepower. More rates require more pumps and sometimes the size of the pad (well site) and many other factors do not allow the operator to have as many pumps as necessary for the job.

Another limiting factor in achieving the necessary rate is *pressure*. There are various limitations on pressure such as surface equipment and casing burst pressure ratings. For example, the maximum allowable surface-treating pressure when fracturing in Marcellus Shale is usually 9500 psi based on 5½ inches, 20 lb/ft, P-110 production casing. This pressure is determined from the casing, surface, and wellhead pressure ratings used for the job. For example, if during a frac stage treatment the surface-treating pressure is about 9500 psi at 60 bpm (BBLs of frac fluid per minute) the rate will be limited to 60 bpm. Rate can only be increased if pressure decreases below the maximum allowable treating pressure during the frac stage. Rate is basically the most important parameter in water frac; however, sometimes the size of the pad, cost, and pressure limitations can prevent achieving the designed rate during frac jobs.

nt lesson that is crucial to emphasize is that a higher
ore surface area. George Mitchel is the pioneer of the
.d he spent several years designing the best practice to
_ally produce from Barnett Shale. The introduction of high-rate
slick water frac was the key to his success. Many shale plays across the
United States are full of natural fractures, which are one of the main
sources of transferring fluid into the wellbore. As more natural fractures
and surface areas are contacted by hydraulic fracturing stimulation, better
productivity will be attained. Low-viscosity fluid, such as water frac,
tends to follow natural fractures, contact more surface area, and create
a complex fracture system within the reservoir. The reason a water frac is
called "slick water frac" is because of a chemical additive called friction
reducer (FR). Without FR, a slick water frac cannot be pumped at a
high rate. The addition of the FR to water reduces the friction and makes
the water very slick.

The best type of frac fluid is not necessarily freshwater. Formation
water (flow-back water) is actually believed to be a better frac fluid in
some areas since it contains the Earth's minerals. Using freshwater for fracing could cause a filter cake along the created fractures, which causes a
reduction in permeability and conductivity. The majority of companies
use a mixture of treated or untreated flow-back and freshwater to obtain
the volume of water needed for the job. Some companies have even tested
100% produced water for frac with significant advancements in proper FR
selection that can handle high total dissolved solids (TDS), irons, etc. Each
slick water hydraulic fracturing stage typically uses about 4000—11,000
BBLs of water (168,000—462,000 gallons of water) depending on the size
of the job (sand volume), treatment complexity, production results, etc.
For example, if the designed sand volume in a stage is 200,000 lb, it will
take less volume of water to pump the stage compared to the designed
sand volume of 500,000 lb. Higher sand volume basically requires more
water to be placed into the formation. Some stages could take more water
because the stage is very difficult to treat when higher sand concentrations
are run. For example, some stages do not like higher sand concentration,
such as 3 ppg (3 lb of sand per gallon of water); therefore, the stage
is treated at a lower sand concentration to put all the designed sand
into the formation. In a water frac, if the stage is hard to treat, it is
more important to place the designed sand into the formation at a lower
concentration (to achieve more surface area) compared to running higher
sand concentration and cutting the stage short of design.

Another important candidate for a water frac is formations with high brittleness. When a rock is brittle, it helps to keep the fractures open after breaking down. For example, glass is a brittle material and when a glass is broken, it is scattered. The main application for a water frac is in formations with high Young's modulus and low Poisson's ratio. High Young's modulus and low Poisson's ratio is basically an indication that the rock is brittle and slick water frac can be used to break the rock. In a water frac, a maximum sand concentration of 3—3.5 ppg (lb per gallon) can be pumped in the best case scenario. Given a healthy, high rate and ease of the formation, 4 ppg sand concentrations can also be achieved with slick water (very rare). Pumping higher sand concentrations (more than 4 ppg) is not possible with a slick water frac and can lead to sanding off the wellbore (screening out). Achieving higher sand concentration can be performed using other frac fluid systems such as cross-linked gel, which will be discussed.

As previously mentioned, the main objective of a water frac treatment is to create a complex, but not a dominant fracture network. In general, in a slick water frac, low-viscosity fluid leads to a complex fracture network, while converting to higher-viscosity frac fluid (e.g., linear gel, cross-linked gel) tends to create dominant hydraulic fractures. The essence of a water frac in naturally fractured reservoirs is to follow natural fractures, while creating multiple flow paths as a result of applying high energy to the rock. This energy is only achieved with a higher rate. The combination of a higher rate and low-viscosity fluid slick water will cause the sand to be placed farther into the formation and result in better long-term productivity. One of the major issues with our industry is being so dependent on short-term production. Some companies only look at the initial production of the well and ignore the long-term production. This is another recipe for failure when designing and comparing the performance of a well. It is also crucial not to make decisions based on a single well's production data. Instead, a field of production data can be used to make critical economic decisions.

Average treating rate in slick water frac varies from stage to stage based on the pressure limitations discussed. The goal is to achieve the maximum designed rate, which is typically 70—100 bpm. In Barnett Shale, the average surface-treating rate is even higher and pump rates as high as 130 bpm have been achieved. Rate also overcomes leak-off and fracture-width problems during the frac stage. Leak-off refers to the fracturing fluid getting lost in the formation. Having a high rate during the frac stage eliminates the concern of having high leak-off. Having high leak-off could lead to sanding off the well. One of the main reasons rate should not be sacrificed in low-permeability unconventional reservoirs is because of not creating the surface area needed for long-term production. When hydraulic fracturing is performed on a low-permeability reservoir with a limited rate, only limited surface area is achieved. Once the reservoir is drained in that particular surface area, the production will be decreased significantly. In high-permeability conventional reservoirs, surface area is not the only deciding factor, because even after creating so much surface area the permeability beyond the stimulated reservoir volume region to transmit fluid into the created hydraulic fractures is still high.

Over time, naturally fractured reservoirs become the best refrac candidates to enhance oil and gas recovery. For example, if after 20 years of production, the production rate declines below the economic limit, it is highly recommended to restimulate the well due to natural fractures that exist in the reservoir to enhance the recovery. In addition to having a natural fracture system, reservoirs should have high initial gas in place (IGIP), high pore pressure, and superior reservoir properties to be the most successful candidates for refrac. It is also very important to select candidates based on their original completion design. Having all of the conditions discussed above, poor initial completion design wells are better candidates for refrac. Fig. 5.1 shows the schematic of typical hydraulic fracture and complex fracture network interactions in slick water frac.

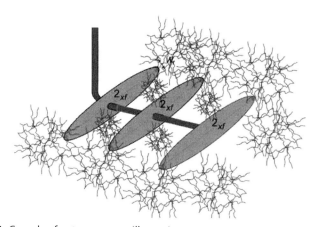

Figure 5.1 Complex fracture system illustration.

CROSS-LINKED GEL FLUID SYSTEM

This type of fluid system is used in conventional and unconventional reservoirs to achieve the so-called biwing fracture system. Cross-linked gel is a heavy viscous fluid. In this type of frac fluid system, viscosity (not velocity) is used to place proppant into the formation. Cross-linked gel is typically used in ductile formations with higher permeability (e.g., oil windows of Eagle Ford and Bakken Shales). Cross-linked gel is also heavily used in oil windows of various shale plays to be able to obtain the necessary fracture width for optimum oil production. The goal in this type of frac is to achieve the maximum sand concentration near the wellbore (higher conductivity) through the use of a viscous fluid. As opposed to a water frac that uses velocity to carry proppant, cross-linked gel uses heavy viscous fluid (cross-linked gel) to place the proppant into the formation. High rate is not required in this type of frac fluid system and usually 25−70 bpm is used to place the proppant into the formation. Higher sand concentration up to 10 ppg can be obtained if and only if a great cross-linked gel is obtained. If for any reason throughout the stage cross-linked gel is cut (not pumped) due to an equipment malfunction, the first thing to do is to cut sand and flush the well to prevent sanding off the well. The reason being is that viscosity carries the high sand concentrations into the formation and without the heavy viscous fluid the well could be sanded off easily at such high sand concentrations. One of the most common mistakes with using cross-linked fluid is pumping the job at lower sand concentrations (<6 ppg). The advantage of using a cross-linked fluid system is utilization of heavy, viscous fluid to pump very high sand concentrations, which will create a dominant fracture with a large proppant pack near the wellbore.

Another advantage of cross-linked fluid system is fluid leak-off reduction. In high-permeability reservoirs where fluid leak-off is significant, the cross-linked fluid system is known to reduce fluid leak-off and keep the proppant suspended until closure. In addition, very high viscosities (thousands of centipoise) can be reached using the cross-linked fluid system.

Another main criterion for choosing cross-linked gel is ductility. Formations with very low Young's modulus and very high Poisson's ratio that have higher permeability are the best candidates for cross-linked frac. It is extremely important to use production data to come up with the best frac fluid system in any formation. Sometimes after taking all

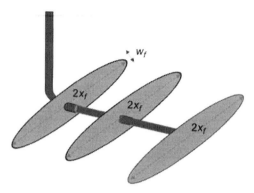

Figure 5.2 Biwing fracture system illustration.

the necessary parameters into account, one type of frac does not yield the best production results. For example, if after hydraulically fracturing a formation using cross-linked gel, the well produces below expectation, a different technique ought to be utilized to maximize production. Theories are good to know and understand, however, the main deciding factor in choosing the frac fluid system is production data. In both Marcellus and Barnett Shale, the reason a water frac is chosen as the main frac fluid system by the majority of operating companies is because of successful production results. If production using slick water frac was not promising, many companies would have tried a different frac fluid system.

As previously mentioned, in a cross-linked frac, viscosity (as opposed to rate or velocity) is used to place high proppant concentrations into the formation. One of the biggest concerns with cross-linked fluid is the gel residue that this type of fluid system leaves in the formation. Cross-linked residue, if not broken properly at reservoir conditions, can cause serious damage to the created fractures by reducing the permeability and fracture conductivity (to be discussed). Fig. 5.2 illustrates the schematic of a biwing fracture system using cross-linked gel.

HYBRID FLUID SYSTEM

In this type of fluid system, slick water is used to pump at a lower sand concentration followed by cross-linked or linear gel to pump at a higher sand concentration to maximize near wellbore conductivity.

Some companies use this type of frac in unconventional reservoirs in the event there are severe issues with placing higher sand concentrations into the formation. This is because some formations do not like higher sand concentrations and the only way to put all the designed sand away is either using linear gel (less viscous compare to cross-linked gel) or cross-linked gel at higher sand concentrations. For example, in a Marcellus Shale operation, if all the designed sand is not able to be placed into the formation by using slick water frac fluid, linear gel is used to increase the viscosity of the fluid and as a result increase the fracture width and over-come near wellbore tortuosity. Some stages in a Marcellus Shale formation need to have some type of viscous fluid such as linear gel for all the designed sand to be successfully placed into the formation. Typically if there are issues with establishing a good rate during higher sand concentrations, 5−10 lb linear gel is used at the start to provide more fracture width and better proppant transport. In some stages, if a baseline flow rate (bpm) is not established to start even lower sand concentrations, 15−20 lb linear gel could be used to carry the proppant into the formation without screening out at a lower rate. Once the necessary baseline flow rate is established, linear gel can be reduced or completely eliminated throughout the stage. For example, if only 25 bpm is reached at the maximum allowable surface pressure, a 15−20 lb gel system can be used and 0.1−0.25 ppg (lb of sand per gallon of water) is started. Once a base rate is established, gel concentration can be lowered or cut throughout the stage. Gel concentrations are typically provided in lb of polymer per 1000 gallons of base fluid (water). For example, a 20 lb ABC gel system is prepared with 20 lbs of ABC per 1000 gallons of base fluid. Gel typically comes in two forms: dry and wet liquid gel concentrate (LGC). The LGC is made by mixing high concentrations of dry gel in a solvent to make an LGC. The concentration of dry gel used varies but it is usually 4 lb of gel per gallon of solvent (water). The main equation for linear gel usage is shown in Eq. (5.1).

$$Linear\ gel\ (gpt) = \frac{Linear\ gel\ system\ in\ \text{lb}}{4}$$

Equation 5.1 Linear gel conversion to gpt.

For example, 5 lb linear gel system is equal to 5 divided by 4, which yields 1.25 gpt (gallons of gel per thousand gallons of water). In other words, LGC is made using 4 lbs of gel in a gallon of solvent. To make

Figure 5.3 20 lb linear gel system.

a 5 lb/1000 gallon gel system, 1.25 gpt of LGC will be needed. The 5 lb gel system is not a superviscous fluid and this type of gel system has just enough viscosity to overcome the friction pressure and provide more fracture width to be able to place the sand into the formation. When tortuosity is severe, higher gel concentrations such as 15–20 lb gel systems will be needed. Fig. 5.3 shows a sample of a 20 lb linear gel system that was used during a Marcellus stage treatment.

FOAM FRACTURING

Foam fracturing is not a common frac fluid system in the majority of unconventional shale plays, but this type of fracturing fluid system provides some attributes that others do not provide. Foams are made up of two parts. The first part is gas bubbles, referred to as the internal phase, and the second part is liquids, referred to as the external phase. Nitrogen foam frac is the most commonly used form of foam-fracturing fluid system. In this type of fluid system, nitrogen is typically pumped with water and other additives to form a foam-like fluid. The nitrogen foam-fracturing fluid system is common in coalbed methane, tight sands, and some low-permeability shale reservoirs that are normally less than 5000 feet deep.

Foam-fracturing fluid, just like other types of fracturing fluid systems, has advantages and disadvantages. Since a nitrogen foam frac has less fluid in the system and a big percentage of the fluid system is composed of nitrogen, it is ideal for water-sensitive formations (e.g., clay-containing formations). Due to the fact that less liquid is pumped in a nitrogen foam frac, clay swelling and formation damage are minimized in water-sensitive formations. A nitrogen foam frac is ideal for low-pressured and depleted formations in which the energy of nitrogen is used to help the well cleanup and flow back after the frac job is completed. As foam frac fluid is mostly composed of greater than 60% gas, recovery of fracturing fluid in low-pressured reservoirs is more efficient as compared to nonfoam-fracturing fluid systems. The compressible nature of the foam frac fluid will help recover the liquid due to gas expansion as the fracturing fluid travels to the wellbore. Due to the fast well cleanup after frac, cleanup time will be minimized in a nitrogen foam frac. Without this type of fluid system in depleted and low-pressured formations, the reservoir does not have the energy to recover the frac fluid pumped downhole. Since the foam-fracturing fluid system only contains 5−35% liquid, low liquid percentage foam will have less hydrostatic pressure acting on the formation.

Another advantage of foam-fracturing fluid system is the fluid-loss capability. As previously discussed, less fluid is pumped downhole in this type of fluid system and as a result, a foam frac provides better fluid efficiency, which in turn yields low fluid loss. This fluid-loss capability can be demonstrated by putting some shaving cream on your hand and flipping your hand upside down. The shaving cream does not readily fall off of the hand. This is indicative of the fluid-loss capability of the foam-fracturing fluid system. When fluid-loss additives are not required, any detrimental damage to fracture permeability and conductivity can be reduced. Note that sometimes fluid-loss capability might be required in highly naturally fractured formations with higher permeability. When nitrogen is injected into a liquid such as water, some foaming will occur. However, due to water being thin, some bubbles will rupture. Adding a foaming agent such as soap will cause the bubbles to become more stable. Soap is known to be a type of surfactant that will stabilize the foam when injecting nitrogen. The general rule of thumb is that in formations with permeability greater than 1 md, fluid-loss additives could be beneficial.

Another important advantage of foam-fracturing fluid is proppant transport. As opposed to a slick water fluid system, foam allows proppant transport into the formation without settling out. This will allow for uniform distribution of proppant particles throughout the fractures. The amount of proppant that foaming frac fluid can suspend will depend on the foam quality (to be discussed). When regular sand (SG = 2.65) is used

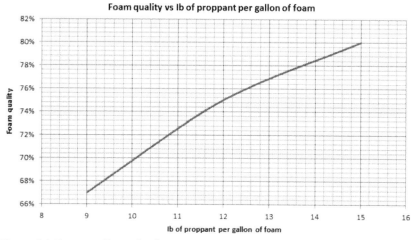

Figure 5.4 Foam quality vs lb of proppant per gallon of foam.

with a foaming agent and 3 lb of sand per gallon of foam is desired, the sand concentration at the blender will be 9 ppg for 67% quality foam, 12 ppg for 75% quality foam, and 15 ppg for 80% quality foam. Fig. 5.4 illustrates this concept for regular sand with specific gravity of 2.65.

FOAM QUALITY

Foam quality is the ratio of gas volume to foam volume (gas + liquid) over a given pressure and temperature. Nitrogen or CO_2 can be used to create foam in liquid status, but nitrogen is typically preferred because CO_2 can be extremely harsh and eroding when water is existent.

$$FQ = \frac{Gas\ volume}{Gas\ volume + liquid\ volume}$$

Equation 5.2 Foam quality.

where FQ = foam quality, %; gas volume = BBLs or gallons; and liquid volume = BBLs or gallons.

When foam quality is between 0% and 52%, gas bubbles do not contact each other and are spherical. Foam viscosity is also low because there is a lot of free fluid in the system, which in turn will affect the fluid-loss capability. When foam quality is between 52% and 96%, the gas bubbles are in contact with each other and as a result an increase in viscosity will occur. Foam qualities of 52% and 60% do not have the proppant-suspension capability. Finally, when foam quality is more than 96%, the foam will degenerate into mist and as a result there will be a loss in viscosity. Note that higher foam quality has a higher viscosity and is better able to suspend proppant. As foam quality increases, more hydraulic horsepower will be needed. This is because an increase in foam quality will decrease the hydrostatic pressure and in turn will increase the surface-treating pressure. An increase in surface-treating pressure will cause an increase in hydraulic horsepower. The most frequently used foam quality is typically 70—75%.

FOAM STABILITY

There are several factors that affect foam stability. Foam quality, surfactant type/concentration, and polymer type/concentration are some examples. One of the most important aspects of foam fracturing is to keep the foam in motion. If foam is not in motion, it will be unstable. When foam stops moving, gravity will cause the free fluid in the foam to drain. This drainage can cause foam instability issues. The rate of foam drainage will depend on many factors such as temperature, viscosity of the liquid phase, and foaming-agent concentration. An increase in temperature can potentially cause a reduction in the viscosity of the fluid. As temperature increases, more foaming agents must be used. Gelling agents are also very important because they can be used to add stability to the fluid. Gelling agents will increase the viscosity (not considerably), but will improve proppant transport and fluid-loss control. Gelling agents must be used in moderation because higher fluid viscosity will be harder to foam and pump, and as a result, will require more hydraulic horse power.

TORTUOSITY

Tortuosity refers to the pressure loss by fracturing fluid between the perforations and main fracture(s). It is basically the restricted pathways between the perforations and main fractures. Tortuosity can be justified as one of the main causes for majority of screen-outs. Tortuosity was not an issue in vertical wells; however, tortuosity seems to be very severe in horizontal wells, wells with moderate-to-severe inclinations, hard rock reservoirs, and wells with dispersed perforations. Pumping conditions and rock properties have direct impact on tortuosity. Tortuosity can be severe in some stages and this is why viscous fluid such as linear gel is run to fight the problem and be able to successfully put all the designed sand in the formation. In general, not being able to obtain sufficient rate during a frac stage could potentially be due to severe tortuosity issues. This problem can be easily solved by pumping higher viscous fluid such as linear gel, which will cause the surface-treating pressure to drop as soon as the linear gel hits the perforations. A drop in the surface-treating pressure is an indication of overcoming the tortuosity issues between the perforations and main fractures. There are various ways to figure out if tortuosity is the problem. The first and the most commonly used method for identifying the tortuosity is pumping a sand slug at low concentrations after the pad. If the sand slug hits the formation and pressure rises, it is an indication of tortuosity. If the sand slug causes an increase in pressure followed by a considerable break in pressure, it means the removal of tortuosity. Finally, if there is no impact when sand hits the perfs, then there are no issues with tortuosity.

When sufficient rate is not established during the pad stage in a slick water fluid system, a low-concentration sand slug (typically $0.1-0.25$ ppg) is run to see the impact on pressure and figure out whether severe tortuosity exists in the formation or not. Another option is to run the sand slug for a second time to attack the tortuosity if pressure permits. Another process for determining if tortuosity exists or not during a frac stage is by subtracting the closure pressure from the instantaneous shut-in pressure (ISIP). If the difference between ISIP and closure pressure is more than 400 psi, there is a high possibility of tortuosity. Some of the most commonly used techniques (as discussed) to combat severe tortuosity are as follows:

1. Pump low-concentration proppant slugs
2. Use high gel loading (>15 lb system)
3. Increase rate (if possible).

It is challenging to determine whether tortuosity or high perforation friction is the cause of not being able to pump into a zone. When dealing with high perforation friction pressure, the following techniques can be used to overcome the issue:

1. Pump low-concentration proppant slugs
2. Spot acid (for the second time)
3. Reperforate.

Sometimes spotting acid for the second time might help resolve the issue in the event all the perforations were not cleaned of cement and debris the first time. When the first five options listed above fail to deal with tortuosity and high perforation friction pressure, reperforating is used to overcome the problem and get into the stage. Some companies do not even try any of the techniques listed above and simply reperforate since the cost of reperforating could possibly be cheaper than trying various techniques. Fig. 5.5 shows the schematic of possible tortuosity between perforations and hydraulic fracture. Fig. 5.6 illustrates the basics of frac fluid design selection based on rock mechanical properties from brittle rocks with high Young's modulus and low Poisson's ratio to ductile rocks. As Fig. 5.6 illustrates, moving from brittle rocks to ductile rocks will need a change in the frac fluid system (slick water to cross-linked gel). This will lead to increase in viscosity of the fluid, better proppant deliverability, lower fracture complexity, and lower flow rate.

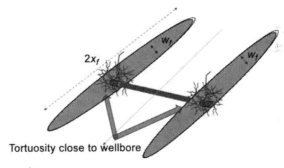

Figure 5.5 Tortuosity.

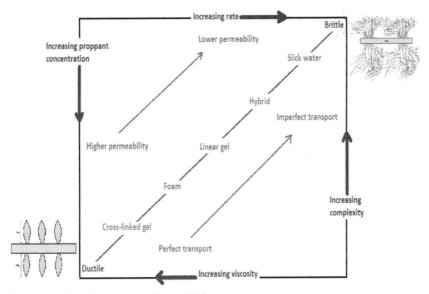

Figure 5.6 Frac design basis (Britt, 2011).

TYPICAL SLICK WATER FRAC STEPS

As previously discussed, slick water is the most commonly used frac fluid system in shale plays. There are four main steps during each slick water hydraulic frac stage that are followed in sequence; they are described in the following sections.

ACIDIZATION STAGE

In this stage, various concentrations of HCl (hydrochloric) or HF (hydrofluoric) acid is pumped downhole to clean the perforations (holes) of any type of debris or cement. The purpose of acid is to clean the perforations of any cement or debris and is not meant to acidize the formation. Acid does help when the formation has limestone streaks or calcite. Different companies have various theories regarding the volume and concentration of acid. Depending on the operating region and formation type, typically 500–4000 gallons of 3–15% acid is pumped downhole to clean the perforations. The acid stage can easily be seen on the surface-treating pressure chart because as soon as the acid hits the perforations (holes in the casing), the surface-treating pressure decreases and a higher rate can be established. Table 5.1 shows specific gravities of HCl acid at various HCl acid concentrations.

Contact time is very important when pumping acid. In general, contact time can be reached by pumping lower acid concentrations. For example, instead of pumping 3000 gallons of 15%, 3000 gallons of 3–7% can be used

Table 5.1 Specific Gravity (SG) of HCl Acid

%	SG
3	1.015
5	1.025
10	1.048
15	1.075
20	1.100
25	1.126
28	1.141
30	1.151
31.45	1.160
36	1.179

in an attempt to have longer contact time with the perforations. This could possibly enhance the cleaning process compared to pumping higher acid concentrations. Typically service companies have acid-blending plants and bring 31.45% acid into their acid-blending plant. It is then mixed as needed in the yard and sampled to confirm the proper mix. Twenty-eight percent acid is typically the maximum concentration that can be hauled to a location in normal oilfield trucks. Since the acid pumped downhole is usually 5−15%, the desired percentage is mixed on the fly. It is a lot easier operationally to haul the acid out hot and mix it on the fly. One of the advantages of mixing the acid on the fly is that more stages can be pumped out of one acid tube. This will be helpful because fewer complications will occur while moving acid in and out after every stage. The acid concentration achieved from mixing on the fly will be close enough to the desired designed concentration. It is the company representative's responsibility to calculate the dilution rates to ensure the proper rates for getting a desired acid percentage are used. Eq. (5.3) can be used to acquire gallons of acid required to convert to the desired acid concentration:

$$Original\ acid\ volume$$

$$= \frac{\%\ of\ acid\ desired \times SG\ of\ desired\ acid}{\%\ of\ original\ acid \times SG\ of\ original\ acid} \times gallons\ of\ desired\ acid$$

Equation 5.3 Original acid volume.

●●●

Example

How much acid and water are needed given 28% hydrochloric acid (hauled to location) in order to obtain 3000 gallons of 5% acid?

$$HCl\ acid\ volume = \frac{5 \times 1.025}{28 \times 1.141} \times 3000 = 481.3\ gallons\ of\ 28\%\ HCl\ acid$$

$$Water\ volume = 3000 - 481.3 = 2518.7\ gallons\ of\ water$$

As can be seen in this example, to make 3000 gallons of 5% hydrochloric acid (28% original trucked to location), only 481 gallons would be acid and the rest would be water. Now, let's do one more calculation: how much acid is required to make 3000 gallons of 15% instead of 5%?

$$HCl\ acid\ volume = \frac{15 \times 1.075}{28 \times 1.141} \times 3000 = 1514.2\ gallons\ of\ 28\%\ HCl\ acid$$

$$Water\ volume = 3000 - 1514.2 = 1485.8\ gallons\ of\ water$$

Therefore, more 28% acid is needed to achieve higher acid concentration.

PAD STAGE

After pumping the designed volume and concentration of acid, pad (which is a combination of only water and chemicals) is pumped downhole to initiate the hydraulic fractures by creating fracture length, height, and width before starting the main proppant stage. In other words, pad is the volume of fluid pumped downhole to create a sufficient fracture network before pumping the proppant stage. It is very crucial to obtain as much rate as possible during the pad stage for a bigger fracture network. Pad volume is extremely important to determine in order to prevent premature sand-off (tip screen-out). Engineers strongly believe that if a sufficient fracture network is not created during this stage, a premature screen-out can be the consequence. The hydraulic fracture network is created throughout the entire treatment; however, the majority of the fracture network is created during pad injection. If not enough pad is pumped, at some point during the treatment the sand will reach the tip of the created fractures causing them to bridge with sand and eventually pack off all the fractures. This will result in sanding off the wellbore if the stage is not ended early by cutting sand. On the other hand, too much pad can be harmful as well. If too much pad is pumped, fracture tips continue to propagate after pumping is stopped resulting in a large unpropped (unpropped means no proppant) region near the tip of the fracture. Propped fracture regions can move toward the unpropped region and essentially leave a poor final proppant distribution inside the main body of the fractures. This underlines the importance of calculating and understanding the pad volume before the main treatment. Pad volume is calculated using Eqs. (5.4), (5.5), and (5.6), which are functions of fluid efficiency in the formation.

$$Nolte\ method = Pad\ volume\ \% = (1 - FE)^2 + 5\%$$

Equation 5.4 Nolte method pad volume.

$$Shell\ method = Pad\ volume\ \% = \frac{1 - FE}{1 + FE}$$

Equation 5.5 Shell method pad volume.

$$Kane\ method = Pad\ volume\ \% = (1 - FE)^2$$

Equation 5.6 Kane method pad volume.

Fluid efficiency is the ratio of stored volume within the fracture to the total fluid injected. Fluid efficiency is inversely related to fluid leak-off. Higher fluid efficiency means lower fluid leak-off and lower fluid efficiency means higher fluid leak-off. Leak-off is the amount of fracturing fluid lost to the formation during or after treatment. Unconventional shale reservoirs in general have lower leak-off because the permeability is very low. Low fluid leak-off in unconventional shale reservoirs indicates that the fluid does not get lost in the formation as much as it would in high-permeability formations. The fluid pumped in low-permeability formations will effectively create fractures because of low fluid leak-off. Generally shale has a high fluid efficiency (low leak-off), therefore, it requires less pad to be pumped. Fluid efficiency can be calculated using the diagnostic fracture injection test, which will be discussed.

Sanding off the wellbore or screening out can be a costly issue when hydraulically fracturing high-permeability formations because the fluid gets lost very quickly to the formation and the amount of pad that was originally pumped is lost in the formation as well. This is the main reason that high-permeability reservoirs, which have higher fluid leak-off, need more pad volume to effectively place all the designed sand into the formation.

Example

Calculate the pad volume % needed for a frac stage with 70% fluid efficiency if 7000 BBLs of frac fluid is designed for the stage:

$$Nolte\ method = pad\ volume\ \% = ((1-0.7)^2 + 0.05) \times 100 = 14\%\ pad\ volume$$

$$Pad\ volume = 7000 \times 14\% = 980\ barrels$$

$$Kane\ method = pad\ volume\ \% = (1-0.7)^2 \times 100 = 9\%\ pad\ volume$$

$$Min\ pad\ volume = 7000 \times 9\% = 630\ barrels$$

$$Shell\ method = pad\ volume\ \% = \frac{1-0.7}{1+0.7} \times 100 = 17.6\%\ pad\ volume$$

$$Max\ pad\ volume = 7000 \times 17.6\% = 1232\ barrels$$

In the example above, a minimum of 9% pad volume will be needed for the job. Another important parameter to keep in mind during the frac job treatment is the pressure chart. If halfway or closer to the end of the frac job, surface-treating pressure starts rising, it could be an indication of high fluid leak-off and losing the pad. In this particular scenario, a minipad or extended sweep can be pumped in the middle of the stage to clear the near wellbore of sand accumulation, create more room for getting back into the stage by pumping more sand, and place the existing sand farther into the formation.

A sweep is essentially when sand is cut and only water and chemicals are pumped downhole as the sand starts packing off. Usually after a hole casing volume sweep, surface-treating pressure begins to decline, which can be an indication of sand accumulation being swept away from the near wellbore area. On the other hand, an *extended sweep*, also referred to as a *minipad*, is when more than one hole casing volume is pumped until the surface-treating pressure shows some relief by having a downward pressure trend. Sweeps can be very common in some areas especially when pumping high volumes of sand. Some sweeps are scheduled in the design, while others are only run as needed. Cutting sand on time and running sweeps as needed is strongly recommended to be able to get back to the stage and put all the designed sand away. If the formation gives out (this can be easily seen on the surface-treating pressure chart) and sweeps are not run, screening out can be the consequence. Experienced frac engineers and consultants are not afraid to cut sand and run a sweep as needed throughout any frac stage.

PROPPANT STAGE

After pumping the calculated pad volume the proppant stage can be started. The proppant stage is the stage during which combinations of proppant, water, and chemicals (called slurry) are pumped downhole. In a slick water frac, it is very important to establish enough flow rate before starting sand. As previously discussed, when using slick water fluid system, rate is the primary mechanism for placing the sand into the formation. If enough rate is not achieved (at least 35 bpm), the proppant stage should not be started because it might result in sanding off the wellbore. Sometimes small concentrations of proppant slugs (such as 500–1000 lb of 0.1–0.25 ppg) are pumped downhole to make sure the formation is able to take in the introduced sand slug before starting the actual pump schedule. The sand stage typically starts with 0.1−0.25 ppg and is gradually increased to higher sand concentrations in a slick water frac. It is important to make sure the current sand concentration hits the perforations before staging up to the next sand concentration to make sure the formation tolerates the amount of sand concentration. For example, if 1.5 ppg of sand stage is being pumped downhole it is crucial to let it reach the perforations before staging up to 1.75 or 2.00 ppg of sand stage.

The entire casing volume capacity (of slurry fluid) must be used for the sand to hit the perforations. This maximum casing volume is calculated to discern when the sand will hit the perforations. In the field operation there is a common saying: "did sand hit the bottom?" This question is asking whether the sand has reached the perforations. In a slick water frac, typically 0.25 ppg jumps are taken to increase sand concentration. However, with more aggressive schedules, 0.5 ppg jumps are attempted as well. It is very crucial to start up sand at very low concentration, such as 0.1 or 0.25 ppg, to erode the perforations in a slick water frac. Starting with higher sand concentrations such as 1 ppg can cause the packing off of all of the perforations and as a result screening out in a slick water frac. A frac stage is very similar to an individual starting to run. Typically, the individual stretches before running and starts at very low speed and gradually gets up to speed. Frac stage follows the same pattern in that it starts with low sand concentrations and gradually stages up throughout the stage.

FLUSH STAGE

After pumping the designed pump schedule, proppant is cut and the well is flushed. Flushing means water and chemicals are only pumped downhole to clear the inside of the production casing of sand until all of the remaining proppant in the casing has been removed/flushed to the formation. Flush volume can be calculated given the casing size, grade, weight, and bottom perforation. The rule of thumb is to pump at least one hole casing volume of water and chemicals to the bottom perforation depth *after* all the surface lines are cleared of sand. There is a densometer (reads sand concentration) at the end of the surface lines and before the entrance to the wellhead. This densometer indicates 0 ppg when all the surface lines are clear of sand. As soon as the densometer shows 0 ppg, the flush count starts. Flush stage is very important to pay attention to because after cutting sand, the hydrostatic pressure increases (due to losing slurry hydrostatic pressure) and pressure needs to be monitored to make sure the maximum allowable pressure is not exceeded. The flush volume is calculated using Eq. (5.7).

$$Flush\ volume = Casing\ capacity \times bottom\ perforation\ measured\ depth$$
Equation 5.7 Flush volume

where casing capacity = BBL/ft and bottom perforation MD = ft.
Casing capacity can also be calculated using Eq. (5.8).

$$Flush\ volume = Casing\ capacity \times bottom\ perforation\ measured\ depth$$
Equation 5.8 Casing capacity

where ID = inside diameter of production casing, feet.

Example
Calculate the flush volume if a 5½ inches, 20 lb/ft, P-110 (ID = 4.778 inches) production casing is used and the bottom perforation of the stage is located at 12,650 feet.

$$Casing\ capacity = \frac{4.778^2}{1029.4} = 0.0222\ bbl/ft$$

Please note that casing capacity can be found from any casing table, which can be found in any service company's standard handbook.

$$Flush\ volume = 0.0222 \times 12650 = 280\ barrels$$

Therefore, 280 BBLs are needed to flush the well to the bottom perf after the densometer reads 0 ppg on the surface lines.

Some operators flush 10—40 BBLs over the bottom perf flush volume (overflush) just to make sure the wellbore is completely clear of any sand. This is just a safety precaution taken by some of the operators to make sure that during a plug-and-perf completion technique, the composite bridge plug and perforation guns can be pumped downhole without any issues. If the wellbore is not fully clear of proppant near the perforations, the settled proppant can sand off some of the perforations, and pressuring out while pumping down composite bridge plug and perforation guns can be the consequence. Overflushing is basically a taboo in vertical wells because by overflushing the well, the sand that was placed near the wellbore will be swept away, which can cause lower conductivity near the wellbore, affecting the production. As discussed earlier, the industry standard practice in multistage hydraulic fracturing in plug and perf technique is to over flush by 10–40 BBL in horizontal wells (sometimes more depending on the operator). This practice has raised concerns about changing the near wellbore conductivity and as a result loss in productivity. Despite this controversial practice, due to satisfactory production and economic results from various shale plays across the United States, this practice is being continued. More experimental and numerical studies must be performed to truly understand the impact of over flushing on horizontal wells with different frac fluid systems and formation properties. Besler et al. (2007) raised concern about over flushing in Bakken Shale when using cross-linked gel fluid system in transverse fracture system. Gijtenbeek et al. (2012) concluded that over flushing in slick water frac fluid system might not be detrimental to production because of poor proppant transport. In addition, formation properties such as brittleness have a large impact on how over flushing affects production results. But it is still recommended not to overflush in horizontal wells as there could be loss in near wellbore conductivity. The impact of overflushing has not been thoroughly studied in horizontal wells. Fig. 5.7 shows an example of a densometer used in one of the slick water frac jobs.

Figure 5.7 Densometer.

FRAC FLUID SELECTION SUMMARY

Frac fluid selection is one of the most challenging aspects of a hydraulic frac design. A comprehensive understanding of formation properties such as Young's modulus, Poisson's ratio, and formation permeability is essential in designing a proper frac fluid system. There are advantages and disadvantages with each frac fluid system and there is not a perfect fluid system out there. Frac fluid selection alters frac geometry, the formation damage, the cleanup, and the ultimate cost of the fracturing treatment.

CHAPTER SIX

Proppant Characteristics and Application Design

INTRODUCTION

Proppant is used to keep the fractures open after the frac job is complete. Proppant provides a high-conductivity pathway for hydrocarbons to flow from the reservoir to the well. After the frac job is completed, proppant prevents the fractures from closing due to overburden pressure. However, unpropped areas will reclose under the overburden pressure and lose their conductivity with time.

One of the most important factors in every frac job is the type of proppant used for the job. Without proppant in the formation, the formation will reclose under the overburden pressure. Pumping only water without proppant downhole might result in good initial production (IP); however, the production will decrease dramatically and the well will not be economical in the long run due to the absence of proppant to keep the fractures open. There are various types of proppant used in hydraulic fracturing; they are discussed in the following sections.

SAND

Sand is the lowest-strength proppant and is highly available and reasonably priced (it is the cheapest). Sand can typically handle closure pressure of up to 6000 psi (closure pressure is the pressure at which the fracture closes). Two of the major sands used in hydraulic fracturing are known as *Ottawa* and *Brady* sands. Ottawa sand (also known as Jordan, White, and Northern) is the type of proppant used in many shale plays across the United States and it comes from the northern United States (Jordan deposits). This type of proppant is high-quality white-colored sand with monocrystalline grains. On the other hand, Brady sand, which comes from near Brady, Texas and mined from the Hickory formation outcrops,

© 2017 Elsevier Inc.
All rights reserved.

is also high-quality sand used for hydraulic fracturing. This type of sand is called "brown sand" because of its color and it is typically cheaper than Ottawa sand due to containing more impurities and having a more angular form than Ottawa sand. The quality of Brady sand is lower compared to Ottawa sand. The specific gravity of sand is typically 2.65.

PRECURED RESIN-COATED SAND

Resin-coated sand is considered to be an intermediate-strength proppant. Resin-coated sand is more expensive than regular sand and, therefore, economic analysis must be performed to determine the economic viability of using this type of sand. The first type of resin-coated sand is called precured resin-coated sand (PRCS). PRCS has a hard coating around the sand grains, which causes this sand to have higher conductivity as compared to uncoated sand. This type of sand is used in formations with a closure stress of between 6000 and 8000 psi. Resin-coated sand is designed to encapsulate fines, but will not bond in fractures. It is believed that this type of sand prevents the migration of crushed fines. Sand fines are created after the closure pressure is applied on the sand.

The cost of resin-coated sand could potentially be one of the primary reasons when not utilized in formations with closure pressures of more than 6000 psi. Hydraulic fracturing is not just about pumping any type of sand downhole based on the closure pressure, but it is also about cost per stage and evaluating the economic aspects of the frac job. It is very important to understand both design theory and economic evaluation of the design.

CURABLE RESIN-COATED SAND

Curable resin-coated sand (CRCS) has very similar properties to PRCS. One of the main applications for this type of sand is controlling the flowback. If, after the frac job and during the flowback period, a large amount of the sand that was pumped downhole travels back (i.e., flows back) to the surface, CRCS is pumped (tailed in) at the end of each frac stage to mitigate this problem. This type of sand will bind in the fractures (under closure pressure) preventing flowback of the sand to the

Figure 6.1 Curable resin-coated proppant at standard conditions.

Figure 6.2 Curable resin-coated proppant under reservoir conditions.

surface after the frac job is over. In addition, this type of sand, just like PRCS, typically has a crush resistance of 6000—8000 psi. Fig. 6.1 shows curable resin-coated proppant at standard conditions while Fig. 6.2 shows the same proppant bonded under reservoir conditions. Yuyi et al. (2016) experimentally tested the impact of each proppant type (sand, resin-coated, and ceramic proppant) in three deep dry Utica wells located on the same pad in order to make an economic decision on the type of sand to be used on the future pads. They concluded that based on the 2016 market conditions, about 13% and 26% uplift in EUR (from the base case) are needed to justify the incremental Capex associated with pumping resin-coated and ceramic sand respectively. Therefore, performing such experimental testing and analysis is crucial in making an important decision for creating long-term value for the shareholders.

INTERMEDIATE-STRENGTH CERAMIC PROPPANT

The next type of proppant, which is the best-quality proppant and has a higher quality than resin-coated sand, is called ceramic proppant. This type of proppant has uniform size and shape and is thermally resistant. An example of an intermediate-strength proppant is low-density fused ceramic proppant. Intermediate-strength proppant can withstand closure pressure of between 8000 and 12,000 psi. The specific gravity of intermediate-strength proppant is 2.9−3.3 (could be lower depending on the manufacturer and this variance is due to raw material sources used by different proppant manufacturers) (Economides and Martin, 2007). Ceramic proppant has a very high crush resistance. Ceramic proppant has a crush resistance that is so high that if you pour some of the sand on a flat table and beat it with a hammer as hard as you would like, the proppant will not crush and will disperse over the flat area. This demonstrates the high crush resistance of this type of proppant.

LIGHTWEIGHT CERAMIC PROPPANT

Lightweight ceramic proppant (LWC) is not as strong as intermediate-strength proppant. This type of proppant can withstand closure pressure of 6000−10,000 psi (Economides and Martin, 2007). The specific gravity of LWC is typically 2.72 and can be as close as the specific gravity of regular sand. This type of sand provides better conductivity because of better sphericity and sieve distribution (to be discussed). Lightweight proppant also has uniform size and shape and is thermally resistant.

HIGH-STRENGTH PROPPANT

An example of a high-strength proppant is high-strength sintered bauxite, which is the strongest type of proppant used in the industry. It can handle a closure pressure of up to 20,000 psi and is used in deep high pressured formations where closure pressure exceeds 10,000 psi. This type of proppant has corundum, which is one of the hardest

Figure 6.3 Ceramic proppant.

Table 6.1 Proppant Comparisons

Regular Sand	Resin-Coated Sand	Ceramic Proppant
Cheapest	More expensive (compared to regular sand)	Most expensive
Lowest conductivity	Medium conductivity	Highest conductivity
Lowest strength	Medium strength	Highest strength
Irregular size and shape	Irregular size and shape	Uniform size and shape
Naturally occurring product	Manufactured product	Engineered and manufactured product

materials known and is used in high-pressure and high-temperature environments. High-strength and intermediate-strength sintered bauxite are produced using the same manufacturing process. The main difference between the two is the raw materials used. Intermediate-strength bauxite can typically handle closure pressure of 15,000 psi while high-strength bauxite can handle closure pressure of up to 20,000 psi. Sintered bauxite typically has a specific gravity of 3.4 or greater. Fig. 6.3 shows an example of intermediate strength ceramic proppant.

Table 6.1 is the summary breakdown of the three main types of proppant that were discussed.

PROPPANT SIZE

Now that the concept of proppant type is clear, the next concept that must be discussed is proppant size in unconventional shale reservoirs. There are different proppant sizes that can be used depending on the frac design and production enhancement of each proppant size. The following sizes are the most commonly used in unconventional shale reservoirs.

100 Mesh

100 mesh is very similar to baby powder since the mesh size is very small and is designed to be placed in hairline cracks of the formation. Frac jobs usually start with 100 mesh to seal off microfractures. 100 mesh also effectively decreases leak-off through any encountered cracks. 100 mesh provides a conduit for the upcoming sands by covering small microscopic cracks in the formation and erosion of perforations. Sometimes some engineers consider 100 mesh to be part of the percentage of the pad volume. This type of sand is highly recommended in naturally fractured formations. Although this type of sand is not designed for conductivity it is frequently used for sealing off microfractures, perforation erosion, and obtaining as much surface area as possible by traveling farther into the formation. 100 mesh is typically the smallest sand mesh size used during frac jobs. Fig. 6.4 shows an example of 100 mesh sand size.

40/70 Mesh

40/70 mesh is typically used after 100 mesh. 40/70 mesh is larger in size compared to 100 mesh. Pumping this kind of sand downhole creates the required fracture length for maximum surface area and some

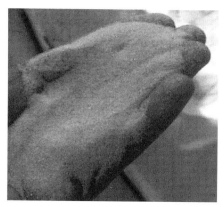

Figure 6.4 100 mesh sand size.

conductivity in the fractures. Combinations of 100 mesh and 40/70 are typically the most common sand sizes used in majority of the unconventional shale reservoirs. It is a known fact that smaller mesh sizes will have a higher crush resistance as compared to the same type of material in a larger mesh. This is because in a fixed fracture width, there are more grains in that width that are able to support the stress. In other words, the stress is more evenly distributed across more grains of proppant with smaller mesh sizes. Therefore, it is crucial to take this concept into consideration when designing proppant size for any frac job.

30/50 Mesh

30/50 mesh is larger than 40/70 and as a result has greater conductivity providing larger flow paths for multiphase flow. Some companies do not run 40/70. Instead, 30/50 mesh proppant is pumped right after 100 mesh for better fracture conductivity near the wellbore especially in liquid-rich and oil windows. Others prefer to run 30/50 after 40/70 for a better transition after 100 mesh. 30/50 mesh is recommended in liquid-rich areas (high BTU). This is because of the multiphase flow effect (to be discussed). Some operators do not believe in pumping 30/50 after 100 mesh because 30/50 mesh does not travel as far as 40/70 mesh into the formation because of its larger size. Stokes' law states that the distribution of proppant inside the fracture depends on its settling velocity in the fracturing fluids. In addition, some operators do not prefer using 30/50 or 20/40 mesh size proppants due to operational issues such as screening out (sanding off) at higher sand concentrations using bigger mesh sizes. Therefore, it is important to perform a risk/reward analysis to see weather the operational risk of pumping bigger sand sizes (if any) is worth the production uplift (if any) or not.

It can be determined that smaller sand particles penetrate deeper into the formation as compared to bigger sand particles. As proppant diameter increases, single-particle settling velocity increases as well. Therefore, 40/70 mesh (with smaller sand particle size) will penetrate more into the formation compared to 30/50 mesh (bigger sand particle). Some operators tail in 30/50 to achieve better conductivity in both dry and liquid-rich areas near the wellbore. Ultimately the final decision on what sand size to use must come from production data and success in each area. If production performance of the wells that only pumped 100 mesh and 40/70 is better than the production performance of the wells that used 100 mesh and 30/50 in the same geologic area, 100 mesh and 40/70 needs to be used on future wells and vice versa. In summary, it is important to have the best design based on theory and simulation (to be discussed);

however, at the end of the day the sand size must be justifed by existing prodution data to design a successful frac job.

20/40 Mesh

20/40 mesh is typically the largest sand size used as compared to all the other sizes discussed thus far. Some operators tail in 20/40 mesh for maximizing near-wellbore conductivity. Some operators do not even run 20/40 mesh and 40/70 or 30/50 mesh is the last sand size pumped down-hole. Production performance must ultimately be the deciding factor on what sand size to use in each area.

Depending on the frac job formation and design, each frac stage requires between 200,000 and 700,000 lb of sand. If the design schedule for a stage is 400,000 lb of sand, the following are some example designs:

Example design # 1 (400,000 lb of sand/stage):
- 50,000 lb of 100 mesh
- 200,000 lb of 30/50 mesh
- 150,000 lb of 20/40 mesh

Example design #2 (400,000 lb of sand/stage):
- 120,000 lb of 100 mesh
- 230,000 lb of 40/70 mesh
- 50,000 lb of 30/50 mesh

Example design #3 (400,000 lb of sand/stage):
- 70,000 lb of 100 mesh
- 330,000 lb of 40/70 mesh

The above designs are just examples underscoring the fact that sand combinations can vary greatly depending on the design, production performance, and economics. The type(s) of sand needed for well optimization is debatable amongst operators, each having preferred recipes for achieving optimal production. There are different hydraulic frac software programs used to run various models to come up with the optimal sand size, type, and volume for the hydraulic frac design.

PROPPANT CHARACTERISTICS

It is important to have a basic knowledge of proppant characteristics and why some proppant types such as resin and ceramic are much more expensive as compared to regular sand. Some characteristics of proppant that are important to understand and monitor are roundness, sphericity,

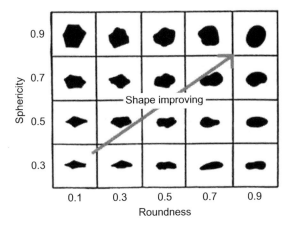

Figure 6.5 Visual estimation of roundness and sphericity (Krumbein and Sloss, 1963). *Modified from Saaid, I.M., Kamat, D., Muhammad, S. 2011. Characterization of Malaysia sand for possible use as proppant. Am. Int. J. Contemp. Res. 1 (1), 37.*

crush resistance, specific gravity, bulk density, acid solubility, sieve size, silt and fine particles, and clustering.

Roundness is the measure of relative sharpness of the grain corners. Improving proppant roundness results in more even stress distribution and potentially improves proppant pack porosity. *Sphericity* is the measure of how round an object is or how closely the grain approaches the shape of a sphere. The American Petroleum Institute (API)-recommended limit for sand in both roundness and sphericity is 0.6 or greater. Fig. 6.5 shows physical roundness and sphericity from Krumbein and Sloss in 1963.

Crush resistance measures the fines created under a given load (exposure to stress). This can be performed in the lab by applying various stresses such as 3000, 4000, 5000 psi, etc. API recommends various percentages of generated fines for different types of sands. K–value testing is an important test that can be performed on various proppant types and sizes in order to understand the % fines generated under each specified stress. K–value is the closure stress (rounded down) under which 10% of the proppant will crush and become fines or out of the standard mesh size. To test the quality of the proppant, it is highly recommended to take a sample from one of the sand haulers onsite and send it to a renowned proppant testing company for K–value and other standard testing. Please note that this type of testing should not be performed by the sand supplier in order to maintain the integrity of the test. The standard API crush-testing procedure typically calls for a loading of 4 lbs/ft^2 in the crush-testing apparatus. However, it is very difficult to obtain such loading in a slick water frac fluid system. Therefore, it is very important to perform this crush testing at various sand loadings such as using the average fracture width near or away from the wellbore to obtain a more realistic view of crushing effect.

Specific gravity is the measurement of absolute density of individual proppant divided by the absolute density of water. The API-recommended maximum specific gravity is 2.65 for sand.

Bulk density is the volume occupied by a given mass of proppant. The API-recommended maximum for proppant is 105 lb/ft^3.

Acid solubility is the solubility of proppant in 12% HCl or 3% HF acids. Acid solubility indicates the amount of contaminants present in the proppant, in addition to relative stability of proppant in acid. The API-recommended maximum acid solubility is 2% for larger sand (30/50 mesh) and 3% for smaller sand (40/70 mesh).

Sieve analysis is a necessary test performed on the proppant throughout the frac job to ensure proper proppant size and quality control of the proppant. It indicates the size distribution of the proppant within the defined proppant size range. In this analysis, which is typically performed by the sand coordinator, 90% of the sample should fall within the designated sieve size. Not more than 0.1% should be larger than the first sieve size and not more than 1% should be smaller than the last sieve size. For example in Table 6.2, if 40/70 mesh is being tested, not more than 0.1% of the sample size test should be larger than 0.0165, and not more than 1% should be smaller than 0.0083. The operating company representative is responsible for verifying the properly tested sieve analysis throughout the frac treatment.

Silt and fine particles measure the amount of silt, clay, and other fine materials (impurities) present in the sample. The API recommendation for silt and fine particles is 250 FTU (formation turbidity unit) or less. Fig. 6.6 shows an example of test sieve shaker used in the laboratory to find proppant size distribution. In addition to laboratory testing, this type of test can also be easily performed in the field.

Table 6.2 Standard Sieve Openings (Ely, 2012)

US Series Mesh	Sieve Opening (in)	US Series Mesh	Sieve Opening (in)
4	0.187	25	0.0280
6	0.132	30	0.0232
8	0.0937	35	0.0197
10	0.0787	40	0.0165
12	0.0661	60	0.0098
14	0.0555	70	0.0083
16	0.0469	100	0.0059
18	0.0394	170	0.0035
20	0.0331		

Figure 6.6 Test Sieve shaker.

Clustering measures the degree of attachment of individual proppant grains to one another. The API–recommended maximum for clustering, which is measured by percentage weight, is 1%. One of the main reasons for this type of test is that during processing the grains were not broken apart (Ely, 2012).

PROPPANT PARTICLE-SIZE DISTRIBUTIONS

The max-to-min ratio in the majority of API sieve designations is approximately 2 to 1. For example, a 20 mesh particle is roughly twice the diameter of a 40 mesh particle as can be seen in Table 6.2. A 20 mesh particle has a diameter of 0.0331 inches compared to a 40 mesh particle with a diameter of 0.0165 inches (~ half of 0.0331 inches). Table 6.2 shows different U.S. sieve and their opening sizes.

PROPPANT TRANSPORT AND DISTRIBUTION IN HYDRAULIC FRACTURE

During hydraulic fracturing, different proppant concentrations are pumped based on the initial frac design and to the extent that reservoir formation permits. The pumped proppants move in both horizontal and vertical directions. In the horizontal direction, proppant follows the fracture tip with the same velocity as fracturing fluid. However, in the vertical direction the proppant velocity, i.e., settling velocity, is different than fluid vertical velocity due to gravitational forces and slippage between proppant particles and fluid. Proppant movement in the direction of the fracture width is commonly neglected due to scale effect (fracture width is much smaller than fracture length and height). As proppant particles settle, they fill up the fracture width and, therefore, increase the proppant concentration in the vertical cross section. There is a critical proppant concentration beyond which screening out (sanding off) occurs. Rate of proppant bank growth or screening out is a function of proppant settling velocity. Settling velocity of a single and perfectly spherical proppant particle can be obtained using Stokes' law assuming infinitely large fracture (boundary effects are neglected). Settling velocity is derived for different flow regimes based on the dimensionless Reynolds number. If the Reynolds number is less than 2, proppant settling velocity can be obtained using Eq. (6.1).

$$V_{ps} = \frac{g(\rho_p - \rho_f)d_p^2}{18\mu}$$

Equation 6.1 Proppant settling velocity $Re \leq 2.0$.

If the Reynolds number falls between 2 and 500, proppant velocity can be obtained using Eq. (6.2).

$$V_{ps} = \frac{20.34(\rho_p - \rho_f)^{0.71}d_p^{1.14}}{\rho_f^{0.29}\mu^{0.43}}$$

Equation 6.2 Proppant settling velocity $(2 < Re < 500)$

For flow regimes with high Reynolds number (i.e., >500), Eq. (6.3) will be used.

$$V_{ps} = 1.74\sqrt{\frac{g(\rho_p - \rho_f)d_p}{\rho_f}}$$

Equation 6.3 Proppant settling velocity $(Re \geq 500)$

In the above equations, ρ_p and ρ_f stand for proppant and fracturing fluid density, μ is the fluid dynamic viscosity, d_p is proppant diameter, and V_{ps} is the uncorrected proppant settling velocity. As mentioned earlier, proppant velocity obtained using Stokes' law neglects the boundary (fracture width) effect by assuming an infinitely large fracture. It also ignores interactions between proppant particles, since it has been developed for a single particle. Gadde et al. (2004) defined a correlation to correct the proppant settling velocity for these two factors as follows:

$$V'_{ps} = V_{ps}\left[0.563\left(\frac{d_p}{w}\right)^2 - 1.563\left(\frac{d_p}{w}\right) + 1\right](2.37c^2 - 3.08c + 1)$$

Equation 6.4 Corrected proppant settling velocity

In Eq. (6.4), V'_{ps} is the corrected proppant settling velocity and c is the proppant concentration. As proppant settling occurs, the frac fluid viscosity will change. The change in frac fluid viscosity as a function of proppant concentration can be obtained using Eq. (6.5).

$$\mu = \mu_0\left\{1 + \left[0.75(e^{1.5n} - 1)e^{-\frac{\gamma(1-n)}{1000}}\right]\frac{1.25c}{1 - 1.5c}\right\}^2$$

Equation 6.5 Frac fluid viscosity

In Eq. (6.5), μ_0 is uncorrected fluid viscosity to proppant concentration and n and γ are the non–Newtonian fluid constants.

Kong et al. (2015) investigated the effect of proppant settling velocity on proppant distribution and fracture conductivity in the Marcellus Shale reservoir and showed that ignoring proppant settling velocity could lead to more than 18% overestimation in dimensionless productivity index. They showed that in tighter formations and using larger proppant size the overestimation in dimensionless productivity index can be as large as 32%. A more realistic prediction of proppant distribution in hydraulic fractures can significantly help operators to design the optimum frac job. In ultralow permeability formations such as shale with permeability less than 1 µD, there is a critical proppant size that can lead to the highest hydraulic fracturing efficiency, as shown in Fig. 6.7. In the hydraulic fracturing process, multisize proppant combinations are injected into the wellbore. Usually a smaller-size proppant is injected first, followed by a larger-size proppant. In ultralow permeability formations such as shales, there is a critical combination of small and large proppant sizes that will result in maximum well productivity index as shown in Fig. 6.8.

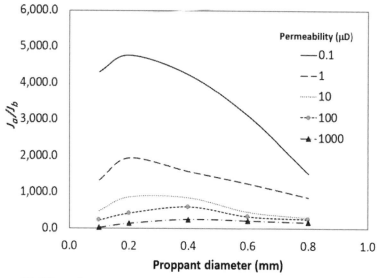

Figure 6.7 Effect of proppant size on dimensionless productivity index for different reservoir permeability. *Modified from Kong, B., Fathi, E., Ameri, S. 2015. Coupled 3-D numerical simulation of proppant distribution and hydraulic fracturing performance optimization in Marcellus shale reservoirs. Int. J. Coal Geol. 147—148, 35—45.*

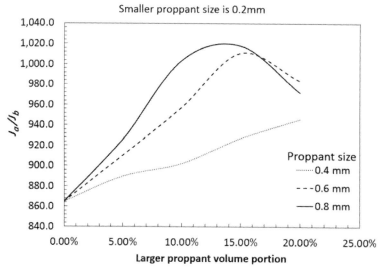

Figure 6.8 Effect of different proppant size and volume combination on well dimensionless productivity index. *Modified from Kong, B., Fathi, E., Ameri, S. 2015. Coupled 3-D numerical simulation of proppant distribution and hydraulic fracturing performance optimization in Marcellus shale reservoirs. Int. J. Coal Geol. 147—148, 35—45.*

FRACTURE CONDUCTIVITY

Fracture conductivity is one of the most important concepts in hydraulic fracturing, and thus is considered in every design. Conductivity is essentially the multiplication of fracture width (ft) and proppant permeability inside the fracture (md). Proppant permeability and conductivity change under different stresses. For example, the permeability of 20/40 mesh (and ultimately conductivity) under 6000 psi of closure pressure is different compared to 10,000 psi. Conductivity is also referred to as *flowback capacity* and its unit is md-ft. Conductivity is the ability of the fractures to transmit reservoir fluid to the wellbore. As closure pressure increases, conductivity decreases. Proppant suppliers typically provide a chart for each proppant type that shows fracture conductivity on the y-axis vs. closure pressure on the x-axis.

Factors that affect fracture width are proppant density, proppant loading, gel filter cake, and embedment. Also, factors that affect proppant permeability are typically proppant size, sphericity, strength, fines and gel damage. Fig. 6.9 shows the schematic of a fracture and fracture width that will be used in fracture conductivity calculation.

$$Fracture\ conductivity = k_f \times W_f$$

Equation 6.6 Fracture conductivity

where K_f = proppant permeability, md, and W_f = fracture width, feet.

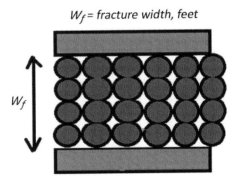

Figure 6.9 Fracture width.

DIMENSIONLESS FRACTURE CONDUCTIVITY

Dimensionless fracture conductivity is the ability of fractures to transmit reservoir fluid to the wellbore divided by the ability of the formation to transmit fluid to the fractures. Dimensionless fracture conductivity is denoted in F_{CD} and is defined as:

$$F_{CD} = \frac{K_f \times W_f}{K \times X_f}$$

Equation 6.7 Dimensionless fracture conductivity

where K_f = fracture permeability in the formation, md, W_f = fracture width, feet, K = formation (matrix) permeability, md, and X_f = fracture half–length, feet.

Fig. 6.10 shows two-stage hydraulic fracturing and reservoir-stimulated volume characteristics used to calculate dimensionless fracture conductivity. Fig. 6.11 illustrates qualitative comparisons of fracture conductivity and closure pressure for different proppant types.

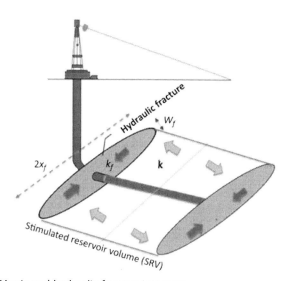

Figure 6.10 Matrix and hydraulic fracture interactions.

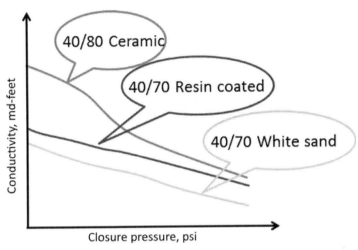

Figure 6.11 Fracture conductivity testing.

INTERNATIONAL ORGANIZATION FOR STANDARDIZATION (ISO) CONDUCTIVITY TEST

The fracture conductivity testing that is performed for each proppant type and size is typically performed under the following conditions:
- Ohio sandstone
- 2 lb/ft^2 proppant loading
- Stress maintained for 50 hours
- 150°F to 200°F
- Extremely low water velocity (2 mL/min) (2% KCl water)

This conductivity test accounts for proppant size, proppant strength/crush profile, some embedment, some temperature effects, and wet system. However, this conductivity test does not account for the following:

a) Non–Darcy flow
b) Multiphase flow
c) Reduced proppant concentration
d) Gel damage
e) Cyclic stress
f) Fines migration
g) Time degradation

NON-DARCY FLOW

As opposed to Darcy's law, which assumes laminar flow in the formation, non-Darcy flow is referred to as fluid flow, which deviates from Darcy flow by having a turbulent flow in the formation and especially near the wellbore. Non-Darcy flow is very common in high-rate gas wells near the wellbore. Therefore, some operators like the idea of tailing in higher-conductivity proppant at the end of each frac stage to accommodate for the non-Darcy flow effect near the wellbore.

MULTIPHASE FLOW

Hydraulic fracturing usually encompasses flow of liquid (water, oil, and condensate) and gas. Fluid flow in hydraulic fracturing highly depends on the relative permeability of the formation to each of these phases. As Fig. 6.12 illustrates, by increasing the saturation of one phase, the relative permeability to that phase increases while the relative permeability to the other phase decreases. Therefore, proppant saturated with liquid is less conducive to flowing gas. This effect is not taken into account in ISO conductivity testing. The importance of relative permeability comes into play in high-BTU gas reservoirs (primarily retrograde condensate reservoirs) and oil reservoirs where fluid exists as liquid at reservoir conditions. Liquid tends to accumulate in the fractures. This will occupy porosity that is not

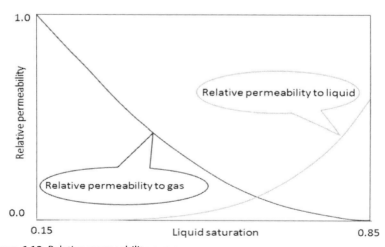

Figure 6.12 Relative permeability curve.

available for gas flow. In wet areas, higher conductivity sand is typically pumped near the wellbore to accommodate the relative permeability effect.

REDUCED PROPPANT CONCENTRATION

An ISO conductivity test is performed using $2\,lb/ft^2$ proppant loading, which can be misleading. Proppant concentration at formation conditions is typically less than $1\,lb/ft^2$. For example, if fracture conductivity for regular sand such as 40/70 mesh under 6000 psi of closure pressure is 400 md–ft at $2\,lb/ft^2$, the conductivity at formation (which is $1\,lb/ft^2$ or less) will be much less than 400 md–ft, roughly 200 md–ft. Therefore, proppant concentration in the formation has been reduced down to 200 md–ft, as opposed to the reported 400 md–ft when the conductivity test is performed.

GEL DAMAGE

Gel damage often occurs in cross-linked jobs, where heavy viscous fluid is used during the frac job. The residual gel effect can have detrimental impact on the fracture conductivity even after using gel breaker. It is important to note that breaker loading can significantly improve the cleanup of distributed gel in the formation. In a slick water frac, gel damage is not as common unless a high concentration of linear gel was used to facilitate placing proppant into the formation. One example of gel damage is fracture face damage, which is caused by filtrate leaking into the rock. Another example of gel damage is the accumulation of residual gel. Gel residue tends to accumulate in very narrow pore throats ultimately affecting flow capacity of the fluid flow. Gel damage can also cause a reduction in effective fracture width due to filter cake buildup. Filter cake forms as frac fluid slurry leaks off into the formation. Filter cake is forced out into the fractures upon closure. Fracture width retention is extremely important to minimize gas velocities in the fracture. Decreasing gas velocity will significantly reduce pressure drop as stated in the Forchheimer equation. Gel is a non–Newtonian fluid. As opposed to Newtonian fluid where the relationship between shear stress and shear rate is linear, in non–Newtonian fluid the relationship between shear stress and shear rate is different and can be time dependent. Another form of detrimental gel damage is referred to as loss of effective fracture length due to gel plug-in tip. Since gel is a non–Newtonian fluid, it must achieve some pressure differential before being able to move. This feature of gel will cause reduction in effective fracture length by plugging the tip of the fracture. Laboratory studies suggest that proppant with better roundness, sphericity, porosity, and permeability facilitate the cleanup of gel as opposed to other types of proppant.

CYCLIC STRESS

When producing a well, there is a pressure referred to as flowing bottom-hole pressure. Flowing bottom-hole pressure is the bottom-hole pressure inside the wellbore at flowing condition denoted as P_{wf}. This pressure is extremely important as proppant stress (stress placed on the proppant) is a function of closure pressure and flowing bottom-hole pressure. Every time flowing bottom-hole pressure changes, the proppant distribution inside the fracture rearranges and some conductivity loss can be the consequence. Fig. 6.13 shows the ideal designed proppant placement and real proppant placement in a hydraulic fracture during the frac job and after the well is in line. In an ideal case, uniform proppant distribution inside the fracture is assumed. After starting to produce from a well, hydraulic fracture width decreases due to an increase in effective stress. However, in reality, proppant distribution will not be uniform due to proppant settling velocity and fluid leak-off to the formation. As a result, there will be propped and unpropped

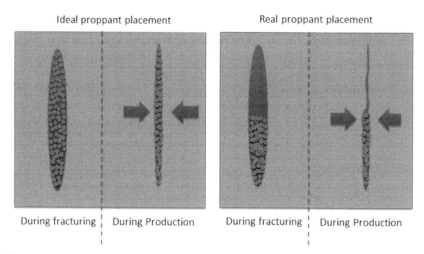

Figure 6.13 Proppant placement in hydraulic fracturing. *Modified from Kong, B., Fathi, E., Ameri, S. 2015. Coupled 3-D numerical simulation of proppant distribution and hydraulic fracturing performance optimization in Marcellus shale reservoirs. Int. J. Coal Geol. 147–148, 35–45.*

regions in a hydraulic fracture. The unpropped region will be closed and fracture width in propped region will decrease due to an increase in effective stress during production. This leads to reduction in fracture conductivity.

FINES MIGRATION

Under closure pressure at downhole conditions, proppant will generate some fines (depending on the type of proppant used), which will reduce conductivity. This will highly depend on the rate of change in effective stress, which is a function of operational conditions. Unfortunately, there is a tendency to achieve high IP when the well first starts producing (initial flowback) in order to impress the investor community. Very high IP can only be achieved through very aggressive flowing bottom-hole pressure drawdown. The practice of "pulling hard" or "rip it and grip it" is common in a lot of the unconventional shale plays. This practice leads to an excessive amount of stress on proppant and as a result some proppant embedment and fines migration that will lead to loss in conductivity. Loss in conductivity is equal to loss in well productivity and ultimately revenue. Intermediate initial flow rates will lead to moderate pressure drawdown, which will result in less damage to the fracture conductivity as shown in Fig. 6.14. Eq. (6.8) shows the importance of minimizing flowing bottom-hole pressure. Minimizing flowing bottom-hole pressure can be achieved by minimizing pressure

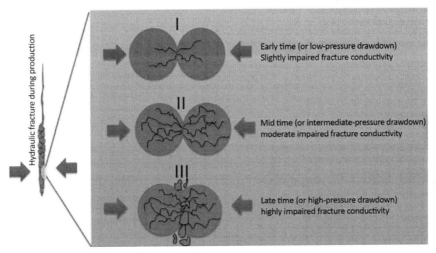

Figure 6.14 Proppant crushing embedment.

drawdown when producing the well. Belyadi et al. (2016) used actual field data from eight studied wells in Utica/Point Pleasant and illustrated that up to 30% improvement in EUR can be achieved using a managed pressure drawdown of 15—20 psi/D casing or tubing pressure drop. In addition, they also showed that reservoir damage was most likely caused by pressure dependency of hydraulic fracture conductivity. Optimum economic rate can be determined based on a company's long-term financial metric as well as gas pricing. In essence, aggressive pressure drawdown can damage a well's performance while conservative pressure drawdown can impact the near-term economic value. Therefore, pressure drawdown schedule and optimum economic rate must be determined for each field based on a company's strategic metric and goal.

$$Proppant\ stress = P_{closure} - P_{wf} + P_{net}$$

Equation 6.8 Proppant stress

where $P_{closure}$ = closure pressure, psi, P_{wf} = flowing bottom–hole pressure, psi, and P_{net} = net pressure, psi.

●●●————————————————————————————

Example

Calculate proppant stress given the two following conditions assuming net pressure is zero:
1. The well is initially producing at 4500 psi flowing bottom-hole pressure and the closure pressure from DFIT is calculated to be around 6500 psi.
2. The flowing bottom-hole pressure is drawn down very aggressively to about 1000 psi after the initial flowback over the course of a 2-day time period (assume closure pressure stays constant after 2 days).

Condition 1: $Proppant\ stress = P_{closure} - P_{wf} = 6500 - 4500 = 2000$ psi
Condition 2: $Proppant\ stress = P_{closure} - P_{wf} = 6500 - 1000 = 5500$ psi

As can be seen in this example, it turns out that proppant stress has increased from the initial 2000 psi to almost 5500 psi in just 2 days. This practice will lower proppant conductivity significantly and will result in a loss in production and ultimately revenue.

TIME DEGRADATION

Fracture conductivity will reduce with time. The rule of thumb is that fracture conductivity will be reduced by 75% with time. This behavior, which leads to early production decline in hydraulically fractured shale reservoirs, is not well understood yet. One of the major physical

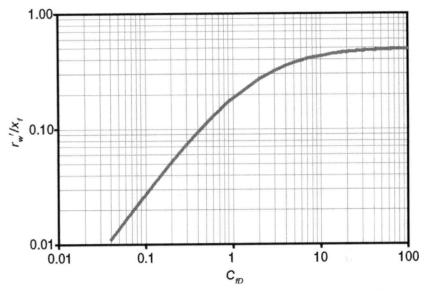

Figure 6.15 Dimensionless frac conductivity vs effective drainage radius (Cinco-Ley and Samaneigo, 1981).

phenomena that has been investigated as the possible source of early production decline is the hydraulic fracture and shale matrix permeability time–dependent characteristics, also known as creep deformation under constant loading. This has mainly been attributed to matrix and hydraulic fracture interaction with fracturing fluid.

FINITE VS INFINITE CONDUCTIVITY

If F_{CD} is greater than 30, it is considered to be *infinite conductivity* and if F_{CD} is less than 30, it is considered to be *finite conductivity*. Cinco-Ley and Samaneigo in 1981 presented Fig. 6.15 that plots dimensionless fracture conductivity (x-axis) versus effective wellbore radius/fracture half-length (y-axis) to define the concept of dimensionless fracture conductivity.

Example

Assume the dimensionless fracture conductivity for a well in the Marcellus Shale is calculated to be 50. From Fig. 6.15, the effective wellbore radius can be found as follows:

$$\frac{r_{w'}}{X_f} = 0.5$$

Assuming a calculated fracture half-length of approximately 300 feet and substituting $X_f = 300$ feet,

$$\frac{r_{w'}}{300} = 0.5 >> r_{w'} = 150 \text{ feet}$$

As long as the dimensionless fracture conductivity is greater than 30, the fracture is considered to be in infinite conductivity and the effective drainage radius does not change. For example, if a dimensionless fracture conductivity of 30 can be achieved using 30/50 mesh sand at downhole condition (taking into account all of the factors that the ISO conductivity test does not take into account), does pumping 20/40 mesh really matter? As long as a dimensionless fracture conductivity of 30 is reached (infinite conductivity) at downhole conditions, pumping a larger sand size is not recommended. The hardest part is figuring out the fracture conductivity after all the discussed effects are taken into account. Since the permeability in unconventional shale reservoirs is very low, it is important to note that achieving infinite conductivity is easier. However, with all the factors discussed that can reduce fracture conductivity, infinite conductivity could potentially be very difficult to obtain in some formations depending on many factors such as type of sand, type of fluid system, type of pressure drawdown, type of reservoir fluid, etc.

Example

A horizontal well with low permeability is going to be hydraulically fractured using a slick water fluid system. The matrix permeability of the reservoir is 0.0003 md (300 nd) with an estimated propped fracture half-length of 300 ft. Fracture conductivity under 6000 psi of closure pressure is estimated to be 400 md-ft from the lab ISO conductivity test (2 lb/ft^2). Calculate fracture conductivity at 1 lb/ft^2 and assume 85% reduction in conductivity due to all of the effects discussed. Calculate dimensionless fracture conductivity and specify whether the fractures are considered to be finite or infinite.

$$Fracture \; conductivity \; at \; 2\frac{lb}{ft^2} = K_f \times W_f = 400 \text{ md-ft}$$

$$Fracture \; conductivity \; at \; 1\frac{lb}{ft^2} \; and \; 85\% \; reduction = \frac{400}{2} \times (1 - 0.85) = 30 \text{ md-ft}$$

$$Dimensionless \; fracture \; conductivity = F_{CD} = \frac{K_f \times W_f}{K \times X_f} = \frac{30}{0.0003 \times 300} = 333$$

Since $F_{CD} > 30 >>>$ infinite fracture conductivity

As can be seen from this example, since the formation permeability is so low even after taking into account some of the effects that can alter fracture conductivity, fractures are considered to be at infinite conductivity.

Unconventional Reservoir Development Footprints

INTRODUCTION

Unconventional reservoir developments encompass activities such as hydraulic fracturing and wastewater deposition in underground reservoirs. These activities introduce manmade stresses that change the in situ stress condition of the underground formations leading to the cases of induced seismicity. The magnitude of induced seismicity is a function of orientation, magnitude, and relative state of the surrounding stress field. Some statistics provided by the U.S. Geological Survey (USGS) suggest exponential growth in the cumulative number of earthquakes in the central and eastern United States since 2005, coinciding with unconventional reservoir developments in these areas. There have also been studies trying to correlate these statistics to hydraulic fracturing, withdrawal, or fluid injection by the oil and gas industry; however, there is no direct evidence and detailed studies that can prove this idea. By taking a closer look at the National Seismic Hazard Map published by the USGS in 2014, one can see that most of the earthquakes higher than a magnitude of 3 occurred in the vicinity of major fault planes that happened to also be very close to the major unconventional shale developments. Having said so, in unconventional reservoir developments, there are many cases of stress field alteration, which may impact the stability of the underground formations, faults, and any discontinuities. This might lead to hydraulic fracture and fault reactivation, or hydraulic fracturing and aquifer interaction. In ultralow-permeability shale reservoirs, which dominate most of the unconventional oil and gas resources in the United States, hydraulic fracturing treatment is absolutely essential to obtain an economic level of production. These hydraulic fracturing activities mainly introduce low-magnitude induced seismicity. These low–magnitude seismic events are used by the oil and gas industry to obtain the geometry of the hydraulic fractures. Often, these low–magnitude seismic events cannot be felt at the

Hydraulic Fracturing in Unconventional Reservoirs
DOI: http://dx.doi.org/10.1016/B978-0-12-849871-2.00007-1
© 2017 Elsevier Inc.
All rights reserved.

surface and will be limited to the treatment zone. However, in extremely rare cases due to unintended interactions between hydraulic and natural fractures, these events might have some impacts at the surface.

Environmental impacts of hydraulic fracturing are not limited to inducing seismicity. Wellbore integrity is also one of the major concerns in the oil and gas industry and is highly regulated by state legislation. Hydraulic fracturing can significantly impact the geomechanical behavior of the wells. These concerns in the oil and gas industry have become a focus of research in areas such as cement behavior and cement bond under confining pressures applied to cement during hydraulic fracturing, and hydraulic fracture pressure communications with old and abundant wells. In some cases during the hydraulic fracturing treatment, the production casing might burst due to exceeding the casing burst pressure, or flaw in the casing manufacture. This opens a great deal of discussion on casing selection and design. Other environmental impacts of unconventional resource developments can be classified in issues related to groundwater protection, wildlife impacts, community impacts, and surface disturbance. On the other hand, unconventional resource development can have an enormous positive social and political impact in terms of providing more jobs, increasing the energy security and sustainability, decreasing the pollution by providing much cleaner energy as compared to conventional developments, and in general increasing the quality of the life in the country.

CASING SELECTION

There are four different types of casings commonly used in the industry for horizontal wells, and they are as follows:
- conductor casing
- surface casing
- intermediate casing
- production casing.

CONDUCTOR CASING

The conductor casing is installed prior to the arrival of the drilling rig. This hole is usually 18−36″ in diameter and 20−50′ long. This casing basically keeps the top of the hole from caving in and additionally it prevents the collapse of loose soil near the surface. Moreover, it is used to help in the process of circulating the drilling fluid up from the bottom of the well. This casing needs to be either cemented or grouted in place.

SURFACE CASING

After placing and cementing the conductor casing, the next hole size needs to be drilled before placing the surface casing. The next hole size is drilled using a drilling rig to the desired depth, which is usually anywhere between a few hundred feet and 2000 feet. This is the most crucial casing as far as the Environmental Protection Agency (EPA) is concerned since the water source is usually located in that range. As a result, to protect the water source from contamination, the EPA typically requires setting the surface casing and cementing it to at least 50′ deeper than the deepest fresh groundwater zone. In some parts of Pennsylvania, the Department of Environmental Protection (DEP) requires two surface casings to protect the coal seams as well. The main purpose of surface casing is to protect freshwater from contamination. Freshwater contamination can be caused by leaking hydrocarbon or salt water from the producing formation *if and only if* the casing or the cementing operation is not done properly. Please note that this crucial process is highly regulated by the EPA and violators are heavily fined and could be suspended from drilling and completion processes if not in compliance. In addition, the environmental agency (varies depending on the state) has to be notified 24 hours before and after the cement job is started and completed to ensure a proper seal between the freshwater zone and the well. Oftentimes (depending on the state), a representative from the state will be present during the cement job to ensure the quality of the cement job and compliance with all laws and regulations. If after cementing the surface casing, cement is not received at surface, the operation cannot continue until a course of action is summarized and submitted to the state for their review and approval in order to make sure the problem is completely resolved before moving forward. Another purpose of surface casing is to make sure the drill hole is not being damaged or collapsed when drilling the next hole section of the well. If proper casing is not placed, the drilled hole could be damaged or even collapsed for many reasons that exist downhole (pressure, temperature, water invasion, etc.). Another important reason for installing the surface casing is to provide primary well-control equipment to be rigged up (examples of primary well control equipment are blow-out preventers).

Surface casing is the first casing to provide the necessary means of installing primary well-control equipment. A typical surface casing size is 13 ⅜″.

INTERMEDIATE CASING

After placing the surface casing, cementing it in place, and getting a confirmation from the environmental agency to continue operations, the next section of the hole is drilled. After drilling this section, intermediate casing is placed in the hole for many reasons. The primary reason for using intermediate casing is to minimize the hazards associated with abnormal underground pressure zones or formations that might otherwise contaminate the well, such as underground salt water deposits. This casing is often used in longer laterals so the hydrostatic pressure of the drilling fluid remains between formation and fracture pressures. Even if none of the above conditions are present, this casing is very important as insurance for any type of unexpected abnormal pressure downhole. The intermediate casing size that most of the operating companies use is 9 ⅝″ casing.

PRODUCTION CASING

Finally, after placing the intermediate casing and cementing it all the way to the surface, the next section of the hole is drilled. The kickoff point (KOP) is the point at which the curve section of the wellbore is started and built. KOP is the point at which a well starts to incline using the predetermined engineering plan to get to the desired zone of interest. After building the curve, the landing point is reached. The landing point is the point at which the target formation is reached and from that depth the well can be horizontally drilled to total depth (TD). Once the TD of the well is reached, typically 5½″ production casing can be run all the way to the surface and cemented in place. Depending on the formation and design, 4½″ or 7″ production casings could also be run. A production casing, which is also called "long string," is the deepest section of the casing of a well since it goes from the desired depth all the way to the surface. This casing is basically a conduit from the surface of the well to the actual producing formation. The size of production casings depends on various considerations including lifting equipment to be used, types of completion processes required, and the possibility of deepening the well

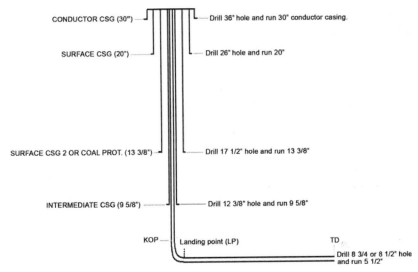

Figure 7.1 Various casing string illustrations.

at a later date. For example, if the well is expected and designed to be deepened at a later time, the production casing should be large enough to allow passage of the drill bit later to drill the next designed section. Fig. 7.1 shows various casing string illustrations.

HYDRAULIC FRACTURING AND AQUIFER INTERACTION

This is a controversial topic amongst environmentalists concerned with the chemicals used in hydraulic fracturing and whether or not they impact drinking water sources. Hydraulic fracturing itself does not cause drinking water contamination. Hydraulic fracturing started in 1947 without a single case of drinking water contamination. The main issue that causes water contamination is bad cement jobs in the casings. After running thousands of feet of casing (steel pipe), there is always a possibility of a microannulus leakage in the casing because of a bad connection between casing joints. However, if there is not a cement bond behind the surface casing, there are oftentimes two, or in some cases three additional casings (coal seam, intermediate, and production casings) that are run and cemented to protect the surface water from any type of contamination. Cementing is a crucial part of the drilling operation. This is why the cementing operation is so highly regulated by the Environmental Protection Agency (EPA) to protect freshwater from any types of contaminates.

When hydraulic fracturing is performed, the fractures created by hydraulic horsepower (HHP) do not extend all the way to the surface. Later on in this chapter, the concept of fracture height is discussed. For example, in Marcellus Shale the true vertical depth (TVD) of an average well is anywhere between 6500′ and 8000′ (depending on the area) and hydraulic fracturing is performed at that depth. Water sources are located between 50′ and a maximum depth of 1000′. Based on current frac microseismic data, it is highly unlikely that the fractures created during hydraulic fracturing could grow to a length of 6000′ to upwards of 7000′, thereby contaminating the local drinking water. Fractures are naturally limited due to natural formation barriers, stresses in the rock (vertical, minimum horizontal, and maximum horizontal stresses), leak-off limits, and height growth. If we had no stresses in the earth, the fractures would easily grow to the surface when hydraulic fracturing. Operators have run microseismic in various basins and formations to identify fracture azimuth, fracture height growth, fracture length, fracture width, etc. The frac microseismic data demonstrated that the average height could be up to 1000′. Therefore, the maximum height that fractures can grow based on seismic data is still thousands of feet away from drinking water sources.

One of the main reasons for water contamination is a bad cement job, and a bad cement job nowadays would never be approved by the state's environmental agency. The industry is heavily regulated and careful regarding this matter, as it is a very sensitive subject. Having said so, introducing manmade stresses to the prestressed formation during hydraulic fracturing can cause induced seismicity. This induced seismicity might reactivate faults and discontinuities. The problem with fault reactivation or slippage is that it can extend all the way to the surface as shown in Fig. 7.2. If hydraulic fracturing causes fault reactivation or slippage, the fault will work as a flow path that can transfer frac fluid all the way to the surface. This leads to frac job failure due to huge frac fluid leak-off to the fault and can lead to severe environmental impacts.

HYDRAULIC FRACTURING AND FAULT REACTIVATION

During hydraulic fracturing, the in situ stress condition of the reservoir will change. The magnitude of the change in in situ stress is directly related to the formation mechanical properties, induced hydraulic gradient through fracture initiation and propagation, and properties of the

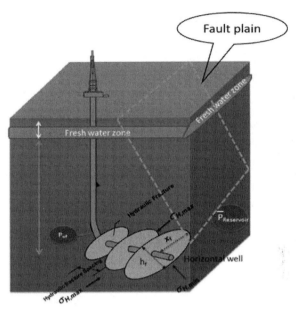

Figure 7.2 Fracture growth limitation.

possible fault or discontinuities. Therefore, faults or any discontinuities in the region affected by stress change due to hydraulic fracturing might be reactivated. Fault slippage or fault motion is directly related to the coefficient of friction or friction factor. Experimental measurement of friction factor of different rock types under different stress conditions showed that the friction factor is changing in a small range between 0.6 and 1.0. Friction factor is a contact property that is measured along preexisting faults and fracture planes. As the in situ stress field is modified due to hydraulic fracturing, the friction factor along deactivated fault and fracture planes changes, which can lead to fault reactivation, instability, and rock failure.

Gao et al. (2015) performed analytical and numerical studies to investigate the stability of the identified and unidentified faults around the multistage hydraulic fracture in shale reservoirs. They showed that the stability of the fault depends on its position with respect to the hydraulic fractures. Assuming fault is in a critically stressed condition with an initial slip tendency of 0.6, there is a critical angle ($\theta = 50°$) and distance r, below which there is a great possibility of fault reactivation. In other words, for a fault in a critically stressed condition with slip tendency of 0.6, if the angle between fractures and fault plain becomes less than 50 degrees ($\theta \leq 50°$) and distance between the fault plain to hydraulic fracture initiation point becomes less than 2.5 times the fracture height ($r \leq 2.5H$), there is a high

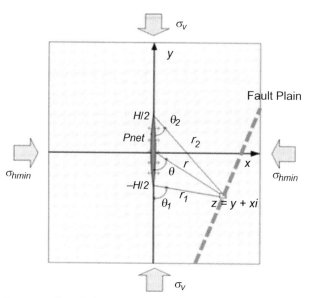

Figure 7.3 Geometry of single hydraulic fracture and fault plain. *Modified after Gao, Q., Cheng, Y., Fathi, E., Ameri, S., 2015. Analysis of stress-field variations expected on subsurface faults and discontinuities in the vicinity of hydraulic fracturing. SPE-168761, SPE Reservoir Evaluation & Engineering Journal.*

potential for fault reactivation as shown in Fig. 7.3. Fig. 7.4 illustrates the impact of pressurized hydraulic fractures on stability of different regions around hydraulic fractures. High pressure on the hydraulic fracture surface leads to a decrease in slip tendency perpendicular to the hydraulic fracture plain and an increase in slip tendency parallel to the fracture plain, especially at the fracture tip. Therefore, perpendicular to the hydraulic fracture, the stability of the region increases while regions in the direction of hydraulic fracture propagation become unstable. The stable and unstable regions around multistage hydraulic fracture are shown in Fig. 7.4.

As discussed earlier, stability or failure of the fault is determined using slip tendency, which is defined as the ratio of shear to normal stress acting on fault plain as shown in Eq. (7.1).

$$T_s = \frac{\tau}{\sigma_n} \geq \mu_s$$

Equation 7.1 Slip tendency.

where T_s is slip tendency, τ is shear stress, σ_n is normal stress and μ_s is static friction factor.

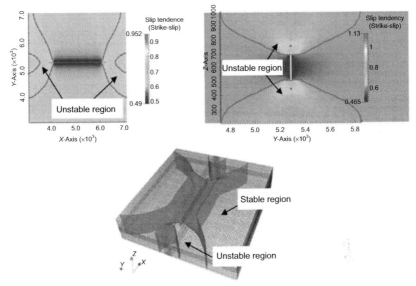

Figure 7.4 Numerical simulation of change in slip tendency around pressurized multistage hydraulic fractures using finite element technique (Gao et al., 2015).

HYDRAULIC FRACTURING AND LOW-MAGNITUDE EARTHQUAKES

Fault reactivation or slippage can lead to earthquake activities commonly below 1.0 in magnitude. However, in very rare cases, earthquakes with magnitude around 3 are also reported (Ellsworth, 2013). USGS recently released the map of earthquakes in the United States with magnitude of 3 or higher from 1973 to 2014. They showed that the cumulative number of earthquakes with a magnitude of 3 or higher has significantly increased since the beginning of the 20th century. The media in the United States has quickly pointed to the oil and gas industry without any scientific and sound justifications that could show that these events are actually correlated to the oil and gas activities, such as hydraulic fracturing or disposal of contaminated water in wastewater injection wells. The oil and gas industry is heavily regulated on tracking these events using microseismic studies and there has not been any relationship between the oil and gas industry activities and large magnitude earthquakes.

Having said so, there is still a critical need for further research in this area due to the potential consequences that are associated with large-magnitude seismic events. Determining the cause and influencing factors in the occurrence of large-magnitude seismic events is essential in preventing hazards associated with these events. The main objective of these kinds of studies should be preventing harm to the public health and infrastructure by reducing or eliminating the major causes of these unintended events. The hydraulic fracturing and post-hydraulic fracturing activities such as disposal of contaminated flowback water are not the only causes of induced seismicity. Hydrocarbon production from these reservoirs could also trigger seismic activities (Soltanzadeh and Hawkes, 2009). In this case, the study of complex rock and fluid interactions that influence the formation stress behavior is required. To provide an effective prediction of change in state of stress, an accurate model representation must be made using a coupled hydromechanical numerical solution. Verification of this solution shall be obtained through analytical means and progressively refined through experimental results, using real-time downhole data such as microseismic, fiber optics, and advanced imaging technologies.

CHAPTER EIGHT

Hydraulic Fracturing Chemical Selection and Design

INTRODUCTION

Chemical selection and design is another important aspect of a hydraulic fracture design. As opposed to public perception, i.e., that many toxic chemicals are being pumped downhole, it is important to note that the industry has done a tremendous job developing new chemicals that are environmentally friendly and do not cause any harm to the public health and safety. Each chemical used in the hydraulic fracturing process has a very specific purpose. For instance, the slick water frac fluid system uses friction reducer (FR) to reduce the friction pressure when pumping at high rates, whereas the cross-linked frac fluid system uses linear gel and cross-linker to create the viscosity needed to place higher sand concentration proppant into the formation. Exploration and production (E&P) companies can save hundreds of thousands of dollars on chemical selection and design. An optimized chemical package including types and concentrations of each needed chemical is crucial in a successful and cost-effective frac job. Therefore, various field and laboratory tests must be performed to find the optimum design. Since chemicals are part of each frac stage cost and E&P companies are responsible for paying for the type and amount of each chemical used, it is very important to perform various field and laboratory tests to find the optimum chemical design, as will be discussed in this chapter. Chemicals used during hydraulic frac jobs are costly and running the chemical at higher concentrations than needed can add a significant amount of expenditure to each frac stage, which can add up rapidly. There are a limited number of chemicals used in hydraulic fracturing. The most commonly used chemicals in hydraulic fracturing are discussed in the following sections.

Hydraulic Fracturing in Unconventional Reservoirs
DOI: http://dx.doi.org/10.1016/B978-0-12-849871-2.00008-3
© 2017 Elsevier Inc.
All rights reserved.

FRICTION REDUCER

FR is the most important chemical used during slick water frac jobs. FR is a type of polymer used to reduce the friction inside the pipe significantly in order to successfully pump the job under the maximum allowable surface-treating pressure. FR reduces the friction between fracturing fluids and tubular. Without FR, it is impossible to pump slick water frac jobs because of very high friction pressure inside the pipe. The high friction pressure is due to the high flow rate that is used during slick water frac jobs. The concentration of FR used varies from 0.5 gpt to 2.0 gpt depending on the quality of FR and water. The unit for FR concentration is gpt, which stands for gallons of FR per thousand gallons of water. A measurement of 1.0 gpt means there is 1 gallon of FR in 1000 gallons of water. The quality of water has a significant impact on FR. For example, if freshwater is used for the frac job, lower concentrations of FR are needed to control the friction pressure; however, if reused water is used for the frac job, the FR needs to be run at higher concentrations to reduce the friction pressure. Another important factor that necessitates running FR at higher concentrations is the quality of FR. It is very important that operating companies discuss the type and quality of FR that a service company provides. One of the main reasons that the type and quality of FR must be monitored is the cost. All fracturing chemicals are expensive and most frac service companies typically make most of their money from chemicals. Therefore, not monitoring the quality and type of FR can cost the operator lots of money by running the FR at higher concentrations, which might not be necessary. The most commonly used FR in the industry is polyacrylamide, of which there are nonionic, cationic, and anionic types. FR comes in dry powder and liquid form with a mineral oil base. Polyacrylamide is also used for soil stabilization and children's toys. FR selection depends on:

- chemistry of source water for fracturing;
- high salinity versus fresh water = different products;
- quality of FR (supplier or service company provider).

FR FLOW LOOP TEST

Economically attractive oil and gas production rates from unconventional shale reservoirs highly depend on the effectiveness of hydraulic fracturing stimulation that can provide maximum reservoir contact. This can be achieved by establishing high pumping rates to inject millions of gallons of fracturing fluid in these tight formations. However, there are technical and environmental problems associated with this technique that must be resolved. To meet the high pumping rates, the main problem is to overcome the tubular friction pressure that can reduce the hydraulic horsepower demand by 80% (Virk, 1975). This can be achieved by adding FRs to the fracturing fluid. The postfracturing flowback fluids have extremely high salinity and different concentrations of dissolved mineral and chemicals that cannot be just discharged to the environment. Treating the produced water is also extremely expensive so that most of the operators decide to recycle the produced water by directly using that for subsequent fracturing applications. The performance of FRs as salinity and total dissolved solid content of flowback fluid increases still remain as unsolved problems.

Different shale plays have different temperature and salt content and there is no one-size-fits-all formula for the amount or composition of each additive of fracturing fluids. Salt affects the functions of additives including surfactants, polymers, and gels in a complex way. The most efficient way to test its effects and get the most suitable formula is physical experiment, such as via dynamic flow loop experiment. Fig. 8.1 shows the schematic of the dynamic flow loop experiment. This setup includes 13′ of ½″ stainless steel in the direct and 13′ of ¾″ straight tubing in the return direction, 16 gallon reservoir tank, 7 HP electric motor, variable capacity pump, flow meter, overhead mixer, thermocouple, insulation, and band heater. We also think FRs can impact the

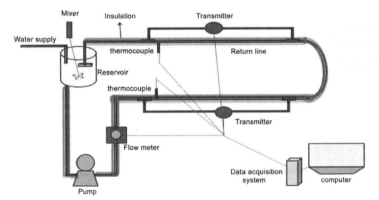

Figure 8.1 Schematic of flow loop apparatus.

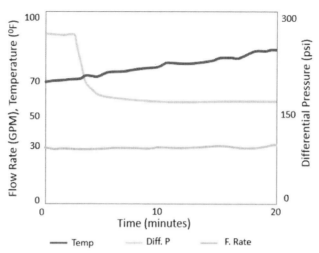

Figure 8.2 Flow loop test results.

proppant transport and settling, and since the effectiveness of hydraulic fracturing significantly depends on proppant displacement, this also needs to be investigated. Numerical simulation is a powerful tool for simulation of proppant transport and displacement in hydraulic fractures as discussed earlier.

FR concentration can be determined by performing lab tests such as FR flow loop test. This test can be performed by taking a water sample that is being used during the frac job to the lab. This water sample is then run through a loop and various FR types and concentrations are tested to find the best FR type and optimum FR concentration of that particular FR. To find the best FR type, different FRs at the same concentration are tested and the one with the maximum friction reduction is selected. Next, the FR concentration of the preselected FR type is increased gradually until the addition of FR does not have a significant impact on pressure drop. This concentration is then recorded and reported to the operating company for their optimum FR concentration design. Fig. 8.2 shows an example of flow test results, where the flow and temperature are measured by flow meter and thermocouples as a function of time and are plotted against average pressure drop measured by two differential pressure transducers in the direct and return lines. During the experiment, the flow rate is kept constant and pressure drop is monitored as FR is added to the fluid system.

PIPE FRICTION PRESSURE

Before discussing the next chemical, it is important to understand friction pressure in slick water jobs without FR. Friction pressure inside the pipe is impacted by rate, fluid viscosity, pipe diameter, and fluid density. A smaller pipe diameter causes the friction pressure to increase. For example, if a 4 ½″ production casing is used instead of a 5 ½″ casing for the hydraulic fracturing treatment, friction pressure inside the pipe will increase. Fluid viscosity and density are very important parameters as well. Various concentrations of FRs are used by different operating companies in an attempt to lower the pipe and perforation friction pressure. Water quality during the frac job is an important factor to consider when designing FR concentration. If heavy flowback water is used without any treatment, more FR has to be run to reduce the friction inside the pipe.

Rate is another important parameter affecting pipe friction pressure. Rate has a proportional relationship with pipe friction pressure. This implies that by increasing the rate, pipe friction pressure will increase as well. Rate not only increases pipe friction pressure, but it also increases perforation friction pressure, which will be discussed.

Friction pressure is calculated using Eq. (8.1):

Pipe friction pressure

$$= \frac{11.41 \times \text{fanning friction factor} \times \text{length of pipe} \times \text{fluid density} \times \text{flow rate}^2}{\text{inside pipe diameter}^5}$$

Equation 8.1 Pipe friction pressure

Length of pipe = ft
Fluid density = ppg
Flow rate = bpm
Inside pipe diameter = inches.

Fanning friction factor is the most difficult parameter to find and there are various methods that can be used to come up with the fanning friction factor. There are two parameters that must be known in order to calculate fanning friction factor, including Reynolds number and relative roughness of pipe.

REYNOLDS NUMBER

The Reynolds number needs to be calculated in order to come up with the fanning friction factor. Slick water is considered to be a Newtonian fluid; therefore, the following equation is used to calculate the Reynolds number for Newtonian fluid:

$$N_R = \frac{1.592 \times 10^4 \times \text{rate} \times \text{fluid density}}{\text{Inside diameter of pipe} \times \text{fluid viscosity}}$$

Equation 8.2 Newtonian fluid Reynolds number

RELATIVE ROUGHNESS OF PIPE

Relative roughness is the amount of surface roughness that exists inside the pipe. The relative roughness of a pipe is known as the absolute roughness of a pipe divided by the inside diameter of a pipe.

$$\text{Relative roughness} = \frac{\varepsilon}{D}$$

Equation 8.3 Relative roughness

$\varepsilon =$ absolute roughness in inches

$D =$ inside diameter of pipe in inches.

Once relative roughness and Reynolds numbers are calculated, fanning friction factor can be obtained depending on whether the flow is laminar or turbulent. There are two equations to calculate the fanning friction factor. Eq. (8.4) is for laminar flow (which means the Reynolds number is less than 2300) and is as follows:

$$F = \frac{16}{N_R}$$

Equation 8.4 Fanning friction factor for laminar flow

If the Reynolds number is more than 4000, then Eq. (8.5) is used to calculate the Darcy friction factor:

$$f(D) = \left(\frac{1}{-1.8 \times \log_{10}\left[\left(\frac{\text{relative roughness}}{3.7} \right)^{1.11} + \left(\frac{6.9}{N_R} \right) \right]} \right)^2$$

Equation 8.5 Darcy friction factor for turbulent flow

Please note that:
Darcy friction factor $= 4 \times$ fanning friction factor $>>>$ therefore:

$$\text{Fanning friction factor for turbulent flow} = \frac{f(D)}{4}$$

Equation 8.6 Fanning friction factor for turbulent flow

Once the fanning friction factor is obtained, the pipe friction pressure can be calculated. This is just one method of pipe friction calculation, which can be very tedious and time consuming. However, there are various handbooks and software available that calculate friction pressure inside the pipe considering the impact(s) of FR concentration. Please note that this calculation does *not* take the use of FR into account. This is just to demonstrate that if FR is not used during high-rate slick water jobs, it is practically impossible to pump the job. Table 8.1 shows the relative roughness of different pipe materials.

Table 8.1 Relative Roughness (Binder, 1973)

Pipe Material	Absolute Roughness (inches)
Drawn brass	0.00006
Drawn copper	0.00006
Commercial steel	0.0018
Wrought iron	0.0018
Asphalted cast iron	0.0048
Galvanized iron	0.006
Cast iron	0.0102
Wood stave	0.0072–0.036
Concrete	0.012–0.12
Riveted steel	0.036–0.36

●●●

Example

Calculate pipe friction pressure if 11,000″ of 5 ½″, 20 lb/ft, P-110 pipe (ID = 4.778″) is run in the hole without the use of FR (assume fresh water is used for frac). Assume a relative roughness of zero inside the pipe.

Rate = 100 bpm, Freshwater density = 8.33 ppg, Freshwater viscosity = 1 cp.

Step 1) The Reynolds number needs to be calculated in order to come up with the fanning friction factor:

$$N_R = \frac{1.592 \times 10^4 \times 100 \times 8.33}{4.778 \times 1} = 2,775,504$$

Step 2) Since the relative roughness inside the pipe is assumed to be zero and the Reynolds number confirms that the flow is turbulent, the Darcy friction factor for turbulent flow can be calculated:

$$f(D) = \left(\frac{1}{-1.8 \times \log 10 \left[\left(\frac{0}{3.7} \right)^{1.11} + \left(\frac{6.9}{2,775,504} \right) \right]} \right)^2 = 0.009826$$

$$\text{Fanning friction factor} = \frac{\text{darcy friction factor}}{4} = \frac{0.009826}{4} = 0.00246$$

Step 3) Now that the fanning friction factor is calculated, pipe friction pressure can be calculated using the following equation:

$$\text{Pipe friction pressure} = \frac{11.41 \times 0.00246 \times 11,000 \times 8.33 \times 100^2}{4.778^5} = 10,314 \text{ psi}$$

Without running FR during the frac treatment, there will be 10,314 psi of pipe friction pressure, and pumping a high-rate frac job will be virtually impossible. This example shows the importance of running FR at predetermined concentrations.

FR BREAKER

FR breaker is used to reduce the viscosity of FR. An example of a FR breaker is hydrogen peroxide.

BIOCIDE

Biocide is another important chemical used in hydraulic fracturing. The primary duty of biocide is killing and controlling bacteria. Bacteria can cause instability in viscosity. The concentration of biocide typically varies between 0.1 and 0.3 gpt. Prejob water testing is performed to measure preexisting bacteria present in the water. This test introduces an effective agent with frac source water. A change in the bottle sample directly relates to the bacteria count. Results are then used to determine the biocide concentration (gpt) required for the frac job. The most commonly used biocide product is called glutaraldehyde and this product is typically pumped as a liquid additive with hydraulic fracturing fluid. The basic types of oilfield bacteria are:

- sulfate-reducing bacteria, which is the oldest known bacteria, and which creates H_2S (hydrogen sulfide, poisonous gas) and sulfide, which can form FeS (iron II sulfide) scale;
- acid-producing bacteria, which produces corrosive acid and can adapt to aerobic or anaerobic conditions;
- general heterotrophic bacteria, which are usually formed in aerobic conditions.

The consequences of not using biocide are:

- H_2S creation in the formation (a safety hazard for producing wells)
- microbiological influenced corrosion
- production restriction due to microbial growth.

Now that the different types of bacteria have been discussed, we will take a look at the types of biocide used in hydraulic fracturing and other applications:

- Oxidizing biocide causes irreversible cell damage to the bacteria. Put simply, this type of biocide burns the cell. Examples of oxidizing biocides are chlorine, bromine, ozone, and chlorine dioxide.
- Nonoxidizing biocide alters the cell wall permeability, interfering with biological processes. This type of biocide essentially gives the bacteria cell cancer, which can result in bacteria either dying or surviving. Examples of nonoxidizing biocides are aldehydes, bronopol, DPNPA, and acrolein.

SCALE INHIBITOR

Scale inhibitor is another commonly used chemical in hydraulic fracturing. Scale inhibitor prevents iron and scale accumulation in the formation and wellbore. In addition, scale inhibitor enhances permeability by eliminating scale in the formation and casing. Scale is a white material that forms inside the pipe (casing) and restricts the flow. The concentration of scale inhibitor is usually 0.1−0.25 gpt. Scale is formed by temperature changes, pressure drops, the mixing of different waters, and agitation. An example of a commonly used scale inhibitor is ethylene glycol (commonly used in antifreeze). The most common types of scale in the oil and gas field are as follows:

• calcium carbonate, which is the most common type of scale and, as opposed to most forms of scale, is less soluble in higher temperatures;
• barium sulfate, which forms a very hard and insoluble scale that has to be mechanically removed;
• iron sulfide, which is the most common sulfide type, and is formed as sulfate-reducing bacteria reduces sulfate;
• sodium chloride, also known as salt, is another self-explanatory type of scale.

LINEAR GEL

Linear gel is sometimes used with slick water frac to facilitate placing proppant into the formation. Linear gel increases the viscosity of frac fluid, adds friction reduction, and eases the proppant transport into the formation. Higher fluid viscosity increases fracture width, and proppant can be transported more easily into the formation, especially at higher sand concentrations. In addition, gelling agents such as linear gel increase fluid-loss capabilities. Typical polymer types for gelling agents are guar (G, raw guar contains 10−13% insoluble residue), hydroxypropyl guar (HPG, 1−3% insoluble residue), carboxymethyl hydroxypropyl guar (CMHPG, 1−2% insoluble residue), hydroxyethyl cellulose (HEC, minimal residue), and polyacrylamide (FR, minimal residue). Typically, the less-residue gelling agents are associated with more refined gelling agents, and as a result are more expensive.

Guar is the most common linear gel that is currently being used in the industry. Guar is considered the cheapest polymer compared to the other polymers discussed above because it leaves much more insoluble residue behind. Guar is primarily grown in India and Pakistan. Guar is often harvested by hand as a secondary crop by subsistence farmers and can be used for human and cattle food. Guar seeds can be grounded into powder. Guar is typically used as slurry concentrate; however, it can also be used as dry powder that is mixed on the fly.

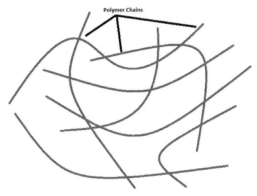

Figure 8.3 Linear base gel (polymer chains).

As previously mentioned, linear gel increases fracture width and bigger sand size and higher sand concentrations can be eased into the formation. When linear gel hits the perforations during a slick water frac, the surface-treating pressure will decrease. The decrease in surface-treating pressure allows surface-treating rate to increase. For example, if a stage is being treated at 9500 psi (maximum allowable surface treating pressure for 5 ½″, P-110, 20 lb/ft casing) and 30 bpm, linear gel is used with slick water to overcome tortuosity, increase fracture width, decrease surface-treating pressure, and be able to get into the stage. The reason surface-treating rate can be increased when using linear gel is because linear gel increases the width of the fracture and the viscosity of the fluid, which in turn improves the proppant transport into the fractures. The concentration of linear gel used during the stage varies and is typically 5−30 lb systems depending on the severity of tortuosity. Since guar concentrate is commonly mixed at ∼4 lb/gal, a 5-lb system means 5 divided by 4, which is 1.25 gpt (gallons of gel per thousand gallons of water). Fig. 8.3 illustrates the schematic of polymer chains in a linear base gel system.

GEL BREAKER

Gel breaker is pumped along with gelling agents so the gel will break once it has been placed into the formation. Gel breaker reduces the viscosity of the gel in the formation. Gel breaker causes the gel to break (reduces viscosity) at certain temperatures at downhole conditions. The degree of gel reduction is controlled by gel breaker type, gel concentration, breaker concentration, temperature, time, and pH. Gel breaker can be tested at the surface by heating it to the formation temperature (using a bath) to ensure proper reduction in viscosity after breaking. It is strongly recommended to do a gel breaker test to visualize the reduction in viscosity. If gel is not completely broken in the formation, it can cause serious formation damage, such as a reduction in conductivity.

BUFFER

Buffer is another chemical that is used when linear gel is used. Buffer is run at predetermined concentrations based on the lab test analysis. Buffer essentially adjusts and controls the pH of the gel for maximum effectiveness. The only time it is necessary to run buffer in conjunction with gel (with slick water frac) is if the base fluid has a basic pH (pH from 8 to 14 is considered basic). If the base fluid has a basic pH, buffer must be run to bring the pH down to neutral or slightly acidic pH (6.5–7). It is the service company's quality assurance/ quality control (QA/QC) responsibility to measure the pH of the fracturing fluid to determine whether buffer is needed or not. There are two types of buffer used in the industry. The first one is referred to as *acidic buffer*, which is used to speed up the gel hydration time. The second type of buffer is called a *basic buffer*, and is used with cross-linked fluid to create delayed cross-linked. The *delayed cross-linked* delays the process of cross-linked fluid to ensure less friction inside the pipe when the fluid passes through thousands of feet of pipe. Having passed through the pipe, the cross-linked fluid starts working normally in an attempt to overcome the tortuous path after the perforations and before the main body of fractures. pH measures how acidic or basic a substance is, and ranges from 0 to 14. A pH level of 7 (e.g., distilled water) is neutral. A pH of less than 7 (e.g., black coffee and orange juice) is acidic. Finally, a pH of more than 7 (e.g., bleach and baking soda) is considered to be basic. Examples of two buffers used in frac jobs are potassium carbonate and acetic acid.

CROSS-LINKER

Cross-linker is the chemical used to create a cross-linked fluid system. When cross-linker is combined with a 20—30 lb linear gel system, cross-linked fluid is created. Cross-linker increases the viscosity of gelling agents by connecting the separate gel polymers together. Cross-linker significantly increases the viscosity of linear gel by increasing the molecular weight of the base polymer by linking multiple molecules together. Cross-linker increases molecular weight without additional polymers. From an economic perspective, it is far more expensive to create heavy viscous fluid with linear gel than a cross-linked fluid system. For example, when linear gel is used to create a 150-centipoise fluid system, it is considerably more expensive than using a cross-linked fluid system to create the same viscosity. This feature of the cross-linked fluid system is considered to be the biggest advantage of using this type of fluid system. One of the disadvantages of cross-linkers is the potential increase in friction pressure. On the other hand, cross-linkers improve the fluid's ability to carry proppant and create viscosity for wider fracture geometry. Common cross-linkers are borate (high pH and moderate temperatures) and zirconate (low pH and high temperatures). Fig. 8.4 illustrates how cross-linkers link multiple molecules together.

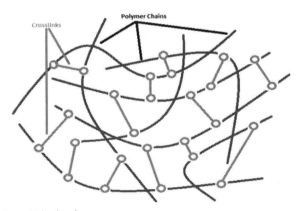

Figure 8.4 Cross-linked gel.

SURFACTANT

Surfactants have different applications. The main application of surfactant is to reduce the surface tension of a liquid. Surface tension is the tendency of a liquid surface to resist an external force. There are various surfactants available in the industry for different usages. The most commonly used surfactants in hydraulic fracturing operations are as follows:

- Microemulsion is a type of surfactant that changes the contact angle, which results in reducing surface tension. Reducing the surface tension results in more fluid recovery during flowback. This type of surfactant was used in the early development of Marcellus Shale to gain more fluid recovery; however, many operators stopped using surfactants as fracturing fluid in dry gas regions of the Marcellus and Barnett Shales.
- Nonemulsifiers minimize or prevent the formation and treatment fluids from emulsion. This type of surfactant is typically used in formations with oil or condensate in an attempt to separate oil or condensate from an aqueous emulsion. Some companies use nonemulsifier in liquid-rich areas of Utica Shale.
- Foamers create stable foam and allow for effective proppant transport.

Surfactants have many more applications and selection depends upon the desired goal. Examples of surfactants are methanol, isopropanol (common use: glass cleaner), and ethoxylated alcohol.

IRON CONTROL

Iron control is used to control and prevent dissolved iron in frac fluid. Iron control prevents the precipitation of some chemicals, such as carbonates and sulfates, which can plug off the formation. Examples of iron controls are ammonium chloride, ethylene, citric acid (food additive), and glycol.

CHAPTER NINE

Fracture Pressure Analysis and Perforation Design

INTRODUCTION

The next step in hydraulic fracturing stimulation is to understand the basic pressure concepts for a successful fracture design, treatment, and execution. Understanding pressure is one of the key aspects of a safe and successful frac operation. One of the most important concepts is the calculation of surface-treating pressure, which is used for production casing design by completion engineers and is discussed in further detail in this chapter. Casing design is very important in new exploration areas because some operators are not able to successfully initiate hydraulic fracturing due to underestimating the expected surface-treating pressure and using a low-burst casing pressure size and grade. Therefore, understanding the basic pressure concepts that will be used in casing design calculation is crucial to a successful frac job. Perforation design is another important parameter in completion design. In this chapter, a special emphasis will be placed on designing limited entry for optimum production enhancement in unconventional reservoirs.

PRESSURE (psi)

Pressure is defined as force divided by area. The unit of pressure in the oil and gas field is psi, which stands for pounds per square inch. For example, 3000 psi means 3000 lbs of force over a square inch of area.

$$P = \frac{F}{A}$$

Equation 9.1 Pressure.

Hydraulic Fracturing in Unconventional Reservoirs
DOI: http://dx.doi.org/10.1016/B978-0-12-849871-2.00009-5
© 2017 Elsevier Inc.
All rights reserved.

P = Pressure in psi
F = Force in lbs
A = Area in square inches.

HYDROSTATIC PRESSURE (psi)

Hydrostatic pressure is the pressure of the fluid column exerted in static condition. Hydrostatic pressure is one of the most important concepts that must be learned by heart. Hydrostatic pressure depends on the weight of fluid (ppg) and true vertical depth (TVD) of the well. In addition, 0.052 is a constant for conversion to psi. One of the most common mistakes that beginners make is using measured depth (MD) instead of TVD to calculate hydrostatic pressure in the wellbore. Measured depth can be used for volume calculation; however, TVD has to be used for hydrostatic pressure calculation. The hydrostatic pressure can be calculated using Eq. (9.2).

$$P_h = 0.052 \times \rho \times TVD$$

Equation 9.2 Hydrostatic pressure.

P_h = Hydrostatic pressure, psi
ρ = Fluid density, ppg
TVD = True vertical depth, ft.

HYDROSTATIC PRESSURE GRADIENT (psi/ft)

Hydrostatic pressure gradient refers to the pressure exerted by the column of fluid per foot of TVD. For example, freshwater has a hydrostatic pressure gradient of 0.433 psi/ft, which means 0.433 psi of fluid column acts on 1 ft of TVD. Hydrostatic pressure gradient is the multiplication of 0.052 constant by fluid density (ppg). Hydrostatic pressure gradient can be calculated using Eq. (9.3).

$$P_h \text{ gradient} = 0.052 \times \rho$$

Equation 9.3 Hydrostatic pressure gradient.

P_h gradient = Hydrostatic pressure gradient, psi/ft
ρ = Fluid density, ppg.

●●●

Example

Calculate hydrostatic pressure and hydrostatic pressure gradient for a well with the following properties:

$$TVD = 10,500 \text{ ft}, \quad MD = 19,500 \text{ ft}, \quad \rho = 8.55 \text{ ppg}.$$

Please make sure to use TVD and not MD when calculating hydrostatic pressure.

$$P_h = 0.052 \times \rho \times TVD = 0.052 \times 8.55 \times 10,500 = 4668 \text{ psi}$$

$$P_h \text{ gradient} = 0.052 \times \rho = 0.052 \times 8.55 = 0.4446 \text{ psi/ft}$$

INSTANTANEOUS SHUT-IN PRESSURE (ISIP, psi)

ISIP stands for instantaneous shut-in pressure, and is the pressure at which all of the pumps come offline following a hydraulic fracturing stage treatment or diagnostic fracture injection test (DFIT). ISIP can be obtained using a surface-treating pressure graph after each hydraulic fracture stage treatment. ISIP is extremely important to calculate for new exploration areas where hydraulic fracturing will take place in order to ultimately calculate the estimated surface-treating pressure. Fig. 9.1 illustrates surface

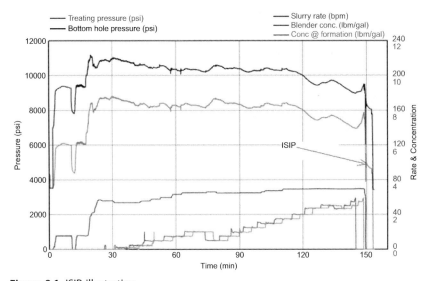

Figure 9.1 ISIP illustration.

treating pressure, calculated bottom hole pressure, slurry rate, blender, and formation sand concentrations. ISIP in Fig. 9.1 is the pressure as soon as all of the pumps are offline (i.e., the slurry rate goes to 0). In this figure, ISIP is approximately 4900 psi. ISIP can also be calculated using Eq. (9.4).

$$ISIP = BHTP - P_h$$

Equation 9.4 Instantaneous shut-in pressure.

ISIP = Instantaneous shut–in pressure, psi
BHTP = Bottom–hole treating pressure, psi
P_h = Hydrostatic pressure, psi.

At the end of a hydraulic fracturing stage when all of the frac pumps are offline, pressure drops significantly and a water hammer effect can be seen on the pressure signal. Pressure continues to drop afterward due to fluid leak-off to the formation. The amount of pressure drop is directly related to the permeability of the formation and frac fluid viscosity. Fig. 9.2 illustrates the surface-treating pressure and surface-treating rate (pump injection rate) versus time. To determine the ISIP, one must draw a vertical line at the point at which surface-treating rate goes to zero and fit a straight line to the pressure fall-off after the shut-in. The intersection of the two lines yields the ISIP.

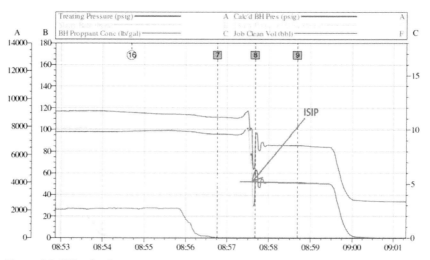

Figure 9.2 ISIP selection.

Example

Calculate ISIP with the following data:

Bottom-hole treating pressure = 10,000 psi, TVD = 7550', Fluid density = 8.9 ppg

$$\text{ISIP} = \text{BHTP} - P_h = 10{,}000 - (0.052 \times 8.9 \times 7550) = 6506 \text{ psi}$$

FRACTURE GRADIENT (FG, psi/ft)

Fracture gradient, also known as frac gradient, is the pressure gradient at which the formation breaks. Frac gradient is crucial to understand in order to calculate the expected bottom-hole treating pressure (BHTP) before the start of a frac job. Eq. (9.5) can be used to calculate the frac gradient.

$$FG = \frac{\text{ISIP} + P_h}{\text{TVD}}$$

Equation 9.5 Frac gradient.

FG = Frac gradient, psi/ft

ISIP = Instantaneous shut-in pressure, psi

P_h = Hydrostatic pressure, psi

TVD = True vertical depth, ft.

Example

ISIP at the end of a DFIT job is obtained to be around 4500 psi. If TVD of the formation is 7500' (assuming fresh water was used during DFIT), calculate the frac gradient.

$$\text{Hydsrostatic pressure} = 0.052 \times 8.33 \times 7500 = 3249 \text{ psi}$$

$$\text{Frac gradient} = \frac{4500 + 3249}{7500} = 1.033 \text{ psi/ft}$$

BOTTOM-HOLE TREATING PRESSURE (psi)

Bottom-hole treating pressure (BHTP) is the amount of pressure required at the perforations to cause fracture extension during hydraulic fracture stimulation. BHTP is the pressure along the fracture face that keeps the fractures open. BHTP is also referred to as bottom-hole frac pressure (BHFP). Correct estimation of BHTP is essential when preparing the estimates of surface-treating pressure and ultimately a frac job. BHTP can be calculated using Eq. (9.6).

$$BHTP = FG \times TVD$$

or

$$BHTP = ISIP + P_h$$

Equation 9.6 Bottom-hole treating pressure equations.

Please note that the second equation can be derived by rearranging the first equation as follows:

$$BHTP = FG \times TVD \to \frac{ISIP + P_h}{TVD} \times TVD = ISIP + P_h$$

●●●

Example
Calculate estimated BHTP if ISIP (obtained from DFIT) is 6427 psi and TVD is 8500′ (assuming fresh water).

$$\text{Frac gradient} = \frac{ISIP + P_h}{TVD} = \frac{6427 + (0.052 \times 8500 \times 8.33)}{8500} = 1.189 \text{ psi/ft}$$

$$BHTP = FG \times TVD = 1.189 \times 8500 = 10,109 \text{ psi}$$

or

$$BHTP = ISIP + P_h = 6427 + (0.052 \times 8500 \times 8.33) = 10,109 \text{ psi}$$

TOTAL FRICTION PRESSURE (psi)

There are various types of friction pressures that must be considered and calculated before and after treatment to derive perforation efficiency and optimum design. Friction pressures during a frac job are pipe friction pressure, perforation friction pressure, and tortuosity pressure. Total friction pressure after each frac stage can be calculated using Eq. (9.7).

$$FP_T = \text{Avg surface treating pressure} - ISIP$$

Equation 9.7 Total friction pressure.

FP_T = Total friction pressure, psi

Avg surface-treating pressure = psi

ISIP = Instantaneous shut-in pressure, psi.

As the name indicates, average surface-treating pressure is the average surface-treating pressure during each hydraulic frac stage. ISIP can also be obtained after each hydraulic frac stage treatment.

Example

A frac stage was completed in Barnett Shale and the data listed below was obtained at the end of the frac stage. Calculate total friction pressure for this stage.

Average surface-treating pressure for the stage = 8650 psi, ISIP = 4500 psi

FP_T = Avg surface treating pressure − ISIP = 8650 − 4500 = 4150 psi

From this example, 4150 psi indicates total friction pressure which consists of pipe, perforation, and tortuosity pressures during the stage. This basically indicates that out of 8650 psi of average surface-treating pressure, 4150 psi is the total friction pressure. In this example, total friction pressure is about 48% of the average treating pressure. This illustrates the importance of calculating and understanding pipe, perforation, and tortuosity friction pressures. Please note that 4150 psi does include the impact of FR used during the frac job. As previously discussed, without the use of FR pumping a slick water frac stage at a high rate would be impossible.

PIPE FRICTION PRESSURE (psi)

Pipe friction pressure can be calculated excluding FR impacts. However, it is much more important to obtain the pipe friction pressure after FR is added to the fracturing fluid pumped in the well. This calculation depends on the type of FR provided by the service company. There are various tools that can be used to approximate pipe friction pressure depending on the type of FR used. Service companies typically perform lab tests to understand the impact of their particular FR product on pressure, and to quantify the pressure reduction caused by the FR. The pressure reduction of each friction reducer varies depending on the type and manufacturer of the product.

PERFORATION FRICTION PRESSURE (psi)

In addition to pipe friction pressure, which is one of the main considerations in hydraulic fracturing treatment design, perforation friction pressure is another important parameter in hydraulic fracturing design that needs to be calculated and considered. Perforation friction pressure can be calculated using Eq. (9.8) if optimum perforation friction pressure for a particular area is known.

$$\text{Perforation friction pressure} = \frac{0.2369 \times Q^2 \times \rho}{C_d^2 \times D_p^4 \times N^2}$$

Equation 9.8 Perforation friction pressure.

Q = Flow rate, bpm
ρ = Fluid density, ppg
C_d = Discharge coefficient, coefficient of roundness of jet perforation
Assume C_d of 0.8−0.85
D_p = Perforation diameter (hole diameter), inches
N = Numbers of perforations (holes).

In Eq. (9.8), as pump rate (flow rate) and fluid density increase, perforation friction pressure will also increase. On the other hand, as the number of perforations (holes) and perforation diameter increases, perforation friction pressure will decrease. Perforation diameter is also referred to as entry-hole diameter (EHD). Discharge coefficient is the measure of perforation efficiency when fluid passes through the perforations. Typically a discharge coefficient of 0.6 is assumed for new perfs and a discharge coefficient of 0.85 is assumed for eroded perfs.

●●●

Example

Calculate the perforation friction pressure with the following data:
  Q = 85 bpm, ρ = 8.5 ppg, discharge coefficient of 0.8, D_p = 0.42", N = 40

$$\text{Perforation friction pressure} = \frac{0.2369 \times Q^2 \times \rho}{C_d^2 \times D_p^4 \times N^2} = \frac{0.2369 \times 85^2 \times 8.5}{0.8^2 \times 0.42^4 \times 40^2} = 457 \text{ psi}$$

OPEN PERFORATIONS

Open perforations refer to the number of perforations that are actually open during a frac stage treatment. At the beginning of unconventional shale development, some companies used up to 90 perforations (holes in the casing) per treatment stage. Does this mean that all of the perfs are open during the treatment? Absolutely not. This is primarily why the industry lowered the number of perfs in an attempt to improve the number of perfs that remain open during the treatment. Each single hole can take up to $1-3$ bpm depending on the formation. Designed pumping rates for slick water frac jobs are usually anywhere between 70 and 100 bpm. Therefore, completion engineers perform various calculations to derive the optimum design and perf efficiency so as to have as many holes open as possible during a frac stage treatment.

The perforation friction-pressure equation can be rearranged and the number of open holes (perfs) can be calculated using Eq. (9.9) if optimum perforation friction pressure for a particular area is known.

$$\text{Open perfs (holes)} = \sqrt{\frac{0.237 \times \rho_f \times Q^2}{\text{Perf friction pressure} \times C_d^2 \times D_p^4}}$$

Equation 9.9 Number of open perfs (holes).

PERFORATION EFFICIENCY

Perforation efficiency refers to the percentage of open holes either before or after a frac job. Typically the perforation efficiency during hydraulic frac jobs ranges from 30% to 80%. If 80% perforation efficiency can be obtained from a frac stage, it is considered to be an outstanding perforation design. If the designed number of perforations per stage is 45 holes, on average $50-60\%$ of the holes could possibly be open during the frac job. This means hydraulic frac stimulation will have only taken place through $23-27$ holes out of the original 45 holes. Therefore, it is important to understand that perforation efficiency is highly dependent on the perforation design, formation type/heterogeneity, natural fracturing, and stresses around the stimulated zones. These factors could all impact the perforation efficiency that can be obtained from a well. Perforation efficiency can be calculated using Eq. (9.10).

$$\text{Perforation efficiency (\%)} = \frac{\text{Number of open holes}}{\text{Designed number of holes}} \times 100$$

Equation 9.10 Perforation efficiency.

PERFORATION DESIGN

Another important concept in hydraulic fracturing design is the number of holes per stage. Designing the number of holes per stage in a conventional reservoir is completely different than in an unconventional shale reservoir. *Limited entry* is a term of art used in the industry and is referred to as the practice of limiting the number of perforations (holes) in a completion stage to help the development of perforation friction pressure during a frac stimulation treatment. The "choking" effect produces back pressure in the casing, which allows simultaneous entry of fracturing fluid into multiple zones of varying in situ stress states. Treatment distribution among zones can be controlled to a degree. Limited entry is known to increase perforation efficiency, and as a result, production in unconventional shale reservoirs (Cramer, 1987). Limited entry can be achieved using the following steps:

1. Determine the friction pressure of a single perforation for the limited entry design. A value of at least 200−300 psi is recommended since a value of this magnitude should be noticeable in the total surface-treating pressure.
2. Once the friction pressure is chosen, solve for rate per perforation to determine the rate per perf (Q/N). This new equation provides the rate per perforation needed to develop the friction pressure of a single perforation.

$$\text{Original perforation friction pressure} = P_f = \frac{0.2369 \times Q^2 \times \rho}{C_d^2 \times D_p^4 \times N^2}$$

Eq. (9.11) can be obtained by rearranging the perforation friction pressure and solving for Q/N.

$$\frac{Q}{N} = \frac{D_p^2 \times C_d \times \sqrt{\frac{P_f}{\rho}}}{0.487}$$

Equation 9.11 Q/N in limited entry design.

Please note that the value of friction pressure in a single perforation must be chosen based on the production success in an area (knowing the history of injection rates and number of perfs in different wells).

Table 9.1 Limited-Entry Design Example

Number of Perfs	Rate (bpm)
20	32
25	40
30	49
35	57
40	65
45	73
50	81

Example

Calculate the desired rate per perf (Q/N) if 260 psi is the desired perforation friction pressure per perf assuming the following data:

$D = 0.42"$, $C_d = 0.80$ (coefficient of roundness of jet perforation, 1.0 is round), $P_f = 260$ psi, $\rho = 8.33$ lb/gal

$$\frac{Q}{N} = \frac{D_p^2 \times C_d \times \sqrt{\frac{P_f}{\rho}}}{0.487} = \frac{0.42^2 \times 0.8 \times \sqrt{\frac{260}{8.33}}}{0.487} = 1.6 \text{ bpm/perf}$$

Based on the calculated rate per perforation, the injection rate for the limited-entry fracturing treatment can be determined by taking into account the maximum allowable surface-treating pressure. From this example, Table 9.1 can be constructed. This table shows the number of perforations required at various injection rates if 260 psi per perforation is chosen to be the desired perforation friction pressure.

NUMBERS OF HOLES (perfs) AND LIMITED ENTRY TECHNIQUE

Holes in fracing are also referred to as perfs (perforations). The number of holes is important in a frac design. It was the industry's belief that more holes would result in better productivity by having more reservoir entries in unconventional shale reservoirs. However, time and actual production data have proven otherwise. Limited entry is believed to result in better perf efficiency and production. Limited entry means obtaining roughly 2 bpm (rate) or more from each hole. In the limited entry technique, EHD in the casing acts as a choke. During a frac stage, the choked flow rate through a limited number of holes produces back pressure. As a result, back pressure impacts the fracture propagation pressure. To a certain degree, this will yield a controlled treatment distribution among fractured zones. The number of holes per cluster depends on the length of the perf gun. For example, if a 1′ perf gun is used and 6 shots per foot are the designed shot density, 6 shots (holes) will be used in each cluster. If there are 6 clusters in one frac stage, 36 holes are used for one particular stage. The length of the perf gun varies between 1′ and 3′ depending on the operator.

PERFORATION DIAMETER AND PENETRATION

A perforation diameter frequently used in shale plays can range between roughly 0.42″ and 0.58″. A 0.42″ EHD means each hole created in the casing has a diameter of 0.42″. In addition, the nominal penetration depends on the type/size, manufacturer of perforation gun, and the amount of explosives used in each gun. A common nominal penetration obtained in shale formations varies between 7″ and 45″. Deep penetration shots are believed to help bypass the near-wellbore damage (e.g., skin damage from drilling) and be closer to the virgin rock in order to establish the initial fractures.

PERFORATION EROSION

Another important topic in the hydraulic fracturing world is perforation erosion. Does each hole that has a certain diameter (e.g., 0.42″) stay the same size after pumping thousands of pounds of proppant? The answer is no because perforations will erode and get bigger. Perforation friction pressure is dependent on erosion rate. As perforations erode, perforation friction pressure will decrease. As previously discussed, the discharge coefficient in the perforation friction-pressure equation takes into consideration whether the perfs are new or eroded when calculating perforation friction pressure.

NEAR-WELLBORE FRICTION PRESSURE (NWBFP)

Near-wellbore friction pressure (NWBFP) is another term used to indicate the total pressure loss near the wellbore. NWBFP is the sum of perforation friction pressure and tortuosity. Eq. (9.12) is used to calculate NWBFP.

Near wellbore friction pressure = perforation friction pressure + tortuosity

Equation 9.12 Near-wellbore friction pressure.

So far, total friction pressure has been discussed as follows:

Total friction pressure = pipe friction pressure + perforation friction pressure + tortuosity

FRACTURE EXTENSION PRESSURE

Fracture extension pressure is referred to as the pressure inside the fracture(s) that makes the fractures grow as pumping continues. In other words, fracture extension pressure is the pressure required to extend the existing fractures. In order to keep the fractures open while gaining length, height, and width, the fracture extension pressure must be greater than the closure pressure of the formation. Fracture extension pressure can be thought of as bottom hole treating pressure (BHTP). These terms are used interchangeably.

Fracture extension pressure = Frac gradient × TVD

Equation 9.13 Fracture extension pressure.

CLOSURE PRESSURE

Closure pressure is the minimum pressure required to keep the fractures open. In other words, closure pressure is the pressure at which the fracture closes without proppant in place. For example, during a hydraulic fracturing treatment, closure stress in the pay zone must exceed the BHTP in order to grow an existing fracture. This means that BHTP has to be greater than the pay zone's closure stress. Difficulty starting a stage during frac jobs could be due to not exceeding closure stress because of high closure stress in the zone of interest. As a result, it is highly recommended to land the wellbore in a zone that has higher closure pressure above and below in an attempt to keep the fractures contained. Closure pressure can be assumed to be the same as the minimum horizontal stress. Determining the closure pressure is extremely important in a hydraulic fracturing design because it helps the engineers determine the type of sand needed for the job. Closure pressure can be determined from a DFIT or step-rate test.

A step rate test is performed before the frac job and is used to determine the fracture extension pressure (P_{EXT}). Fracture extension pressure is normally slightly higher than fracture closure pressure. The first method in determining the closure pressure is a step-up test, which is part of a step-rate test. Therefore, this test is useful in figuring out the upper boundary of closure pressure.

Procedure

1. Water is typically pumped using the lowest possible rate a pump can handle (usually 0.5–1 bpm). Once the desired rate has been reached, wait for pressure stabilization and then record the exact pressure and rate.
2. After getting the exact pressure and rate, step up the rate to 1.5, 2, 3, 5, and 10 bpm and record the stabilized pressure at each rate.

If done correctly, this test is very simple to perform. Please note that this test needs to be done before starting treatment for accurate results. Conducting this test after a frac stage treatment will yield inaccurate results that cannot be used for determining the closure pressure. Fig. 9.3 illustrates how to determine the fracture extension pressure, which can be used to estimate an upper boundary value for the closure pressure. This test usually takes 15 minutes to conduct.

The second method that can be used to determine closure pressure is an injection falloff test. In this test, the fluid is injected at a constant rate and the well is then shut in. The pressure will naturally fall below the closure pressure, and eventually the fractures close. The permeability of the formation will determine the time to closure. The lower the permeability, the longer it takes to reach closure pressure. The time to reach closure is a

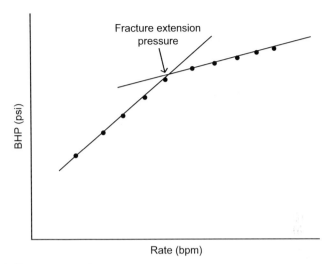

Figure 9.3 Fracture extension pressure.

function of pump time and estimated permeability. Since this type of test requires longer shut-in time for unconventional shale reservoirs due to low permeability, the practical application is limited to higher perm reservoirs. The time to reach closure can be approximated using Eq. (9.14).

$$\text{Time to reach closure} = \frac{0.3 \times \text{pump time}}{\text{estimated perm}}$$

Equation 9.14 Time to reach closure (Barree, 2013).

Pump time = minutes
Estimated permeability = md.

●●●

Example
Calculate the time to reach closure if 5 minutes of pump time was conducted in 0.003 and 0.03 md rock: Time to closure @ 0.003md $= \dfrac{0.3 \times \text{pump time}}{\text{estimated perm}} =$

$\dfrac{0.3 \times 5}{0.003} = 500 \text{ min} = 8.33 \text{ h}$

Time to closure @ 0.03 md $= \dfrac{0.3 \times \text{pump time}}{\text{estimated perm}} = \dfrac{0.3 \times 5}{0.03} = 50 \text{ min} = 0.833 \text{ h}$

Shut-in bottom-hole pressure (y-axis) is plotted versus square root of time (x-axis) to determine the fracture closure pressure. The fracture closure pressure is the point at which the flow deviates from a straight line.

Example

Estimate the fracture closure pressure using Table 9.2:

Table 9.2 Bottom-Hole Pressure (BHP) Versus Sqrt(Time) Example

Shut-in time min	BHP psi	Sqrt(Time) min
0	4300	0.00
1	4073	1.00
4	3840	2.00
6	3740	2.45
8	3605	2.83
10	3470	3.16
12	3350	3.46

BHP (*y*-axis) versus square root of time (*x*-axis) needs to be plotted. The closure pressure is illustrated in Fig. 9.4 where the plotted line deviates from the linear line. In this example, closure pressure is approximately 3600 psi.

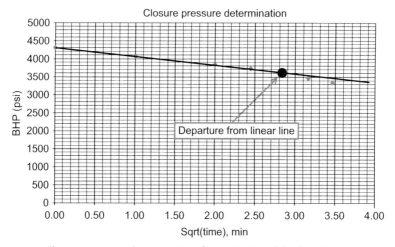

Figure 9.4 Closure pressure determination from injection fall-off test.

Another commonly used method to calculate closure pressure is from the DFIT analysis, which will be covered in detail later in the book.

NET PRESSURE

Net pressure is one of the most important pressures to consider in hydraulic fracturing. Net pressure is the energy required for propagating fractures and creating width during the frac job and refers to the excess pressure over the frac pressure required to extend the fractures. Net pressure is essentially the difference between the fracturing fluid pressure and the closure pressure and is the driving mechanism behind fracture growth. The more pressure inside a fracture, the more potential there is for growth. The term net pressure is only used when the fracture is open. If the fracture is closed, net pressure is equal to 0. Net pressure depends on various parameters such as Young's modulus, fracture height, fluid viscosity, fluid rate, total fracture length, and tip pressure. Net pressure is also referred to as process zone stress and can be calculated using Eq. (9.15) or Eq. (9.16).

$$P_{net} = BHTP - P_c$$

or

$$P_{net} = BH\ ISIP - P_c$$

Equation 9.15 Net pressure, equation 1.

P_{net} = Net pressure, psi
BHTP = Bottom-hole treating pressure, psi
P_c = Closure pressure (approximately minimum horizontal stress), psi
BH ISIP = Bottom-hole ISIP, psi, BH ISIP = ISIP + P_h.

$$P_{net} = \frac{E^{3/4}}{h}(\mu \times Q \times L)^{1/4} + P_{tip}$$

Equation 9.16 Net pressure.

E = Young's modulus, psi
h = Fracture height, ft
Q = Rate, bpm
L = Total fracture length, ft
P_{tip} = Fracture tip pressure, psi.

As can be seen from Eq. (9.16), Young's modulus is raised to the power of 3/4 while the fluid rate, viscosity, and total fracture length are only raised to the power of 1/4. This shows that Young's modulus has more impact on net pressure compared to viscosity, rate, and length. As a result, the Young's modulus measurement of a formation is a key parameter in fracture propagation. Fracture-tip pressure is a quantity that is not easy to find, however, different numerical simulations depending on a wide range of assumptions (e.g. fracture tip with or without fluid lag) will provide estimates of the fracture tip pressure, Bao et al. (2016). In hydraulic fracturing, a dynamic gap zone between fracture tip and fracturing fluid following the tip exists which can impact the fracture tip pressure.

When net pressure (y-axis) versus time (x-axis) is plotted on the log—log plot during a live frac stage treatment, a net pressure chart can be constructed. A net pressure chart is also referred to as a Nolty chart, and is used during the hydraulic frac treatment to follow various pressure trends throughout the stage. Net pressure charts are used to estimate various fracture propagation behavior at different points in time. As previously discussed, since net pressure is the driving mechanism behind the fracture growth, it can be used to predict the fracture dimension. Company representatives in the field rely heavily on the Nolty chart during the treatment since it is very accurate in conventional reservoirs. In unconventional reservoirs, this chart is still a useful tool to determine the fracture propagation, but it is not as accurate as it is in the conventional reservoirs. Fig. 9.5 shows the concept of net pressure during the treatment, which can be used to make critical decisions.

- If the pressure response in the Nolty chart is similar to trend #1, it is an indication of contained height and unrestricted length extension during the treatment (slightly positive slope).
- If the slope of the net pressure line is zero (trend #2), it represents contained height and possibly openning up more fractures with fluid loss. It indicates a less-efficient length extension.

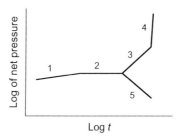

Figure 9.5 Net pressure interpretation.

- Trend #3 pressure response during treatment is bad news because the formation is giving up and there is a high possibility of a tip screening-out (sanding-off) if sand is not cut on time.
- Trend #4 is basically a full screen-out and the pump room needs to be ready to come offline as soon as the pressure starts rising dramatically to avoid exceeding the pressure limitations on the casing and equipment.
- Trend #5 illustrates uncontrolled fracture height growth.

Net pressure typically ranges between 100 and 1400 psi. In some instances, net pressure could be higher. If net pressure is much higher than 1400 psi, this could be due to near-wellbore restriction or large tip plasticity.

Example

Estimate net pressure if closure pressure is obtained from a step-rate test to be 6500 psi and ISIP is 4700 psi. The well has a TVD of 6800′ and used an 8.8-ppg frac fluid to pump the job.

$$\text{BH ISIP} = \text{ISIP} + P_h = 4700 + (0.052 \times 8.8 \times 6800) = 7811 \text{ psi}$$

$$P_{net} = \text{BH ISIP} - P_c = 7811 - 6500 = 1312 \text{ psi}$$

SURFACE-TREATING PRESSURE (STP, psi)

Surface-treating pressure (STP), also known as wellhead treating pressure (WHTP) is the pressure at the surface during a hydraulic fracturing treatment. STP during a hydraulic fracturing treatment is the real-time pressure obtained from the surface pressure transducer on the main line. A transducer uses pulsation to get the real-time pressure during a hydraulic fracture treatment. Surface-treating pressure can be estimated using Eq. (9.17).

$$\text{STP} = \text{BHTP} + P_f - P_h + P_{net}$$

Equation 9.17 Surface-treating pressure.

BHTP = Bottom-hole treating pressure, psi
P_f = Total friction pressure, psi
P_h = Hydrostatic pressure, psi
P_{net} = Net pressure, psi.

It is important to estimate the surface-treating pressure in order to have enough hydraulic horsepower (HHP) on site during the frac job. The HHP needed for the job is a function of surface-treating rate and surface-treating pressure. Once surface-treating pressure is estimated and the designed rate is known, HHP can be calculated using Eq. (9.18).

$$HHP = \frac{WHTP \times R}{40.8}$$

Equation 9.18 Hydraulic horsepower.

WHTP = Wellhead-treating pressure, psi
R = Surface-treating rate, bpm.
By rearranging the surface-treating pressure equation, BHTP can be solved as shown in Eq. (9.19).

$$BHTP = STP - P_f + P_h - P_{net}$$

Equation 9.19 Rearranged BHTP.

Example

You are the completion engineer responsible for determining the anticipated surface-treating pressure for a hydraulic frac treatment in a low-permeability field with the following data. Assuming a designed rate of 80 bpm, how much hydraulic horsepower is needed for the job? If each pump has 2250 HHP, how many pumps will be needed for the job?

ISIP = 7500 psi (from DFIT test), TVD = 10,500′, water density = 8.6 ppg, pipe friction pressure = 4221 psi (calculated assuming 1 gpt FR, 80 bpm, 20,000′ pipe MD, and 4.778″ casing ID), $D_p = 0.42″$, N = 36 perfs, $C_d = 0.8$, $\Delta P_{net} = 0$ psi(Assume net pressure is 0)

Step 1. Calculate hydrostatic pressure:

$$P_h = 0.052 \times \rho \times TVD = 0.052 \times 8.6 \times 10,500 = 4696 \text{ psi}$$

Step 2. Calculate frac gradient:

$$FG = \frac{ISIP + \text{hydrostatic pressure}}{TVD} = \frac{7500 + 4696}{10,500} = 1.16 \text{ psi/ft}$$

Step 3. Calculate BHTP:

$$BHTP = \text{Frac gradient} \times TVD = 1.16 \times 10,500 = 12,180 \text{ psi}$$

Step 4. Calculate perforation friction pressure:

$$\text{Perf friction} = \frac{0.2369 \times Q^2 \times \rho}{C_d^2 \times D_p^4 \times N^2} = \frac{0.2369 \times 80^2 \times 8.6}{0.8^2 \times 0.42^4 \times 36^2} = 505 \text{ psi}$$

Step 5. Calculate STP:

$$STP = BHTP + P_f - P_h + P_{net} = 12{,}180 + (4221 + 505) - 4696 + 0 = 12{,}210 \text{ psi}$$

Step 6. Calculate HHP needed for the job:

$$HHP = \frac{WHTP \times R}{40.8} = \frac{12{,}210 \times 80}{40.8} = 23{,}941 \text{ psi}$$

Step 7. Calculate total number of pumps if each pump is 2250 HHP:

$$\text{Total \# of pumps} = \frac{23{,}941}{2250} = 10.6$$

Typically a 20% safety factor is added to the calculated number to make sure enough HHP is available in the event that some of the pumps malfunction during the job. Therefore, 13 pumps will be needed for this job. Please note that perforation friction pressure calculated above assumes that all of the perforations are open and taking fluid during the treatment. It is recommended to take some precaution and assume that only a percentage of the total designed perforations will be taking fluid (e.g. 60%) and as a result estimate the new perforation friction pressure.

To give some perspective on HHP used during a hydraulic frac job by assuming 16 pumps for the job with 2250 HHP for each pump, this is equivalent to about 72 Corvettes.

CHAPTER TEN

Fracture Treatment Design

INTRODUCTION

Now that the concept of various pressures has been discussed, the next topic that will be discussed is fracture treatment design. In this chapter of the book, various frac schedule concepts and calculations will be presented to design a frac treatment schedule that can be used in the field to pump a frac job. This chapter will primarily focus on designing a slick water and foam fracture treatment schedule with example problems that can be followed and applied. The workflow presented in this chapter can be used and applied to generate various fracture treatment schedules for testing various completions designs.

ABSOLUTE VOLUME FACTOR (AVF, GAL/LBS)

Absolute volume factor (AVF) refers to the absolute volume that a solid occupies in water. For example, pouring 1 lb of Ottawa sand (2.65 specific gravity) into 1 gallon of water will displace 0.0453 gallons of water. The absolute volume factor depends on the density of the frac fluid and the specific gravity of the proppant used, and is calculated using Eq. (10.1).

$$\text{Absolute volume factor} = \frac{1}{\text{Absolute density}} = \frac{1}{\rho_f \times SG}$$

Equation 10.1 Absolute volume factor.

AVF = Absolute volume factor, gal/lb
ρ_f = Fluid density, ppg
SG = Specific gravity of proppant.

As can be seen from the AVF equation, as specific gravity and fluid density increase, AVF decreases.

Hydraulic Fracturing in Unconventional Reservoirs
DOI: http://dx.doi.org/10.1016/B978-0-12-849871-2.00010-1
© 2017 Elsevier Inc.
All rights reserved.
143

Example

Calculate absolute volume factor of Ottawa sand with SG of 2.65 considering freshwater density of 8.33 ppg.

$$\text{Absolute volume factor} = \frac{1}{8.33 \times 2.65} = 0.0453 \text{ gal/lb}$$

Note: 0.0453 gal/lb is commonly used for hydraulic fracturing design schedules when regular sand is utilized.

Calculate absolute volume factor of sintered bauxite with a specific gravity of 3.4 (assuming freshwater).

$$\text{Absolute volume factor} = \frac{1}{8.33 \times 3.4} = 0.0353 \text{ gal/lb}$$

DIRTY (SLURRY) VERSUS CLEAN FRAC FLUID

In hydraulic fracturing operations, two terms are frequently used. The first one is referred to as *clean volume*, which means only water and chemicals make up the volume. The second commonly used term is *dirty (slurry) volume*, which means combinations of water, sand, and chemicals make up the volume. In addition, *clean rate* refers to the rate of the clean side (water and chemicals) and the *dirty rate* refers to the rate of the dirty side (water, sand, and chemicals). The slurry rate is typically read from a flow meter located on the blender and the clean rate is normally calculated using Eq. (10.2).

$$\text{Clean rate} = \frac{\text{Slurry rate}}{1 + (\text{sand concentration} \times \text{AVF})}$$

Equation 10.2 Clean rate.

Slurry and clean rate = bpm
Sand concentration = ppg
AVF = Absolute volume factor, gal/lb.

Example

Calculate clean rate during a 3-ppg sand concentration if the slurry rate is 94 bpm (from flow meter) and Ottawa sand (SG = 2.65) is being used.

$$\text{AVF for ottawa sand} = \frac{1}{2.65 \times 8.33} = 0.0453 \text{ gal/lb}$$

$$\text{Clean rate} = \frac{\text{Slurry rate}}{1 + (\text{sand concentration} \times \text{AVF})} = \frac{94}{1 + (3 \times 0.0453)} = 83 \text{ bpm}$$

The clean rate is *always* *less* than the slurry rate because if *only* water is being pumped downhole, the rate will be less as compared to the mix of water and sand.

SLURRY (DIRTY) DENSITY (PPG)

Slurry density is the density of water and sand that are being pumped downhole. Slurry density has a direct impact on the hydrostatic pressure inside the casing during a frac job. As slurry density increases, hydrostatic pressure increases as well. If only water is being pumped downhole, the hydrostatic pressure of the column of water can be calculated using the hydrostatic pressure equation. However, when various sand concentrations are added to the water at various stages of hydraulic fracturing, slurry density must be considered in the hydrostatic pressure calculation.

$$\text{Slurry density} = \frac{\text{Base fluid density} + \text{sand conc.}}{1 + (\text{sand conc.} \times \text{AVF})}$$

Equation 10.3 Slurry density.

Base fluid density = ppg
Sand concentration = ppg
AVF = gal/lb.

Assuming every parameter in the surface-treating pressure equation stays constant during a hydraulic fracturing stage, as slurry density (sand concentrations) and hydrostatic pressure increase, the surface-treating pressure must decrease. This is because surface-treating pressure is inversely related to hydrostatic pressure. During the flush stage, after all of the designed sand volume is put away and only fluid is being pumped, surface-treating pressure is usually increased (pressure increase depends on the sand concentration). This is due to hydrostatic pressure decreasing (when sand is no longer being pumped), and as a result surface-treating pressure increasing.

Example

Calculate slurry density and hydrostatic pressure of 2.5 ppg Ottawa sand mixed with freshwater at a true vertical depth (TVD) of 7450′. How much hydrostatic pressure will be attained if sand is cut and the well is flushed with only freshwater?

$$\text{Slurry density} = \frac{\text{Base fluid density} + \text{sand conc.}}{1 + (\text{sand conc.} \times \text{AVF})} = \frac{8.33 + 2.5}{1 + (2.5 \times 0.0453)} = 9.73 \text{ ppg}$$

Hydrostatic pressure of the slurry fluid can be calculated:

$P_{h,\text{slurry fluid}} = 0.052 \times 7450 \times 9.73 = 3769 \text{ psi}$

Hydrostatic pressure of *only* fresh water can also be calculated:

$P_{h,\text{fresh water}} = 0.052 \times 7450 \times 8.33 = 3227 \text{ psi}$

Therefore:

Surface treating pressure increase = 3769 − 3227 = 542 psi surface pressure increase

This example shows the importance of sand concentration in relation to surface-treating pressure monitoring during the frac job. As soon as the extra hydrostatic pressure created by the various sand concentrations is cut, the surface-treating pressure will increase.

STAGE FLUID CLEAN VOLUME (BBLS)

Clean volume refers to the volume of water and chemicals. Stage fluid clean volume is the amount of clean volume for every proppant stage concentration. For example, after finishing the acidization and pad stages, the proppant stage is started. In slick water frac, proppant stage concentration starts with low proppant concentration of 0.1−0.25 ppg. Each proppant stage concentration can have varying clean volume. For example, 500 BBLs of frac fluid can be the designed clean volume for a 0.25 ppg proppant stage. After staging up to a 0.5 ppg proppant stage, clean volume can now be 450 BBLs depending on the job design. The amount of clean volume for each proppant stage concentration is determined from the contact surface area that would like to be created and is typically obtained using a hydraulic frac software or from an optimized schedule designed using production data. The amount of water to pump is also a function of water availability and ease of transportation, well spacing (interlateral spacing), formation properties, and distance from adjacent producing wells. Sometimes in an attempt to mitigate fracture communication between new wells and producing wells in an area, the amounts of sand and water are both reduced to avoid fracture interference (also called frac hit). For example, if a horizontal well has been producing for the last four years and a pad consisting of six horizontal wells will be hydraulically fractured right next to the producing well located 750′ apart, it is vital to adjust the schedule accordingly to mitigate fracture communication with the depleted nearby well. If the sand schedule is not properly designed and altered to facilitate this concern, this could have a detrimental production consequence to the depleted producing well.

STAGE FLUID SLURRY (DIRTY) VOLUME (BBLS)

As previously mentioned, slurry volume refers to the volume of water, proppant, and chemicals. Stage fluid slurry volume at different proppant concentrations can be calculated and is provided to the field personnel as part of the frac schedule. Slurry volume is always more than the clean volume since sand is considered part of the volume. Stage fluid slurry volume can be calculated using Eq. (10.4).

$$\text{Dirty volume} = \text{Clean volume} + (\text{sand conc.} \times \text{clean volume} \times \text{AVF}) \quad (10.4)$$

Equation 10.4 Slurry volume.

Clean volume = BBLs
Sand concentration = ppg
AVF = gal/lb.

●●●

Example

Calculate the dirty volume needed at 2 ppg Ottawa sand if 250 BBLs of clean volume is used.

$$\text{Dirty volume} = \text{Clean volume} + (\text{sand conc.} \times \text{clean volume} \times \text{AVF})$$
$$= 250 + (2 \times 250 \times 0.0453) = 273 \text{ BBLs}$$

As can be seen in the above calculation, the dirty volume is 23 BBLs more than the clean volume. This is because the 2 ppg Ottawa sand is used in that stage of treatment. As sand concentration increases throughout each stage, dirty volume increases too.

STAGE PROPPANT (LBS)

The next step in creating a frac schedule is to calculate stage proppant at different concentrations. Stage proppant is basically the amount of proppant that needs to be calculated at various proppant stage concentrations. For example, at 1 ppg proppant concentration, stage proppant might be 20,000 lbs of sand depending on the clean volume. Stage proppant is a function of proppant concentration and stage fluid clean volume. Stage proppant can be calculated using Eq. (10.5).

$$\text{Stage proppant} = 42 \times \text{proppant conc.} \times \text{stage fluid clean volume} \quad (10.5)$$

Equation 10.5 Stage proppant.

Stage proppant = lbs
Proppant concentration = ppg
Stage fluid clean volume = BBLs.

●●● ———————————————————————————————

Example

Calculate stage proppant at 2 ppg proppant concentration if 340 BBLs of clean volume is designed for this particular proppant stage.

$$\text{Stage proppant} = 42 \times \text{proppant conc.} \times \text{stage fluid clean volume}$$

$$= 42 \times 2 \times 340 = 28,560 \text{ lbs}$$

SAND PER FOOT (LB/FT)

Sand per foot is the amount of sand per foot that can be calculated on both stage and well levels (assuming geometric design).

$$\text{Sand per foot} = \frac{\text{total sand per stage}}{\text{stage length}} = \frac{\text{total sand per well}}{\text{perforated lateral length}}$$

Equation 10.6 Sand per foot.

WATER PER FOOT

Water per foot is the amount of water per foot that can be calculated on both stage and well levels (assuming geometric design).

$$\text{Water per foot} = \frac{\text{total water per stage}}{\text{stage length}} = \frac{\text{total water per well}}{\text{perforated lateral length}}$$

Equation 10.7 Water per foot.

Example

In a particular region of the Barnett Shale, 800 lbs/ft of sand and 40 BBL/ft of water has been determined as the optimum sand and water per foot based on actual production data in 400′ frac stage spacing. Calculate total sand and water per stage.

Total sand per stage $= 800 \times 400 = 320,000$ lbs of sand per stage
Total water per stage $= 40 \times 400 = 16,000$ BBLs of water per stage

SAND-TO-WATER RATIO (SWR, LB/GAL)

Sand-to-water ratio is another important metric in frac design. Total sand divided by total water per stage will yield the sand-to-water ratio. A lower sand-to-water ratio means a higher percentage of water in relation to sand. A higher sand-to-water ratio indicates more aggressive stages by pumping higher amounts of sand in relation to water. Typically the sand-to-water ratio in slick water fracs ranges from 0.7 to 1.7. A sand-to-water ratio in more viscous fluid type systems such as cross-linked jobs could be much higher.

$$\text{Sand to water ratio} = \text{SWR} = \frac{\text{total sand}}{\text{total water}}$$

Equation 10.8 Sand-to-water ratio (SWR).

Total sand $=$ lbs
Total water $=$ gal

Example

Calculate sand-to-water ratio for a well with 400,000 lbs of sand per stage and 8500 BBLs of water.

$$\text{SWR} = \frac{\text{total sand}}{\text{total water}} = \frac{400,000}{8500 \times 42} = 1.12 \text{ lb/gal}$$

SLICK WATER FRAC SCHEDULE

Completions engineers design hydraulic frac jobs. Different hydraulic fracturing software that can be used to design an optimum job is available in the industry. The purpose of this section is not to dig deep into the derivation of equations and calculations, but to understand the basic concepts of a frac schedule provided to the field personnel for execution. The idea behind optimum fracture design is to spend the least amount of money and get the most out of the reservoir by stimulating and contacting as much reservoir rock as possible. The best and most comprehensive design is obtained by investigating completed wells and comparing this data to the production performance of the wells. Computer modeling can be run to solve for optimum design. However, the production performance of the well should dictate the completions design that is chosen for the well.

Every horizontal well is divided into many stages. The number of stages for each well depends on the lateral length. Normally, as lateral length increases, the number of stages increases as well. For example, a well with a 4000' lateral length could have 20 frac stages (depending on the design) but a well with an 8000' lateral length could have 40 frac stages. Therefore, hydraulic fracturing occurs throughout multiple stages to stimulate and contact as much reservoir rock volume as possible. During slick water frac jobs, every stage can use anywhere from 150,000 to 800,000 lbs of proppant. The amount of proppant pumped downhole is massive. For example, an 8000' horizontal well that has 40 stages and uses 400,000 lbs of proppant per stage will need 16 million lbs of sand. In addition to proppant, water will be needed. The average amount of water per stage depends on many design factors such as stage length, amount of sand, treatment difficulty, etc. Typically a stage can use anywhere between 4000 BBLs and 14,000 BBLs of water in slick water jobs. In cross-linked jobs, less water is required since high viscous fluid carries the slurry fluid into the formation. For example, an 8000' horizontal well that has 40 stages and uses 8000 BBLs of water per stage will need 320,000 BBLs (13.44 million gallons) of water. These examples are discussed to give perspective on the total amount of sand and water that must be used to stimulate these low–permeability reservoirs during slick water frac jobs.

Example

Fill in Table 10.1 using the following assumptions, and calculate the sand-to-water ratio, sand/ft, water/ft, pad %, % 100 mesh, and % 40/70 mesh.

Proppant type = Ottawa sand, SG = 2.65, Stage length = 350'
Stage fluid slurry volume sample calculation @ 0.25 ppg and 600 BBLs:

$$\text{Dirty volume} = \text{clean volume} + (\text{sand conc.} \times \text{clean volume} \times \text{AVF})$$
$$= 600 + (0.25 \times 600 \times 0.0453) = 607 \text{ barrels}$$

% of total clean volume sample calculation @ 0.25 ppg and 600 BBLs:

$$\% \text{ of total clean volume} = \frac{\text{stage fluid clean volume}}{\text{total clean volume}} \times 100 = \frac{600}{8236} \times 100 = 7.3\%$$

Stage proppant sample calculation at 0.25 ppg and 600 BBLs:

$$\text{Stage proppant} = 42 \times \text{proppant conc.} \times \text{stage fluid clean volume}$$
$$= 42 \times 0.25 \times 600 = 6300 \text{ lbs}$$

% of total proppant sample calculation at 0.25 ppg and 600 BBLs:

$$\% \text{ of total proppant} = \frac{\text{stage proppant}}{\text{total proppant}} \times 100 = \frac{6300}{396,554} \times 100 = 1.6\%$$

Stage time sample calculation at 0.25 ppg and 600 BBLs:

$$\text{Stage time} = \frac{\text{stage fluid slurry volume}}{\text{pump rate}} = \frac{607}{85} = 7.14 \text{ minutes}$$

Sand per foot calculation:

$$\text{Sand per foot} = \frac{\text{total sand per stage}}{\text{stage length}} = \frac{396,554}{350} = 1133 \text{lb/ft}$$

Water per foot calculation:

$$\text{Water per foot} = \frac{\text{total water per stage}}{\text{stage length}} = \frac{8236}{350} = 24 \text{ barrels/ft}$$

Sand-to-water ratio calculation:

$$\text{SWR} = \frac{\text{total sand}}{\text{total water}} = \frac{396,554}{8236 \times 42} = 1.15 \text{ lb/gal}$$

Pad % calculation:

$$\text{Pad \%} = \frac{\text{pad volume}}{\text{total slurry volume excluding acid and ball}} \times 100 = \frac{410}{7876} \times 100 = 4.94\%$$

Please note that some completions engineers do include acid volume as part of the pad volume calculation but this example excludes acid volume as a percentage of the pad volume. Table 10.2 shows the completed slick water schedule for this problem. This slick water example format is very similar to the provided treatment schedule for job execution in the field. As previously discussed, the designed sand and water volumes are

Table 10.1 Slick Water Schedule Example

85 bpm, 396,554 lbs

Stage Name	Pump Rate bpm	Fluid Name	Stage Fluid Clean Vol BBLs	Stage Fluid Slurry BBLs	% of Total Clean Vol %	Prop Conc. ppg	Stage Proppant lbs	% of Total Prop. %	Cumulative Prop. lbs	Stage Time min
Pump ball	15	Slickwater	300			0				
5% HCl acid	85	Acid	60			0				
Pad	85	Slickwater	410			0				
100 mesh	85	Slickwater	600			0.25				
100 mesh	85	Slickwater	550			0.5				
100 mesh	85	Slickwater	375			0.75				
100 mesh	85	Slickwater	550			1				
100 mesh	85	Slickwater	450			1.25				
100 mesh	85	Slickwater	500			1.5				
40/70 mesh	85	Slickwater	450			0.5				
40/70 mesh	85	Slickwater	365			0.75				
40/70 mesh	85	Slickwater	365			1				
40/70 mesh	85	Slickwater	455			1.25				
40/70 mesh	85	Slickwater	350			1.5				
40/70 mesh	85	Slickwater	379			1.75				
40/70 mesh	85	Slickwater	389			2				
40/70 mesh	85	Slickwater	380			2.25				
40/70 mesh	85	Slickwater	360			2.5				
40/70 mesh	85	Slickwater	299			2.75				
40/70 mesh	85	Slickwater	299			3				
Flush	85	Slickwater	350			0				
		Total clean volume	8236	BBLs						

Table 10.2 Completed Slick Water Schedule Answer

85 bpm, 396,554 lbs

Stage Name	Pump Rate bpm	Fluid Name	Stage Fluid Clean Vol BBLs	Stage Fluid Slurry Vol BBLs	% of Total Clean Vol %	Prop Conc. ppg	Stage Proppant lbs	% of Total Prop. %	Cumulative Prop. lbs	Stage Time min
Pump ball	15	Slickwater	300	300	3.64	0	0		0	20.0
5% HCl acid	85	Acid	60	60	0.73	0	0		0	0.7
Pad	85	Slickwater	410	410	4.98	0	0		0	4.8
100 mesh	85	Slickwater	600	607	7.29	0.25	6300	1.6	6300	7.1
100 mesh	85	Slickwater	550	562	6.68	0.5	11,550	2.9	17,850	6.6
100 mesh	85	Slickwater	375	388	4.55	0.75	11,813	3.0	29,663	4.6
100 mesh	85	Slickwater	550	575	6.68	1	23,100	5.8	52,763	6.8
100 mesh	85	Slickwater	450	475	5.46	1.25	23,625	6.0	76,388	5.6
100 mesh	85	Slickwater	500	534	6.07	1.5	31,500	7.9	107,888	6.3
40/70 mesh	85	Slickwater	450	460	5.46	0.5	9450	2.4	117,338	5.41
40/70 mesh	85	Slickwater	365	377	4.43	0.75	11,498	2.9	128,835	4.44
40/70 mesh	85	Slickwater	365	381	4.43	1	15,330	3.9	144,165	4.49
40/70 mesh	85	Slickwater	455	481	5.52	1.25	23,888	6.0	168,053	5.66
40/70 mesh	85	Slickwater	350	374	4.25	1.5	22,050	5.6	190,103	4.40
40/70 mesh	85	Slickwater	379	409	4.60	1.75	27,857	7.0	217,959	4.81
40/70 mesh	85	Slickwater	389	424	4.72	2	32,676	8.2	250,635	4.99
40/70 mesh	85	Slickwater	380	419	4.61	2.25	35,910	9.1	286,545	4.93
40/70 mesh	85	Slickwater	360	401	4.37	2.5	37,800	9.5	324,345	4.71
40/70 mesh	85	Slickwater	299	336	3.63	2.75	34,535	8.7	358,880	3.95
40/70 mesh	85	Slickwater	299	340	3.63	3	37,674	9.5	396,554	3.99
Flush	85	Slickwater	350	350	4.25	0	0	0.0	0	4.12

Total clean volume	8236	BBLs
Sand/water ratio	1.15	27%
Pad percentage	4.94	% 73%

100 mesh (lbs) 107,888 OR 27%
40/70 mesh (lbs) 288,666 OR 73%
Total (lbs) 396,554

Total stage time (min) 118.4

Stage length (ft)	350	
Water/ft	24	BBL/ft
Sand/ft	1133	lb/ft

heavily dependent on the success in each area as well the optimum economic sand schedule that yields the highest net present value (NPV). For instance, if pumping higher sand and water loadings in a particular field would yield better well performance results and justifies spending additional capital expenditure on higher sand and water loadings, more sand and water loadings will be used for that particular area. In essence, the incremental gain obtained from production results must justify spending the additional capital on higher sand and water loadings in order to economically justify pumping such schedules.

FOAM FRAC SCHEDULE AND CALCULATIONS

Nitrogen gas measurements are reported in standard cubic feet (SCF). When pressure is exerted on nitrogen gas, the volume of nitrogen gas will decrease. In contrast, when heat is applied to nitrogen gas, the volume of nitrogen gas will increase. Nitrogen that is brought on location during foam frac jobs is in liquid form. When nitrogen is pumped downhole, it will be exposed to both pressure and temperature at downhole conditions. Since temperature and pressure affect this gas in opposite ways, it is very important to use bottom-hole conditions to calculate nitrogen volume. Volume factor tables or charts can be used to calculate how many SCF of nitrogen gas is equal to one barrel of liquid. To obtain the volume factor of nitrogen at downhole conditions, bottom-hole treating pressure (BHTP) and bottom-hole static temperature (BHST) must be available. The volume factor of nitrogen can be approximated using Eq. (10.9).

$$\frac{\text{SCF}}{\text{BBL}} = \left(\frac{Z \text{ factor at standard condition} \times \text{standard temperature} \times \text{BHTP}}{Z \times (\text{BHST} + 460) \times \text{atmospheric pressure}} \right) \times 5.615$$

Equation 10.9 Standard cubic feet of nitrogen per barrel of liquid.

SCF/BBL = Standard cubic feet of nitrogen per barrel of volume
Z factor at standard conditions = 1
Standard temperature = $520°R$ $(60 + 460)$
BHTP = Bottom-hole treating pressure, psi
Z = Compressibility factor at downhole condition
BHST = Bottom-hole static temperature, °F
Atmospheric pressure = 14.7 psia.

In addition, there are other charts and plots available in various handbooks that can be used to obtain the volume factor of nitrogen at different pressures and temperatures.

FOAM VOLUME

Foam volume can be calculated when clean fluid volume (volume of water) is available. Foam volume is calculated using Eq. (10.10).

$$\text{Foam volume} = \frac{\text{liquid volume}}{(1 - FQ)}$$

Equation 10.10 Foam volume.

Foam volume = BBL
Liquid volume = BBL
FQ = Foam quality, %.

Example
A foam frac requires a total of 600 barrels of foam volume. Calculate the total clean volume (volume of water) required assuming a 70% foam quality.

$$\text{Foam volume} = \frac{\text{liquid volume}}{(1 - FQ)} \rightarrow 600 = \frac{\text{liquid volume}}{1 - 70\%} = 180 \text{ BBLs of liquid volume}$$

NITROGEN VOLUME

Before being able to calculate nitrogen volume, the nitrogen volume factor at downhole pressure and temperature must be known. Once the nitrogen volume factor is calculated, the nitrogen volume for the job can be calculated using Eq. (10.11).

$$\text{Nitrogen volume} = \text{clean foam volume} \times VF \times FQ$$

Equation 10.11 Nitrogen volume.

Nitrogen volume = BBLs
Clean foam volume = BBLs
VF = Nitrogen volume factor, SCF/BBL
FQ = Foam quality, %.

Example

Calculate nitrogen volume assuming 625 barrels of clean foam volume using the following parameters:

BHTP = 2500 psi, BHST = 125 °F, FQ = 70%

Step 1) First, nitrogen volume factor at 125°F and 2500 psi must be obtained using Eq. (10.9). From this equation, nitrogen volume factor is 810 SCF/BBL.

Step 2) Calculate nitrogen volume assuming 70% foam quality:

Nitrogen volume = clean foam volume × VF × FQ = 625 × 810 × 70% = 354, 375 SCF

BLENDER SAND CONCENTRATION

During foam frac jobs, the sand concentration at the blender must be much higher than the sand concentration at downhole conditions. This is because the slurry fluid carrying the sand from the blender will be diluted with nitrogen. Therefore, blender sand concentrations need to be calculated using Eq. (10.12).

$$\text{Blender sand concentration} = \frac{\text{BH sand concentration}}{1 - \text{FQ}}$$

Equation 10.12 Blender sand concentration.

Blender sand concentration = ppg
BH sand concentration = Bottom-hole sand concentration, ppg
FQ = foam quality, %.

Example

The bottom-hole sand concentrations for a foam frac job are designed at 0.5, 1, 1.5, and 2 ppg. Calculate blender sand concentration at these concentrations assuming 75% foam quality.

$$\text{Blender sand concentration @ } 0.5 = \frac{0.5}{1 - 75\%} = 2 \text{ ppg}$$

$$\text{Blender sand concentration @ } 1 = \frac{1}{1 - 75\%} = 4 \text{ ppg}$$

$$\text{Blender sand concentration @ } 1.5 = \frac{1.5}{1 - 75\%} = 6 \text{ ppg}$$

$$\text{Blender sand concentration @ } 2 = \frac{2}{1 - 75\%} = 8 \text{ ppg}$$

SLURRY FACTOR (SF)

Slurry factor is one of the most important calculations that must be performed for designing a foam frac job. Since the proppant concentration at bottom hole and blender are different, slurry factor at surface (blender) and bottom hole must both be calculated. Since adding sand to fluid on foam jobs decreases foam quality, the slurry factor calculation becomes very important. Increasing sand concentration will decrease clean rate when designing a foam job schedule.

$$\text{Slurry factor (SF)} = 1 + (\text{sand concentration} \times \text{AVF})$$

Equation 10.13 Slurry factor (SF).

Sand concentration = ppg
AVF = Absolute volume factor, gal/lb.

Example

Calculate slurry factor at bottom hole and surface (blender) assuming regular sand with SG of 2.65 at various BH sand concentrations of 1 and 2 ppg, assuming 70% foam quality.

$$\text{AVF} = \frac{1}{2.65 \times 8.33} = 0.0453 \text{ gal/lb}$$

Slurry factor @ 1 ppg BH conc. = 1 + (sand conc. × AVF) = 1 + (1 × 0.0453) = 1.0453

Slurry factor @ 2 ppg BH conc. = 1 + (2 × 0.0453) = 1.0906

The 1- and 2-ppg bottom-hole sand concentrations are calculated to be 3.33 and 6.67 ppg sand concentrations at the blender, assuming 70% foam quality.

Slurry factor @ 3.33 ppg blender conc. = 1 + (3.33 × 0.0453) = 1.151

Slurry factor @ 6.67 ppg blender conc. = 1 + (6.67 × 0.0453) = 1.302

CLEAN RATE (NO PROPPANT)

Clean rate (assuming no proppant) during pad has to be calculated when designing a foam frac job. Clean rate during pad and no proppant can be calculated using Eq. (10.14).

$$\text{Clean rate (no proppant)} = \text{Foam rate} \times (1 - \text{FQ})$$

Equation 10.14 Clean rate (no proppant).

Clean rate = Assuming no proppant, bpm
Foam rate = Also known as downhole rate, bpm
FQ = Foam quality, %.

Example

The foam rate for a foam job is designed at 30 bpm. Calculate the clean rate during pad assuming 75% foam quality.

Clean rate (no proppant) = foam rate × (1 − FQ) = 30 × (1 − 75%) = 7.5 bpm

---

CLEAN RATE (WITH PROPPANT)

Once clean rate with no proppant is calculated, clean rate with proppant at different bottom-hole concentrations must also be calculated when designing a sand schedule for a foam frac job. Clean rate (with proppant) is calculated using Eq. (10.15).

$$\text{Clean rate}_{\text{proppant}} = \frac{\text{Clean rate}_{\text{no proppant(pad)}}}{\text{SF}_{\text{BH}}}$$

Equation 10.15 Clean rate (with proppant).

$\text{Clean rate}_{\text{proppant}}$ = Clean rate with proppant, bpm
$\text{Clean rate}_{\text{no proppant}}$ = Clean rate during pad, bpm
SF_{BH} = Slurry factor at bottom-hole sand concentration.

Example

The bottom-hole sand concentration during a foam frac job is at 2 ppg. The design foam rate was 25 bpm. Assuming a 68% foam quality and regular sand with 2.65 SG, calculate clean rate during 2 ppg bottom-hole sand concentration.

Clean rate (no proppant) = Foam rate × (1 − FQ) = 25 × (1 − 68%) = 8 bpm

Slurry factor = 1 + (sand concentration × AVF) = 1 + (2 × 0.0453) = 1.0906

$$\text{Clean rate}_{\text{proppant}} = \frac{\text{Clean rate}_{\text{no proppant(pad)}}}{\text{SF}_{\text{BH}}} = \frac{8}{1.0906} = 7.34 \text{ bpm}$$

SLURRY RATE (WITH PROPPANT)

The next important calculation when designing a foam frac job is the slurry rate with proppant calculation. Slurry rate with proppant can be calculated using Eq. (10.16).

$$\text{Slurry rate}_{\text{proppant}} = \left(\frac{\text{Clean rate}_{\text{no proppant(pad)}}}{\text{SF}_{\text{BH}}}\right) \times \text{SF}_{\text{blender}}$$

Equation 10.16 Slurry rate with proppant.

Slurry rate$_{\text{proppant}}$ = Slurry rate with proppant, bpm
Clean rate$_{\text{no proppant}}$ = Clean rate during pad, bpm
SF$_{\text{BH}}$ = Slurry factor at bottom-hole sand concentration
SF$_{\text{blender}}$ = Slurry factor at blender sand concentration.

Example

Calculate slurry rate with proppant assuming a 72% foam quality and clean rate (no proppant) of 6 bpm during 1.5 ppg bottom-hole sand concentration. Assume regular sand with SG of 2.65.

$$\text{SF}_{\text{BH}} = 1 + (\text{sand concentration} \times \text{AVF}) = 1 + (1.5 \times 0.0453) = 1.068$$

$$\text{Blender sand concentration} = \frac{\text{BH sand concentration}}{1 - \text{FQ}} = \frac{1.5}{1 - 72\%} = 5.36 \text{ ppg}$$

$$\text{SF}_{\text{blender}} = 1 + (5.36 \times 0.0453) = 1.242$$

$$\text{Slurry rate}_{\text{proppant}} = \left(\frac{\text{Clean rate}_{\text{pad}}}{\text{SF}_{\text{BH}}}\right) \times \text{SF}_{\text{blender}} = \left(\frac{6}{1.068}\right) \times 1.242 = 6.98 \text{ bpm}$$

NITROGEN RATE (WITH AND WITHOUT PROPPANT)

The next step in designing a foam frac job is to calculate nitrogen rate with and without proppant. Nitrogen rate without proppant can be calculated using Eq. (10.17).

$$\text{Nitrogen rate (no proppant)} = \text{dirty foam rate} \times \text{VF} \times \text{FQ}$$

Equation 10.17 Nitrogen rate (no proppant).

Nitrogen rate = Assuming no proppant, SCF/min
Dirty foam rate = Designed foam rate, bpm
VF = Nitrogen volume factor, SCF/BBL
FQ = Foam quality, %.
In addition, nitrogen rate with proppant is calculated using Eq. (10.18).

Nitrogen rate (with proppant) = (dirty foam rate − slurry rate) × VF

Equation 10.18 Nitrogen rate (with proppant).

Foam rate = bpm
Slurry rate = bpm
VF = Nitrogen volume factor, SCF/BBL.

Example

A foam frac job is scheduled to have a foam rate of 32 bpm with a slurry rate of 8.2 bpm. Nitrogen volume factor is calculated to be 1001 SCF/BBL and foam quality scheduled for the job is 70%. Calculate the nitrogen rate with and without proppant for this particular stage.

Nitrogen rate(pad) = Dirty foam rate × VF × FQ = 32 × 1001 × 70% = 22, 422 SCF/min

Nitrogen rate(with proppant) = (Dirty foam rate − slurry rate) × VF = (32 − 8.2) × 1001
= 23, 824 SCF/min

Example

You are a completions engineer responsible for designing a foam frac schedule for a coalbed methane (CBM) well. Assuming the following properties and schedule, calculate the rest of the foam frac schedule.

ISIP = 2150 psi, hydrostatic pressure = 1350 psi, BHST = 100 °F, FQ = 70%, SG = 2.65 (regular sand), dirty foam rate (bottom hole) = 30 bpm

Stage Name	BH Proppant Conc. ppg	Dirty Foam Volume BBLs
ACID	0.00	6.0
PAD	0.00	40.0
20/40	1.00	30.0
20/40	1.50	30.0
20/40	2.00	30.0
20/40	2.50	30.0
20/40	3.00	30.0
FLUSH	0.0	45.0

Step 1) Calculate nitrogen volume factor based on BHTP and BHST:

$$BHTP = P_h + ISIP = 1350 + 2150 = 3500 \text{ psi}$$

Nitrogen volume factor at 3500 psi and 100°F is approximately **1139 SCF/BBL** using Eq. (10.9).

Step 2) BH proppant concentration for each proppant stage is provided. Calculate blender proppant concentration for each proppant concentration using the equation below.

$$\text{Blender sand concentration} = \frac{\text{BH sand concentration}}{1 - FQ}$$

Stage Name	BH Proppant Conc. ppg	Blender Proppant Conc. ppg
ACID	0.00	0.00
PAD	0.00	0.00
20/40	1.00	$1/(1-70\%) = 3.33$
20/40	1.50	$1.5/(1-70\%) = 5$
20/40	2.00	$2/(1-70\%) = 6.67$
20/40	2.50	$2.5(1-70\%) = 8.33$
20/40	3.00	$3/(1-70\%) = 10$
FLUSH	0.0	0.00

Step 3) Calculate bottom-hole slurry factor for each proppant stage:

$$AVF = \frac{1}{2.65 \times 8.33} = 0.0453 \text{ gal/lb}$$

$$\text{Slurry factor} = 1 + (\text{sand concentration} \times AVF)$$

Stage Name	BH Proppant Conc. ppg	BH Slurry Factor
ACID	0.00	$1 + (0 \times 0.0453) = 1$
PAD	0.00	$1 + (0 \times 0.0453) = 1$
20/40	1.00	$1 + (1 \times 0.0453) = 1.05$
20/40	1.50	$1 + (1.5 \times 0.0453) = 1.07$
20/40	2.00	$1 + (2 \times 0.0453) = 1.09$
20/40	2.50	$1 + (2.5 \times 0.0453) = 1.11$
20/40	3.00	$1 + (3 \times 0.0453) = 1.14$
FLUSH	0.0	$1 + (0 \times 0.0453) = 1$

Step 4) Calculate clean foam volume simply by taking dirty foam volume (provided) and dividing it by the calculated bottom-hole slurry factor.

Stage Name	Dirty Foam Volume BBLs	BH Slurry Factor	Clean Foam Volume BBLs
ACID	6.0	$1 + (0 \times 0.0453) = 1$	$6/1 = 6$
PAD	40.0	$1 + (0 \times 0.0453) = 1$	$40/1 = 40$
20/40	30.0	$1 + (1 \times 0.0453) = 1.05$	$30/1.05 = 28.7$
20/40	30.0	$1 + (1.5 \times 0.0453) = 1.07$	$30/1.07 = 28.09$
20/40	30.0	$1 + (2 \times 0.0453) = 1.09$	$30/1.09 = 27.51$
20/40	30.0	$1 + (2.5 \times 0.0453) = 1.11$	$30/1.11 = 26.95$
20/40	30.0	$1 + (3 \times 0.0453) = 1.14$	$30/1.14 = 26.41$
FLUSH	45.0	$1 + (0 \times 0.0453) = 1$	$45/1 = 45$

Step 5) Calculate clean fluid volume using the following equation:

$$\text{Clean fluid volume} = \text{Clean foam volume} \times (1 - FQ)$$

Stage Name	Clean Foam Volume BBLs	Clean Fluid Volume BBLs
ACID	$6/1 = 6$	$6 \times (1 - 0\%) = 6$
PAD	$40/1 = 40$	$40 \times (1 - 70\%) = 12$
20/40	$30/1.05 = 28.7$	$28.7 \times (1 - 70\%) = 8.61$
20/40	$30/1.07 = 28.09$	$28.09 \times (1 - 70\%) = 8.43$
20/40	$30/1.09 = 27.51$	$27.51 \times (1 - 70\%) = 8.25$
20/40	$30/1.11 = 26.95$	$26.95 \times (1 - 70\%) = 8.08$
20/40	$30/1.14 = 26.41$	$26.41 \times (1 - 70\%) = 7.92$
FLUSH	$45/1 = 45$	$45 \times (1 - 70\%) = 13.5$

Step 6) Calculate surface slurry factor for each proppant stage:

Stage Name	Blender Proppant Conc. ppg	Surface Slurry Factor
ACID	0.00	$1 + (0 \times 0.0453) = 1$
PAD	0.00	$1 + (0 \times 0.0453) = 1$
20/40	$1/(1 - 70\%) = 3.33$	$1 + (3.33 \times 0.0453) = 1.15$
20/40	$1.5(1 - 70\%) = 5$	$1 + (5 \times 0.0453) = 1.23$
20/40	$2/(1 - 70\%) = 6.67$	$1 + (6.67 \times 0.0453) = 1.30$
20/40	$2.5/(1 - 70\%) = 8.33$	$1 + (8.33 \times 0.0453) = 1.38$
20/40	$3/(1 - 70\%) = 10$	$1 + (10 \times 0.0453) = 1.45$
FLUSH	0.00	$1 + (0.0453) = 1$

Step 7) Calculate dirty fluid volume for each proppant stage as shown below:

Dirty fluid volume = clean fluid volume × surface (blender) slurry factor

Stage Name	Clean Fluid Volume BBLs	Surface Slurry Factor	Dirty Fluid Volume BBLs
ACID	$6 \times (1-0\%) = 6$	$1 + (0 \times 0.0453) = 1$	$6 \times 1 = 6$
PAD	$40 \times (1 - 70\%) = 12$	$1 + (0 \times 0.0453) = 1$	$12 \times 1 = 12$
20/40	$28.7 \times (1 - 70\%) = 8.61$	$1 + (3.33 \times 0.0453) = 1.15$	$8.61 \times 1.15 = 9.91$
20/40	$28.09 \times (1 - 70\%) = 8.43$	$1 + (5 \times 0.0453) = 1.23$	$8.43 \times 1.23 = 10.34$
20/40	$27.51 \times (1 - 70\%) = 8.25$	$1 + (6.67 \times 0.0453) = 1.30$	$8.25 \times 1.30 = 10.74$
20/40	$26.95 \times (1 - 70\%) = 8.08$	$1 + (8.33 \times 0.0453) = 1.38$	$8.08 \times 1.38 = 11.14$
20/40	$26.41 \times (1 - 70\%) = 7.92$	$1 + (10 \times 0.0453) = 1.45$	$7.92 \times 1.45 = 11.51$
FLUSH	$45 \times (1 - 70\%) = 13.5$	$1 + (0 \times 0.0453) = 1$	$13.5 \times 1 = 13.5$

Step 8) Calculate amount of sand for each blender sand concentration as shown below:

Amount of sand = clean fluid volume in gallons × blender proppant concentration

Stage Name	Blender Proppant Conc. ppg	Clean Fluid Volume BBLs	Sand (lbs) Stage	Sand (lbs) CUM
ACID	0.00	$6 \times (1 - 0\%) = 6$	$6 \times 42 \times 0 = 0$	0
PAD	0.00	$40 \times (1 - 70\%) = 12$	$12 \times 42 \times 0 = 0$	0
20/40	$1/(1 - 70\%) = 3.33$	$28.7 \times (1 - 70\%) = 8.61$	$8.61 \times 42 \times 3.33 = 1205$	1205
20/40	$1.5/(1 - 70\%) = 5$	$28.09 \times (1 - 70\%) = 8.43$	$8.43 \times 42 \times 5 = 1770$	$1205 + 1770 = 2975$
20/40	$2/(1 - 70\%) = 6.67$	$27.51 \times (1 - 70\%) = 8.25$	$8.25 \times 42 \times 6.67 = 2311$	$2975 + 2311 = 5286$
20/40	$2.5/(1 - 70\%) = 8.33$	$26.95 \times (1 - 70\%) = 8.08$	$8.08 \times 42 \times 8.33 = 2830$	$5286 + 2830 = 8115$
20/40	$3/(1 - 70\%) = 10$	$26.41 \times (1 - 70\%) = 7.92$	$7.92 \times 42 \times 10 = 3328$	$8115 + 3328 = 11,443$
FLUSH	0.00	$45 \times (1 - 70\%) = 13.5$	$13.5 \times 42 \times 0 = 0$	11,443

Step 9) Calculate nitrogen volume for each stage as shown below:

Nitrogen volume = clean foam volume × VF × FQ

Stage Name	FQ	Clean Foam Volume	Nitrogen (SCF)	
	%	BBLs	Stage	CUM
ACID	0%	$6/1 = 6$	$6 \times 0 \times 1139 = 0$	0
PAD	70%	$40/1 = 40$	$40 \times 70\% \times 1139$ $= 31,892$	31,892
20/40	70%	$30/1.05$ $= 28.7$	$28.7 \times 70\% \times 1139$ $= 22,882$	$31,892 + 22,882$ $= 54,774$
20/40	70%	$30/1.07$ $= 28.09$	$28.09 \times 70\% \times 1139$ $= 22,397$	$54,774 + 22,397$ $= 77,172$
20/40	70%	$30/1.09$ $= 27.51$	$27.51 \times 70\% \times 1139$ $= 21,932$	$77,172 + 21,932$ $= 99,104$
20/40	70%	$30/1.11$ $= 26.95$	$26.95 \times 70\% \times 1139$ $= 21,486$	$99,104 + 21,486$ $= 120,589$
20/40	70%	$30/1.14$ $= 26.41$	$26.41 \times 70\% \times 1139$ $= 21,057$	$120,589 + 21,057$ $= 141,647$
FLUSH	70%	$45/1 = 45$	$45 \times 70\% \times 1139$ $= 35,879$	$141,647 + 35,879$ $= 177,525$

Step 10) Dirty foam rate (designed rate) is given in the problem to be 30 bpm. Clean foam rate can be calculated as follows:

$$\text{Clean foam rate} = \frac{\text{dirty foam rate}}{\text{BH slurry factor}}$$

Stage Name	BH Slurry Factor	Rate (bpm)	
		Dirty Foam Rate	Clean Foam Rate
ACID	$1 + (0 \times 0.0453) = 1$	30	$30/1 = 30$
PAD	$1 + (0 \times 0.0453) = 1$	30	$30/1 = 30$
20/40	$1 + (1 \times 0.0453) = 1.05$	30	$30/1.05 = 28.70$
20/40	$1 + (1.5 \times 0.0453) = 1.07$	30	$30/1.07 = 28.09$
20/40	$1 + (2 \times 0.0453) = 1.09$	30	$30/1.09 = 27.51$
20/40	$1 + (2.5 \times 0.0453) = 1.11$	30	$30/1.11 = 26.95$
20/40	$1 + (3 \times 0.0453) = 1.14$	30	$30/1.14 = 26.41$
FLUSH	$1 + (0 \times 0.0453) = 1$	30	$30/1 = 30$

Step 11) Clean fluid rate for each proppant stage can be calculated as follows:

$$\text{Clean fluid rate} = \text{clean foam rate} \times (1 - FQ)$$

Stage Name	FQ	Rate (bpm)		
	%	Dirty Foam Rate	Clean Foam Rate	Clean Fluid Rate
ACID	0%	30	$30/1 = 30$	$30 \times (1 - 0\%) = 30$
PAD	70%	30	$30/1 = 30$	$30 \times (1 - 70\%) = 9$
20/40	70%	30	$30/1.05 = 28.70$	$28.70 \times (1 - 70\%) = 8.61$
20/40	70%	30	$30/1.07 = 28.09$	$28.09 \times (1 - 70\%) = 8.43$
20/40	70%	30	$30/1.09 = 27.51$	$27.51 \times (1 - 70\%) = 8.25$
20/40	70%	30	$30/1.11 = 26.95$	$26.95 \times (1 - 70\%) = 8.08$
20/40	70%	30	$30/1.14 = 26.41$	$26.41 \times (1 - 70\%) = 7.92$
FLUSH	70%	30	$30/1 = 30$	$30 \times (1 - 70\%) = 9$

Step 12) Dirty fluid rate for each proppant stage can be calculated as follows:

Dirty fluid rate = clean fluid rate × blender (surface) SF

Stage Name	Surface Slurry Factor	Rate (bpm)		
		Dirty Foam Rate	Clean Fluid Rate	Dirty Fluid Rate
ACID	$1 + (0 \times 0.0453)$ $= 1$	30	$30 \times (1 - 0\%)$ $= 30$	$30 \times 1 = 30$
PAD	$1 + (0 \times 0.0453)$ $= 1$	30	$30 \times (1 - 70\%)$ $= 9$	$9 \times 1 = 9$
20/40	$1 + (3.33 \times 0.0453)$ $= 1.15$	30	$28.70 \times$ $(1 - 70\%)$ $= 8.61$	8.61×1.15 $= 9.91$
20/40	$1 + (5 \times 0.0453)$ $= 1.23$	30	$28.09 \times$ $(1 - 70\%)$ $= 8.43$	8.43×1.23 $= 10.34$
20/40	$1 + (6.67 \times 0.0453)$ $= 1.30$	30	$27.51 \times$ $(1 - 70\%)$ $= 8.25$	8.25×1.30 $= 10.74$
20/40	$1 + (8.33 \times 0.0453)$ $= 1.38$	30	$26.95 \times$ $(1 - 70\%)$ $= 8.08$	8.08×1.38 $= 11.14$
20/40	$1 + (10 \times 0.0453)$ $= 1.45$	30	$26.41 \times$ $(1 - 70\%)$ $= 7.92$	7.92×1.45 $= 11.51$
FLUSH	$1 + (0 \times 0.0453)$ $= 1$	30	$30 \times (1 - 70\%)$ $= 9$	$9 \times 1 = 9$

Step 13) Calculate nitrogen rate for each stage using either of the equations listed below:

Nitrogen rate = clean foam rate × FQ × VF or Nitrogen rate = (dirty foam rate − slurry rate) × VF

Stage Name	FQ		Rate	
	%	Dirty Foam Rate	Clean Foam Rate (bpm)	Nitrogen Rate (SCF/min)
ACID	0%	30	30/1 = 30	0
PAD	70%	30	30/1 = 30	30 × 70% × 1139 = 23,919
20/40	70%	30	30/1.05 = 28.70	28.7 × 70% × 1139 = 22,882
20/40	70%	30	30/1.07 = 28.09	28.09 × 70% × 1139 = 22,397
20/40	70%	30	30/1.09 = 27.51	27.51 × 70% × 1139 = 21,932
20/40	70%	30	30/1.11 = 26.95	26.95 × 70% × 1139 = 21,486
20/40	70%	30	30/1.14 = 26.41	26.41 × 70% × 1139 = 21,057
FLUSH	70%	30	30/1 = 30	30 × 70% × 1139 = 23,919

Step 14) The last step is to calculate pump time for each stage as follows:

$$\text{Pump time} = \frac{\text{dirty foam volume}}{\text{dirty foam rate}}$$

Stage Name	Dirty Foam Volume	Rate (bpm)	Time (minutes)	
	BBLs	Dirty Foam Rate	Pump Time	Total
ACID	6.0	30	6/30 = 0.2	0.2
PAD	40.0	30	40/30 = 1.33	0.2 + 1.33 = 1.53
20/40	30.0	30	30/30 = 1	1.53 + 1 = 2.53
20/40	30.0	30	30/30 = 1	2.53 + 1 = 3.53
20/40	30.0	30	30/30 = 1	3.53 + 1 = 4.53
20/40	30.0	30	30/30 = 1	4.53 + 1 = 5.53
20/40	30.0	30	30/30 = 1	5.53 + 1 = 6.53
FLUSH	45.0	30	45/30 = 1.5	6.53 + 1.5 = 8.03

The foam schedule for this example is summarized in Table 10.3.

Table 10.3 Foam Design Schedule Example

Stage Name	Proppant Concentration (ppg)			Foam Volumes (BBLs)		
	BH Proppant Conc.	FQ	Blender Prop Conc.	Dirty Foam Volume	BH SF	Clean Foam Volume
ACID	0.00	0%	0.00	6.0	1.00	6.00
PAD	0.00	70%	0.00	40.0	1.00	40.00
20/40	1.00	70%	3.33	30.0	1.05	28.70
20/40	1.50	70%	5.00	30.0	1.07	28.09
20/40	2.00	70%	6.67	30.0	1.09	27.51
20/40	2.50	70%	8.33	30.0	1.11	26.95
20/40	3.00	70%	10.00	30.0	1.14	26.41
FLUSH	0.0	70%	0.00	45.0	1.00	45.00

Fluid volume (BBLs)			Sand (lbs)		Nitrogen (SCF)	
Clean Fluid Volume	Surface SF	Dirty Fluid Volume	Stage	CUM	Stage	CUM
6.00	1.00	6.00	0	0	0	0
12.00	1.00	12.00	0	0	31,892	31,892
8.61	1.15	9.91	1205	1205	22,882	54,774
8.43	1.23	10.34	1770	2975	22,397	77,172
8.25	1.30	10.74	2311	5286	21,932	99,104
8.08	1.38	11.14	2830	8115	21,486	120,589
7.92	1.45	11.51	3328	11,443	21,057	141,647
13.50	1.00	13.50	0	11,443	35,879	177,525

Rate					Time	
Dirty Foam Rate	Clean Foam Rate	Clean Fluid Rate	Dirty Fluid Rate	Nitrogen Rate	Pump Time	Total
30.0	30.00	30.00	30.00	0	0.20	0.20
30.0	30.00	9.00	9.00	23,919	1.33	1.53
30.0	28.70	8.61	9.91	22,882	1.00	2.53
30.0	28.09	8.43	10.34	22,397	1.00	3.53
30.0	27.51	8.25	10.74	21,932	1.00	4.53
30.0	26.95	8.08	11.14	21,486	1.00	5.53
30.0	26.41	7.92	11.51	21,057	1.00	6.53
30.0	30.00	9.00	9.00	23,919	1.50	8.03

CHAPTER ELEVEN

Horizontal Well Multistage Completion Techniques

INTRODUCTION

Multistage hydraulic fracturing, along with drilling longer horizontal wells, has tremendously helped the industry to make unconventional shale resources, which used to be uneconomical, economically profitable. This requires a vast acknowledgment to the industry that has been effortlessly testing various concepts in the unconventional shale reservoirs in order to make the process safer, cost effective, and environmentally friendlier. This is just the beginning and there are so many more new advancements in technology and science that the industry will see for the years to come in these resources.

There are two commonly used completion (frac) methods in the industry. The first one is referred to as "conventional plug and perf," which is the most used completion method. The second type is called "sliding sleeve," which is less frequently used and can be seen more often in shale oil plays such as the Bakken shale (although lots of operators have stepped away from using the sliding sleeve technique in the Bakken). The choice of which type of frac technique to use depends on the operator's success along with the economics of each particular technique and technology. If sliding sleeve works better in certain areas from economical, operational, and production perspectives, sliding sleeve should be used. However, if plug and perf causes an increase in production by a big proportion without any operational issues and economical concerns, plug and perf must be used. This is driven by each company's success and philosophy on each technique. One important lesson that the industry has learned since the late 1990s developing the unconventional shale reservoirs is that data should be the biggest driven and deciding factor of every engineering and operational decision. There are substantial amounts of complexity and heterogeneity in the unconventional shale reservoirs that would dictate using data to drive the business forward instead of relying on opinions and theories that may or may not function.

Hydraulic Fracturing in Unconventional Reservoirs
DOI: http://dx.doi.org/10.1016/B978-0-12-849871-2.00011-3

© 2017 Elsevier Inc.
All rights reserved.

CONVENTIONAL PLUG AND PERF

Conventional plug and perf is the most commonly used method in unconventional shale plays. Composite bridge (frac) plugs are used for isolation between frac stages. Plug and perf is a completion system that uses perforation guns with a composite bridge plug using wireline. Once at the desired measured depth, a composite bridge plug is set, and each perforation gun is pulled up to the designed depths until all of the perforation guns are fired. Each perforation gun represents a cluster. After firing all of the perforation guns, wireline is pulled out of the hole (POOH). Conventional plug and perf can be done using cemented or uncemented casings. This method involves multiple perforation clusters per stage. This method is also known to be a slow and repetitive perforation and stimulation process. Conventional plug and perf is slow because after every frac stage, wireline must stab onto the well. The plug and perforation guns are sent downhole to set the composite bridge plug, shoot the guns (clusters), and finally pull out of the hole. The wireline process for stage isolation and perforation can take anywhere between 2 and 4 hours depending on measured depth, crew efficiency, wireline speed, etc. For example, if a well has 40 stages, this process must be performed 40 times. If each wireline run is assumed to be about 3 hours, 120 hours are spent on frac stage isolation and perforation. This is 5 days' worth of frac stage isolation and perforation on a well with 40 stages. The industry uses a zipper frac technique where while one well is being fraced, another well is being perforated in an attempt to improve the operational efficiency when using the conventional plug and perf technique on multiwell pads. The conventional plug and perf technique has been very successful in the industry from a production perspective. Otherwise, given the slow progress of the conventional plug and perf method, the industry would have moved away from this technique. The industry has also developed other efficient techniques such as the dissolvable ball and plug, which will dissolve at downhole conditions eliminating the need for drill-out.

COMPOSITE BRIDGE (FRAC) PLUG

Composite bridge (frac) plugs are used for isolation between frac stages in the conventional plug and perf method. The main reason composite bridge plugs are used is because these types of plugs can be easily and rapidly drilled out after the frac job is over. After setting the plug, a ball is dropped and pumped downhole until seated inside of the composite bridge plug. The ball is typically pumped downhole at 10−15 bpm and as soon as the ball is seated inside of the plug, there is a spike in the surface-treating pressure. The spike in surface-treating pressure confirms the ball sitting in the plug. Once the ball is seated, the previous stage has now been isolated and the treatment for the new stage can commence. Fig. 11.1 shows the composite bridge plug and associated components including burn charge, frac plug setting tool, and pump down ring. Fig. 11.2 displays the perforation guns used in the conventional plug and perf technique. Fig. 11.3 shows the inside view of the same perforation guns presented in Fig. 11.2. Fig. 11.4 shows the frac ball seated in the composite bridge plug. Fig. 11.5 shows the schematic of the one-stage hydraulic fracturing (plug to plug) with four clusters.

Stack Fracing

Stack fracing involves fracing one stage, and then waiting for the wireline to perforate the next stage on the same well before being able to frac again. In this type of frac, one well is completed at a time. Stack fracing is very common in exploration areas where only one well is located on a pad. Therefore, frac crews pump a stage and wait for the wireline to set the plug and perforate the next stage by performing routine maintenance on their equipment. Once the wireline is done setting the plug, perforating, and pulling out of the hole, the frac crew will proceed to pumping the next stage. This continues until all of the stages are completed on the same well. The main disadvantage of stack fracing is the frac waiting time between stages.

Zipper Fracing

Zipper fracing refers to fracing a stage on one well while perforating and setting the plug on another well. Zipper fracing can be performed on multiple wells at one time. One of the main advantages of zipper fracing is saving time and money by continuously fracing and perforating. Zipper fracing is very common in the majority of the shale plays.

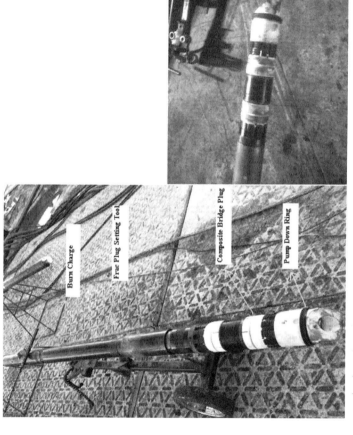

Figure 11.1 Composite bridge plug.

Figure 11.2 Perforation guns.

Figure 11.3 Inside view of perforation guns.

Simultaneous Frac

Simultaneous frac is not as commonly used as zipper or stack frac. In this type of frac, two wells are simultaneously fracked at the same time. This requires a great deal of both coordination and equipment onsite. In addition, the pad has to be large enough to fit all of the frac equipment for this enormous job.

Figure 11.4 Frac ball inside of a composite bridge plug for frac stage isolation.

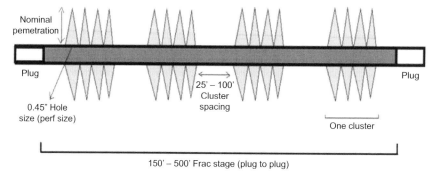

Figure 11.5 Plug and cluster spacing example.

SLIDING SLEEVE

Sliding sleeve is also known as fracturing sleeve. Sliding sleeve is an alternative to plug and perf, and is used to stimulate multistage horizontal wells through holes/ports. This method is operated by a ball and baffle. When the ball lands on the baffle, the inner sleeve is opened and activated. This provides the flow path for the fracturing fluid. This type of frac typically has one opening (cluster) per stage. Multiport technology is also available to mimic plug and perf with multiple clusters. The biggest advantage of sliding sleeve is timing. Since there is no need to send composite plug and perforation guns downhole, it saves a tremendous amount of time, which is equal to saving money. The system can either be cemented or uncemented.

SLIDING SLEEVE ADVANTAGES

An advantage that sliding sleeve is known for is reduction in stimulation cycle times. Given no wireline will be needed for frac stage isolation and perforation, stage after stage can be completed as long as sand and water logistics keep up. Sliding sleeve also reduces water usage and overdisplacement (flush stage). Sliding sleeve is also known for maximizing near-wellbore conductivity and can be used with dissolvable frac balls.

SLIDING SLEEVE DISADVANTAGES

Some of the biggest disadvantages of sliding sleeve are mechanical issues. Anything mechanical can fail, and mitigation processes can be very costly. Another disadvantage is the limited number of stages in cemented applications. In today's conventional plug and perf methodology, clusters (perforation guns) are sometimes placed 30' apart to maximize the contact area. However, depending on the service provider, the number of stages can be limited. Since the sleeves are run downhole with the casing, each joint of casing is usually 40−45' in length. Therefore, the sleeves cannot be placed closer than 40' unless special casing is ordered, which can be expensive. Hole conditioning is another crucial step before running sliding sleeve with casing in the hole. Finally, the industry has limited experience with sliding sleeve as compared to the tried and true conventional plug and perf method. Sliding sleeve can be divided into different types. The most common ones are described in the following sections.

Toe Sleeve/Valve

Toe sleeve is a pressure-operated valve that creates flow path without any intervention of the wireline.

Single Entry-Point Frac Sleeve

The single entry-point frac sleeve system is operated by a ball and baffle. Frac balls are dropped in sequence of smallest to largest in order to activate the sleeve. Ball trailer or pneumatic ball launcher can be used to launch the balls from the surface.

Multi-entry Point Frac Sleeve

As opposed to single entry point, multi-entry point frac sleeve allows multiple entry points in a single stage without the use of plug and perf. The idea is to mimic the plug and perf design by using sliding sleeve technology. One ball can open more than one sleeve. This technique is very similar to the conventional plug and perf and every entry point is similarly referred to as a "cluster."

Hybrid Design

Hybrid design uses a combination of frac sleeves and plug and perf. The first half of the well (toe section) uses sliding sleeve and the second half (heel section) uses plug and perf. The Bakken Shale is an ideal example of the hybrid design. Since the lateral lengths of Bakken wells are typically in excess of 8000' and coiled tubing is limited by depth that can be reached, many operators use the hybrid design to facilitate this process.

FRAC STAGE SPACING (PLUG-TO-PLUG SPACING)

Frac stage refers to the space from plug to plug in a vertical or horizontal well. In many formations across the United States the horizontal lateral length of a well is divided into many stages to optimize production. This is why hydraulic fracturing is often referred to as multistage hydraulic fracturing, i.e., each well has many stages depending on the horizontal length of the well, design, and economic calculations. Therefore, the next interesting subject in hydraulic fracturing is the number of stages necessary to maximize production in horizontal wells. When hydraulic fracturing started, some companies tried to perform a single-stage frac job with no success, thus causing the need for multistage hydraulic fracturing in various formations across the United States.

SHORTER STAGE LENGTH (SSL)

In conventional plug and perf technique, the industry standard for plug-to-plug spacing is anywhere between 150' and 500'. Operating companies have used various frac spacing designs (e.g., 150', 200', 300', etc.) to come up with the optimum production that yields the best economic outcome based on actual production data. Some companies believe in shorter frac stage spacing such as 150–200'. This type of frac is referred to as shorter stage length (SSL) due to shorter plug-to-plug spacing. For example, if the lateral length of a well is 6000' and 200' plug-to-plug spacing is chosen, hydraulic fracturing will take place in 30 stages. This means the process of setting plug, perforating, and fracing needs to be performed 30 times on one well. One of the main factors associated with frac spacing is economics. Every frac stage is very costly and depends on various factors such as service provider, amount of sand, water, chemicals pumped, market conditions, etc. For example, if a stage is pumped using 250,000 lbs of sand with the associated water and chemicals, it will be less expensive compared to a stage that uses 500,000 lbs of sand and the associated water and chemicals. Stage spacing in different areas and formations is ultimately dictated by production success and economic analysis.

The concept of SSL was heavily applied and tested since 2013 because shorter spacing between frac stages often yields higher initial productions (IP) with steeper decline and in some areas shallower decline. In some areas, steeper decline is not noted and decline percentage can sustain itself. For example, the IP from a standard spacing of 300' is about 6 MMSCF/D with an initial annual secant decline of 62%; however, the IP from a well with 150' spacing (SSL) may yield an IP of 8 MMSCF/D with a similar or steeper decline of more than 62%. Sometimes the decline percentage could be shallower than historically observed using SSL depending on the effectiveness of the completions design. The incremental production volume that is initially gained is basically the time value of money and as a result would sometimes be more economically beneficial in certain areas as long as the decline percentage can sustain itself. The time value of money is a concept that means that money available today is worth more than the same amount of money in the future due to its potential earning capacity. SSL does not work everywhere, and the probability of success will depend on the formation properties. In some areas, SSL works very well and a production uplift of 10–40% can be seen. However, in other areas, no production uplift can be seen from SSL. The question then is whether the incremental gain in production offsets the additional capex (capital expenditure) required for any particular design. If so, and the required funding is available, then a more optimized design must be used.

To determine the most economic option, economic analysis must be performed using both methods in each area. The biggest challenge in SSL design is the incremental capital (capex) that must be spent. The additional IP and estimated ultimate recovery (EUR) gained from SSL must be enough to offset the incremental capital spent on the well. The decision to use SSL or standard stage spacing is truly an economic decision and, therefore, economic analysis must be the deciding factor and not the IP or EUR of the well. SSL is truly area dependent.

Why Does SSL Not Work Everywhere?

1. Not enough gas originally in place to be recovered
2. Complex geology may add complications
3. Heavily naturally fractured regions
4. Higher permeability of the rock
5. Lower pore pressure and pumping too much water can be detrimental in some areas
6. Lower pore pressure means less gas in place (GIP), particularly in dry gas areas
7. Hydraulic fracture stage interaction and competing fractures.

CLUSTER SPACING

There are various clusters in each frac stage. A cluster is referred to as a perforating gun. If there are five clusters in one frac stage, there are five perforation guns in that stage that are usually evenly distributed (geometric design). The industry average for the number of guns (e.g., clusters) in the Marcellus Shale formation is anywhere between three and eight clusters that are equally spaced in each stage. For example, if 6 clusters are used for a 300' frac stage, the cluster spacing is 50'. The industry average for cluster spacing is 25—100'. Every operator has its own theory regarding the number of clusters and holes in a frac stage. The major deciding factor in choosing the number of clusters is formation permeability, GIP, and perforation efficiency. A general rule of thumb is that if the formation permeability is higher than usual, fewer clusters will be needed. In contrast, if the formation permeability is lower, more clusters will be necessary. The goal is to achieve the maximum surface area between clusters. In addition, if GIP in a particular area is not significant, fewer clusters and stages will be needed to release the hydrocarbon.

Some operating companies believe that the spacing between clusters needs to be minimized to gain the most surface area out of each zone. On the other hand, others believe a lower number of clusters is necessary to achieve longer fracture networks by forcing the hydraulic fracturing energy to go to a limited number of clusters. Every operator justifies its theory with its production results. Having shorter cluster-to-cluster spacing has shown to maximize initial production in certain areas. Hydraulic fracturing operation in shorter cluster-to-cluster spacing has sometimes shown to be more difficult due to competing fractures or communication between clusters.

REFRAC OVERVIEW

Refrac refers to a second fracture stimulation on a well with existing production data and is another important topic that the industry has been experimenting with in various shale plays since 2011. Discussions on refrac are very common in a low commodity-pricing environment where plenty of time is available for analyzing the previous frac jobs with poor completion designs. In addition, instead of investing more funds into drilling and completing a new well, refrac could potentially offer better economics in areas with excellent reservoir quality and pressure. Refrac has caused a substantial production increase in many of the shale plays including and not limited to Marcellus, Haynesville, Eagle Ford, Barnett, and Bakken. The primary reasons behind refrac are as follows:

1. To implement new or enhanced completion design

    Many wells were completed using old completion designs such as 400−500' stage spacing, low sand/ft, high number of perforations per stage, high or low number of clusters per stage, etc. The combination of the above designs has caused a large percentage of unstimulated (virgin) rock that has not been touched yet. Refracing and implementing new completion designs such as reduced stage spacing, limited entry design, more sand/ft, etc., could potentially improve the production performance in some areas.

2. Contact more surface area by adding diversion, perforation, and reorientation

    One of the most common refrac methods is using diversion (special bimodal degradable particulate), which is offered by various service companies. The basic concept behind using diversion is to pack off the currently open perforations to effectively stimulate the unstimulated perforations, allowing breakdown into new areas of the reservoir. In addition to diversion, adding new perforations and reorientation could aid in increasing the contact area.

3. Bypass skin damage caused by scale, fines migration, and iron/salt deposition

   Skin damage can be caused by scale accumulation in the pipe and formation, salt and iron deposition, or simply by fines migration. The improper use of a chemical package when hydraulically fracturing a well could cause a detrimental impact to the long-term productivity of a well.

4. Wells that did not use managed pressure drawdown (especially in overpressured reservoirs) and caused proppant crushing, embedment, and conductivity reduction

   Unmanaged pressure drawdown has shown detrimental impact to the productivity of the wells in many different shale formations especially in overpressured reservoirs. Aggressive pressure drawdown can cause proppant crushing, proppant embedment, fines migration, cyclic stress, and pressure-dependent permeability effects. Pressure-dependent permeability is very important to consider in overpressured reservoirs. This is due to the pore volume reduction since the natural compaction process is incomplete. Therefore, the available flow area is reduced and permeability decreases with pressure. Refrac has shown to be successful on some of the wells that did not originally follow a managed pressure drawdown in overpressured formations such as the Haynesville and Eagle Ford Shales.

5. Increase conductivity and restore conductivity loss

   One of the most unknown segments of a hydraulic fracture design and production evaluation is how the conductivity loss with time affects the production performance and the economics of the wells. Almost all exploration and production (E&P) companies are interested in the first 5−10 years of producing life of a well because economically speaking that is when 80% + of the value is returned to the shareholders. Therefore, if conductivity loss in the fracture or near the wellbore during this time period does not severely impact the production, it is not a subject that is often discussed. However, if conductivity loss occurs sooner rather than later within the sensitive economic timeframe, it is very important to understand both the mechanism behind this loss and ways to mitigate this issue on future completion designs. Applying refrac on wells that are believed to have encountered some kind of near-wellbore or fracture conductivity loss due to various factors such as unmanaged drawdown, scale accumulation, non-Darcy effect, proppant crushing and embedment, fines migration, liquid trapping, liquid loading, fracture face skin, convergence skin, etc., has shown to restore conductivity loss and production enhancement in many refracs tested to date.

6. Change in fluid system could be successful refrac candidates

Another important reason behind a successful refrac could be the implementation of a different type of frac fluid system that is more compatible with the formation of interest. For example, if a well was originally hydraulically fractured using a cross-linked fluid system without successful production results, other frac fluid systems such as slick water could cause significant increase in production by performing refrac. It is important to note that if the area is not a rich area from both reserve and geologic perspectives, performing refrac is not recommended. Rodvelt et al. (2015) analyzed seven Marcellus wells that were refractured in Greene County, Pennsylvania, and noted 65−123% increase in reserve from refracing these Marcellus wells located in a geologically superior area with high reservoir pressure and excellent reservoir properties by using diversion material for the refrac.

When evaluating wells for refrac, keep the following guidelines in mind:

- Select wells with high remaining reserves and excellent geologic areas.
- Focus on the wells with old completion designs such as wells with larger stage spacing and minimal proppant mass.
- Stay away from wells with mechanical integrity issues, as this can get very costly.
- Select a great refrac candidate first in an attempt to add value to the entire prospect in the event it is successful and can be repeated on all of the remaining wells in the area. Some E&P companies also assign a present value on their refrac candidates and potentials when divesting assets.
- Poor wells often make bad refrac candidates, unless there is solid evidence that the original frac design, materials, or implementation was a failure.
- Stay away from low-pressure or depleted reservoirs where frac fluid recovery will be very challenging.
- Stay away from wells with excellent original design and implementation and focus on the ones with poor designs first, as they are many wells with poor designs that must be fixed first.
- Stay away from poor reservoir quality wells as refracing might not economically generate any additional value due to the poor area.
- Always run economic analysis using the existing refrac wells as analogous wells.

CHAPTER TWELVE

Completions and Flowback Design Evaluation in Relation to Production

INTRODUCTION

After obtaining sufficient production data for analysis, one of the most important aspects of completions optimization is evaluating the completions design. Typically, 6 months to 1 year of data (depending on data quality) is needed to evaluate each completions test in unconventional shale reservoirs. There are various tools that can be used to evaluate the productivity of a well in unconventional shale reservoirs. Calculating estimated ultimate recovery (EUR) using various types of decline curve analysis (DCA) or using rate transient analysis (RTA) is widely used to determine the flow capacity and strength of a well in conjunction with one another to tie back to completions design. One of the most important plots used to determine the flow capacity of a well is referred to as a superposition plot, from which flow capacity or $A\sqrt{K}$ of each well is determined. The flow capacity of a well can be determined by plotting pseudo $\Delta P/q$ on the y–axis versus material balance square root of time (CUM/q) on the x–axis to determine the slope of the linear portion of the plot, which is inversely proportional to $A\sqrt{K}$. $A\sqrt{K}$ is one of the most important parameters that can be used to determine the strength of a well in unconventional shale reservoirs based on their completions design, reservoir quality, pressure drawdown management, and other variables. $A\sqrt{K}$ is basically the contacted surface area multiplied by the effective permeability of the contacted rock. $A\sqrt{K}$ in unconventional reservoirs is the equivalent of kh in conventional reservoirs. $A\sqrt{K}$ is obtained from square root or superposition plots and it is a function of initial reservoir pressure, flowing bottom–hole pressure (with time), production rate (with time), porosity, gas viscosity, total compressibility, and reservoir temperature. The industry has found out that there is a direct correlation between $A\sqrt{K}$ and EUR from all of the analyses that have been

Hydraulic Fracturing in Unconventional Reservoirs
DOI: http://dx.doi.org/10.1016/B978-0-12-849871-2.00012-5
© 2017 Elsevier Inc.
All rights reserved.
183

performed in the past. As opposed to DCA, which assumes constant flowing bottom-hole pressure, drainage area, permeability, skin, and existence of boundary dominated flow, $A\sqrt{K}$ analysis takes pressure and rate with time as well as other reservoir properties into account for accurate determination of a well's strength. $A\sqrt{K}$ is also used to rank the best to worst performers in each field and make important completions design decisions for the company. There are many important parameters that must be taken into consideration when evaluating the production aspect of a completions test. They are described in the following sections.

LANDING ZONE

The landing zone of a well is extremely important to evaluate in an attempt to find the optimum target zone for each field. In theory, the best landing zone would have high resistivity, low water saturation, low formation density, high total organic contents (TOC), low clay content, high effective porosity, high Young's Modulus, and low Poisson's ratio (brittle from a fracability perspective). It is very challenging to find a formation that has all of the aforementioned properties. Therefore, in new exploration areas, different landing zones (keeping all of the other variables constant) must be tested to understand the production performance of each landing zone. In addition, it is important to understand in situ stress around each landing zone from various logging suites such as sonic log. Target zones are typically 5−15'. The ideal landing zone should have excellent barriers above and below the target zone to stay in the rich target rock for as long as possible. Frac barriers are very important in staying within the target area for the maximum production optimization. Susquehanna County, Pennsylvania is known for having excellent frac barriers below and above the Marcellus, with outstanding production performance as a result. The landing zone is heavily area dependent and numerous logs and testing must be performed to understand the best landing zone in each area. The best optimizing technique is to pick two or three different zones in the rich rock and test each landing zone to determine the target zone that yields the best production performance. It is well known that picking the proper landing zone has a substantial impact on the production performance and economic viability of a well in unconventional shale reservoirs. Therefore, special emphasis should be placed on choosing the proper target zone for optimum production enhancement.

STAGE SPACING

Stage spacing is another important completions design parameter that must be evaluated and understood from production analysis and evaluation. The idea is to contact as much surface area as possible within the clusters, while minimizing fracture interference (competing fractures). Stage spacing, just like any other parameter in unconventional shale reservoirs, is area dependent. For instance, if 150' stage spacing is optimum in one area, it does not necessarily mean that it will be optimum in other areas. Therefore, stage spacing needs to be obtained depending on the formation properties. Tighter stage spacing requires more capital expenditure; therefore, it is important to determine the percentage uplift in production from tighter stage spacing to justify the additional capital. In a commodity-pricing environment where completions capital is cheaper than usual, no significant uplift is required to justify tighter stage spacing. However, in an environment where completions capital is expensive, higher percentage uplift in production is needed to economically justify tighter stage spacing. Therefore, economic analysis must be performed at various points in time based on different oil/condensate/NGL/gas pricing and capital expenditure associated with each design. For example, if 150' stage spacing is the optimum design based on $2.5 MM completions capital expenditure given a well with 7000' lateral length, this stage spacing might not be optimum when compared to a completions capital of $5 MM because a higher percentage uplift in the production curve will be needed to justify the higher capital, depending on the commodity pricing. Therefore, production performance and economic analysis in each area must be the sole deciding factor behind choosing the optimum stage spacing.

CLUSTER SPACING

Cluster spacing is another important factor that needs to be considered when evaluating the production capacity of a well. In a conventional plug-and-perf technique, the distance between clusters should be optimized in a manner that yields the highest production result. There are typically 3—10 clusters per stage depending on stage spacing, formation properties, and company philosophy. The number of clusters must be tested in each area to understand the impact of cluster spacing on production performance.

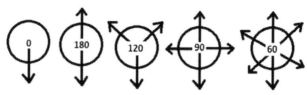

Figure 12.1 Perforation phasing.

NUMBER OF PERFORATIONS, ENTRY-HOLE DIAMETER (EHD), AND PERFORATION PHASING

From a perforation design perspective, the number of perforations, entry-hole diameter (EHD), and perforation phasing are three of the most important perforation design parameters that should be tied back to production results, if any testing is being performed. A rule of thumb that is used in the industry is that the EHD must be at least six times the maximum sand grain size used during the job to prevent sand grains from bridging and screening out during the fracture treatment. It is also known that the perforations should be within 30 degrees of the maximum principle stress orientation to reduce near-wellbore tortuosity and potential treatment issues throughout the frac job. Therefore, a perforating gun with 60-degree phasing will orient the perforations to be within 30 degrees of the maximum principle stress direction, especially since the exact perforation orientation is rarely known. Some companies use other phasing angles such as 0-, 90-, 120-, and 180-degree phasing. As previously discussed, the number of perforations is typically designed based on the limited entry technique for optimum production enhancement in unconventional reservoirs. However, various testing on the number of perforations, EHD, and perforation phasing must be done to understand the outcome of each test. It is very important to test one variable at a time to understand the sole impact of that parameter on production. Fig. 12.1 shows different perforation phasing.

SAND AND WATER PER FOOT

The amount of sand and water per foot are other important design parameters that must be tested to understand their impact on production, and economic viability of determining the optimum sand-and-water-per-foot design. The sand-and-water-per-foot design, just like all of the other parameters that have been discussed, are also area dependent. In areas with higher pore pressure and exceptional formation properties, higher sand and water per foot could potentially help the production performance of a well. On the other hand, in areas with lower pore pressure and poorer formation properties, pumping high volumes of sand and water might not be the ideal production enhancement solution. Therefore, various sand-and-water-per-foot designs must be tested to determine the optimum economic design for each area. For instance, 1000 #/ft, 1500 #/ft, 2000 #/ft, 2500 #/ft, 3000 #/ft, etc. (sand per foot) designs must be tested in new areas with limited data. One exercise that can be performed before pumping higher sand and water loadings is to determine the percentage uplift in production needed to justify the additional capital that will be spent on higher sand and water loadings. Once production data is available, percentage uplift for each design can be easily determined, and the optimum economical sand-and-water-per-foot design can be found for each area.

Another important factor in a downturn market with minimal frac activity is the water disposal cost. In a continuous frac environment, flowback water along with stored water can be continuously used on future frac jobs. This eliminates the cost spent on water disposal. Depending on water infrastructure and trucking fees, the water can be continuously used instead of disposing of the water. However, in a downturn market, where plenty of water is available and the water storage capacity is full, water will have to be disposed of. Water disposal might exceed the trucking fees. Therefore, many companies will continue to frac in a downturn market just to avoid paying a significant amount of money for water disposal, which can become costly depending on the area.

PROPPANT SIZE AND TYPE

Proppant size and type are other important factors when analyzing production results. The decision to choose between different proppant schedules such as pumping 100 mesh and 40/70 versus pumping 100 mesh, 40/70, and tailing in with 30/50 or 20/40 mesh should solely depend on the production performance (including economic analysis) of each proppant design. Sometimes, the availability and price of each proppant size can have a direct impact on the sand size selection. Some fields have shown that pumping a large percentage of 100 mesh yields the best production results. This could be due to the area being heavily naturally fractured and 100 mesh proppant traveling farther into the formation due to its size as compared to

40/70, 30/50, or 20/40. In theory, it might not make sense to pump a large percentage of 100 mesh due to higher closure pressure in a particular area, but if pumping a large percentage of 100 mesh is justified based on production data while saving money, more 100 mesh must be pumped to acknowledge the production results and more studies need to be done to understand the physical mechanism that results in excessive production.

Proppant type is another important factor that must be determined. The proppant type to be used in formations that have closure pressure of less than 6000−7000 psi is straightforward as regular sand is typically used in those formations. However, the discussion on whether ceramic or resin-coated sand is needed in higher closure pressure areas will depend on the closure pressure of the area that can be obtained from diagnostic fracture injection test (DFIT). In theory, if closure pressure is more than 8000 psi, ceramic proppant is recommended to avoid proppant crushing and embedment. However, the economics of pumping ceramic comes into play because ceramic is considered to be very expensive and pumping a large percentage of ceramic proppant might not be economically feasible. In addition, the amount of ceramic needed in overpressured formations with high closure pressure is determined from testing various percentages of ceramic proppant. Proppant-type testing will need to be performed on new exploration areas from day one in an attempt to understand the impact of each proppant type in relation to production. Many operators in different basins have tested various proppant types such as regular sand, resin-coated, and ceramic proppants to understand the impact of each proppant in higher closure pressure formations. Each field is unique due to the heterogeneity of the unconventional plays, in the sense that ceramic proppant might be absolutely necessary in some fields but it might not be economically justified in others, even with higher than 8000 psi closure pressure.

BOUNDED VERSUS UNBOUNDED (INNER VS. OUTER)

Another very important aspect of production evaluation in relation to completions design is whether a well is surrounded by other wells on both sides (bounded) or a well is not bounded by other wells on one or two sides. In Fig. 12.2, the B well is surrounded by A and C wells. Therefore, the B well is considered to be a bounded (inner) well. The A and C wells are both considered to be unbounded from one side, and finally the D well is unbounded from both sides. Production data in some

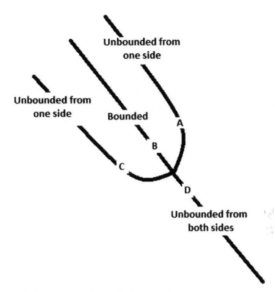

Figure 12.2 Bounded versus unbounded example.

fields has shown that unbounded wells from either one side or two sides have better production performance depending on the sequence in which the wells were hydraulically fractured (frac order). Belyadi et al. (2015) analyzed more than 100 Marcellus wells in the same geologic area and showed that unbounded wells are the best performers in the field. Frac order could also have an impact on fracture propagation by creating a pressure barrier or stress shadow effect around the wellbore and helping fracture propagation outward into the unbounded virgin rock. By assuming that all four wells in Fig. 12.2 have identical completions design, production performance of the wells could potentially be different if the B and D wells are zipper fractured first, followed by zipper fracturing A and C. By creating a stress shadow effect around the B well, the fractures will propagate outward when the A and C wells are hydraulically fractured. This could hypothetically cause better fracture half-length and more contact area on the two unbounded wells. On the other hand, zipper fracturing A and C wells first followed by zipper fracturing B and D wells will create more complexity around the B well. The D well is a standalone well and should not have a direct impact on the production performance of the A, B, and C wells. In an attempt to take advantage of unbounded wells, some companies pump more sand and water per foot on the unbounded wells to contact as much surface area as possible especially if the surrounding area around the unbounded wells will not be

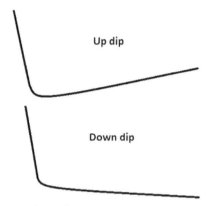

Figure 12.3 Up dip versus down dip.

hydraulically fractured anytime soon. Therefore, it is very important to take all of the aspects of the completions design including frac order, bounded, and unbounded wells into account when testing various completions designs and tying them back to production performance.

UP DIP VERSUS DOWN DIP

Another important parameter that should be taken into consideration when analyzing production data is whether a well was drilled up dip or down dip, especially in liquid-rich fields with undulations. A well with an inclination of more than 90 degrees is called up dip and a well with an inclination of less than 90 degrees is called down dip (Fig. 12.3). Some fields have seen better production results from up-dip wells while others have witnessed better production results from down-dip wells. In some fields, it is very challenging to see the impact of up dip versus down dip wells due to many other completions design changes. Fig. 12.3 illustrates the difference between up dip and down dip inclination.

WELL SPACING

Well spacing, or inter-lateral spacing, is also extremely important when analyzing and comparing the production results from various wells in the same field. Well spacing can range anywhere from 300' to 1500' depending on the rock characteristics (especially permeability), fracture half-length, fracture conductivity, gas pricing, capital expenditure, operating costs (OPEX) and many other parameters. It is crucial to select the optimum well spacing for each area based on production and economic evaluation of each area. Gas pricing has a tremendous impact on well spacing. Higher gas pricing indicates tighter well spacing, while lower gas pricing indicates wider well spacing. Higher capital expenditure dictates wider well spacing but lower capital expenditure indicates tighter well spacing. Various analytical and numerical models can be run to find the optimum economical well spacing for each area. From a production perspective, it is imperative to make sure that well spacing between all of the experimental wells is taken into account when wells with various completions designs are being tested. From a completions design perspective, well spacing should have a direct impact on completions design, i.e., on the amount of sand and water/ft, stage spacing, cluster spacing, etc. Exploration and production (E&P) companies design well spacing for each area based on actual field testing in addition to reservoir modeling techniques such as rate transient analysis (RTA), analytical, or numerical simulators. Belyadi et al. (2016) performed a well-spacing analysis sensitivity in which history matching was done on a dry Utica well in an exploration area and fracture and reservoir simulators along with economic analysis were used to find the optimum well spacing for the area. They concluded that well spacing is heavily influenced by fracture half-length, conductivity, effective permeability, gas pricing, capital expenditure (Capex), and operating costs (OPEX). They also concluded that well spacing is heavily area dependent and a spacing that might be optimum today may not be optimum in the future. In addition to well spacing, lateral length is another important parameter to consider when analyzing production data. The industry has been moving more towards drilling longer lateral wells (8000 + ft) in order to save Capex/ft. As lateral length increases, the economics of the wells typically improve unless severe production impairment is observed from drilling longer lateral wells. Economic analysis should be performed to understand the amount of production that can be lost and still yield better economic results by drilling longer lateral wells. For example, even if drilling a 12,000' lateral length well causes a 5% EUR/ft reduction as compared to a 7000' well due to completions efficiency (and other factors) in longer lateral wells, it might still be more economical to lose insignificant amount of reserve but create a higher value for the shareholders as significant amount of capital can be saved by drilling longer lateral wells. Sometimes a company's acreage position does not allow drilling long lateral wells in some units. Production results in some fields have shown no loss in production by increasing lateral lengths while other places have shown a percentage reduction in EUR/ft as lateral length increases. This must be evaluated from area to area to make the best possible decision for the company.

WATER QUALITY

Water quality used during frac jobs is another debatable topic that many companies are trying to understand. Some of the very important parameters that are used to compare against production performance are total dissolved solids (TDS), water conductivity, and the amount of chlorides that can be measured by taking a water sample prior to every stage. The importance of water analysis becomes more complex when 100% produced water with high TDS ($> 120,000$ ppm) is used for the job. High-TDS water could cause some issues with friction reducer (FR) selection for the job, and various FRs must be tested in the lab and field to determine the best FR type and concentration with high-TDS produced water. Sometimes the best FR selection from the lab analysis might not perform the same when tested in the field. Therefore, a contingency plan should be available with FR selection, especially during high-rate slick water frac jobs.

Other fracture treatment metrics that are recommended to be reviewed are:
- average treating pressure and rate trends
- breakdown pressure and instantaneous shut-in pressure (ISIP) versus % proppant placed
- breakdown pressure versus average treating pressure
- number of clusters and perf diameter versus % proppant placed
- fluid types and volumes versus % proppant placed.

FLOWBACK DESIGN

Flowback design is vital and can be just as important as completions design particularly in overpressured formations. In essence, the way a well is flowed back and produced is just as crucial as the way a well is hydraulically fractured. The notion that unconventional shale wells can be produced just like conventional wells is not correct due to the possibility of losing the integrity of the pumped proppant because of proppant crushing, proppant embedment, geomechanical effect (overpressured reservoirs), fines migration, cyclic stress, near–wellbore conductivity loss, and non–Darcy effects. Therefore, proper care must be taken to prevent proppant damage in the formation, and maintaining the integrity of the well's performance for decades to come. After hydraulic fracturing is finished, the drill–out phase takes place. Drill–out is defined as a poststimulation phase where coil tubing or stick tubing is utilized to clean the wellbore before flowback and production.

New technologies have recently been developed in which dissolvable plugs are used between stages in a conventional plug–and–perf technique. The use of dissolvable plugs eliminates the need for the drill–out phase and flowback through third–party equipment can take place. Sometimes a cleanup run is performed even after using dissolvable plugs to make sure that there is no debris in the hole. After the drill–out or cleanup period (if any), flowback takes place. Flowback is defined as post—drill–out phase (if any) where the well is flowed through the third–party equipment and cleaned up before turning the well over to permanent production equipment. The flowback procedure is typically provided by production or completions engineers with feedback from reservoir engineers. This procedure needs to be provided in a manner that proppant integrity is not sacrificed in any shape or form by following a pressure drawdown, depending on the reservoir characteristics and pore pressure. One rule of thumb that can be used during flowback is to stay within the critical drawdown pressure limit. Critical drawdown pressure is defined in Eq. (12.1).

$$\text{Critical drawdown pressure} = \text{closure pressure} - \text{reservoir pressure}$$

Equation 12.1 Critical drawdown pressure.

The most important part during the life of a well is when a well goes from a full column of fluid to a full column of gas during the flowback period. Therefore, special care must take place to avoid any proppant damage during this period by not exceeding critical drawdown pressure during flowback. The difference between closure and reservoir pressure is defined as critical drawdown pressure because when this pressure is exceeded, stress will start to be applied on the proppant. It is essential to avoid placing stress on the proppant during the flowback period where lots of events, including cleanup/proppant flowback and conversion of the full column of water to gas, occur. As the well cleans up and the gas cut starts occurring, casing pressure builds up until the peak casing pressure is reached. The critical drawdown pressure count begins as soon as the peak casing pressure along with stabilized water and tubing pressure (if there is any tubing in the well) are reached. After deducting the critical drawdown pressure from the peak casing pressure, the pressure drawdown must be in a manner that is controlled with time thereafter. Uncontrolled, sharp pressure drawdown has shown detrimental impact on production in various shale plays such as Utica, Haynesville, and Eagle Ford shale plays. There is a balance between how heavily a well is curtailed back and sacrificing the long-term productivity of the well by pulling hard on the well. Therefore, economic analysis must be run to understand the impact of managed pressure drawdown on the net asset value and obtain the optimum economic rate at which a well should be produced. An analysis that can be performed to make an educated decision for the company is to determine the amount of uplift in EUR needed to justify producing the well at a lower rate. This analysis is heavily influenced by gas pricing. In a low commodity-pricing environment, depending on each company's goal and strategy, it could potentially be economically justifiable to heavily curtail the wells when there is a potential future upside in pricing. However, in a high commodity-pricing environment, it is important to understand and run various economic sensitivities to obtain the optimum economic rate at which a well must be produced. For instance, if the price of gas is $6/MMBTU and there is only a 5% reduction in EUR by producing the well at a higher incremental rate of 5 MMSCF/D, the company might be economically better off to cause 5% damage to the long-term productivity of the well (negligible) and make more gas up front to take advantage of the time value of money. However, if production data shows a 30% reduction in EUR by producing the well at a higher incremental rate of 5 MMSCF/D, economic analysis must be performed to thoroughly understand this impact and determine the optimum economic rate that will create value for the shareholders.

FLOWBACK EQUIPMENT

Flowback equipment is used during flowback and is provided by a third-party flowback company. To save costs, some E&P companies have started flowing back their wells through third-party equipment such as sand traps, choke manifolds, and gas buster tanks when on fluid production and when the well is initially flowed back until the gas cut is reached. Once the gas cut is reached, the well can be diverted to permanent production equipment. Taking a slowback approach, wells must be managed to minimize sand production and stay within the limits of gas production units (GPUs) for water production. During a regular flowback job using a third-party flowback company, the following flowback equipment is typically used.

Choke Manifold

A choke manifold is used to control the flow of a well by providing back pressure. There are two types of chokes used in flowback operations. The first type of choke, which is more commonly used, is called an *adjustable choke*. The adjustable choke has two parts, called the *valve* and *seat*. The valve and seat on an adjustable choke wash out quickly after flowing back lots of sand; however, they can be replaced very easily by diverting the well flow to a different direction and replacing the valve or seat. In addition, adjustable chokes are operated using a wheel and changing the choke size is very easy to perform. The second type of choke is called a *bean choke*. Bean chokes come in various sizes and to achieve the required size, the insert inside the choke must be replaced. The bean choke consists of a replaceable insert (also referred to as the bean) made from steel. The inserts are manufactured with various hole diameters and are available in different sizes. Recently, automated chokes have also become common in the industry to improve efficiency and to remotely operate the wells. Fig. 12.4 shows a choke manifold used during the flowback operation.

Sand Trap

The sand trap is typically located right after the choke manifold on a multi-well pad and is used to prevent erosive materials such as proppant from entering the equipment, in an attempt to prevent washout and damage to the equipment. Flowback fluid (water + gas + sand + oil) coming out of

Figure 12.4 Choke manifold.

the choke manifold typically goes to a sand trap. Since proppant has a heavier density, it will fall down to the bottom of the vessel. Sand is then removed through the outlet located at the bottom of the vessel. The blow-down line located at the bottom of the vessel is used to pump the sand every so often to the flowback tanks depending on the amount of sand that is flowed back. Sand traps typically have pressure ratings of 2.8 to 10 K psi.

In single-well applications, the sand trap is placed upstream of the choke manifold (before the choke manifold). In this scenario, the sand trap should be able to handle 20% above the maximum anticipated well-head pressure. On the other hand, the sand trap is located downstream (after the choke manifold) on multiwell pad applications and lower pressure rating will be required since fluid coming out of the choke manifold has lower pressure. It is better to run the sand trap upstream (before) the choke manifold in any application. This is simply because the differential pressure between wellhead and choke manifold is less, therefore, resulting in slower velocity. However, on multiwell pads, having a sand trap before the manifold would require a sand trap per well, which would be very costly; this is why the sand trap is run after the choke manifold. In addition, sometimes due to the pressure limitations on the sand trap, the sand trap is run after the choke manifold. Another reason for placing the sand trap before the choke manifold is to prevent washing out the choke manifold. Producing substantial sand can wash out the seat and stems (used to control the choke) inside the choke manifold. Seats and stems wash out very quickly after encountering a large amount of sand and to prevent paying a lot of funds for damaged and washed out equipment, a sand trap can be placed upstream of the choke manifold during drill-out and flow operations. Any sand trap used during the operation must have a mechanical pressure-relief system referred to as a pop-off. A pop-off releases the pressure if the pressure exceeds the limitation. In addition, sand traps must be inspected yearly and should also have a bypass system in the event of failure. Fig. 12.5 shows a sand trap used during flow back operation.

High-Stage Separator

High-stage separators are divided into three main types. Vertical, horizontal, and spherical separators are well-known separators throughout various operations. Horizontal separators are more commonly used in a variety of operations. Separators can be two, three, or four phases. A two-phase separator separates fluid from the wells into gas and total liquids. Since water has a heavier density than gas, it leaves the vessel at the bottom while gas

Figure 12.5 Sand trap.

leaves the separator at the top. A three-phase separator separates fluid into gas, water, and oil. The first partition in a three-phase separator is used for water removal. One of the primary differences between two-phase and three-phase separators is that additional weir is used in a three-phase separator to control the oil/water interface. Finally, a four-phase separator (not commonly used) has the ability to separate sand, water, oil, and gas. First of all, the sand falls down into the first partition as soon as it enters the inlet diffuser (because of heavier weight). Oil and water are directed to the second partition (middle portion of the separator). Since water has a higher density, it remains in the second partition and is dumped through a dump line to the flowback tank. Oil on the other hand has a lighter density as compared to water and reaches the third partition, where it is dumped to a low-stage separator that is located right after the high-stage separator. For instance, in Marcellus Shale operations, some areas are known to only produce dry gas and therefore a two-phase separator is used. Sometimes a three-phase separator is used in dry gas windows and two partitions are used to remove water in the event water production is expected to be high. However, if an area is known to produce water, condensate, and gas, it would not be feasible to separate the liquid (condensate and water) from each other in a two-phase separator. In this particular scenario, a three-phase separator would be necessary to efficiently separate water, condensate, and gas. The majority of separation occurs at the inlet diffuser. Adequate settling partitions allow turbulence to fall down and liquid to fall out. Liquid capacity of a separator depends on the retention time of the fluid in the vessel. For a good separation to occur, sufficient time to obtain equilibrium condition between liquid and gas must be met. It is important to note that even after going through the sand trap, there might be some residual sand that was not caught in the sand trap and found its way to the separator. This will be caught in the first partition of the four-phase separator and is basically an added safety against erosion and washout on the vessel.

Every separator has pressure, rate (velocity), and volume limitations depending on the company's manufacturer. The most commonly used units are 720, 1440, and 2000 psi units in different operations. Please note that separator operating pressure must not exceed 75% of its maximum operating pressure for an extended period of time. For instance, if 1440 psi separator is used, it is important not to exceed 1080 psi for an extended period of time as a safety factor. The most important parameters when it comes to units are the rate and the volume that each separator can handle. Different separators can handle various volumes of liquid and gas. The rule of thumb for having sufficient number of separators onsite is that all of the separators must be able to handle 40% over the maximum anticipated production rate (liquid and gas). This safety factor can be increased in multiwell pads and exploration wells. For example, if an eight-well pad is anticipated to produce 64 MMSCF/D, separators should be able to handle at least 90 MMSCF/D or preferably more due to being a multiwell pad. Separators typically have an electronic gas-flow measurement with a backup mechanical meter (Barton meter) in the event the electronic one fails. Another type of safety equipment that must be installed on separators is a check valve, which is rigged up (R/U) right after the exit gas line on the separator. Completions, facilities, or production engineers in the office are typically responsible for designing the necessary flowback equipment for drill-out, production tubing, and flowback operations. A separator has various regulators and regulators are employed to reduce pressure to the areas of the tank. The most common areas that need regulations are as follows:

- *Liquid level controller (LLC)* is a pneumatic controller to evacuate fluid from the separator. When the level of water or oil gets to a predefined limit recommended by the manufacturer and set by the operator, it will automatically dump the water to the flowback tank and it will dump the oil or condensate to the low-stage separator. **Dump valve** is a working valve actuated by a LLC to evacuate fluid. There are two dump valves located on a three-phase separator (water and oil). Dump valve acts like a "toilet flush" when it dumps. The water and oil dumps must have manual bypass in addition to pneumatic.

- The *scrubber pot* removes traces of liquid droplets from gas stream. When gas leaves the vessel at the top, it passes through mist extractor (removes mist from gas stream) followed by scrubber pot, which removes the liquid droplets in the gas stream. The scrubber pot is essentially a water knockout unit used for conditioning fuel gas supply. The supply gas must be dry for separator control or failure can be the consequence if it is not. In the wet gas scenarios, a dryer is run after the scrubber pot to avoid this issue. This regulator prevents the gauge located on the scrubber pot from any moisture as well.

- *Back pressure regulator (BPR)* is located on the separator to hold pressure inside the separator as needed. For instance, if sales line pressure (which goes to the compressor station) is 700 psi, there needs to be a pressure higher than 700 psi on the separator back pressure to send the gas to the sales line. All units must have a pneumatic and manual back pressure regulator.

Other mechanical controls of a separator are as follows:

- *Mechanical pop-off* is used to prevent overpressuring the separator and rupture. A mechanical pop-off is set slightly above the operating pressure. Pop-offs on the separator are inspected annually or if ruptured. If mechanical pop-off gets activated, it will send the gas to the flare to prevent overpressuring the separator.

- *Meter* is used to measure the flow. Flow measurements must be electronic with mechanical backup. There are typically a minimum of two flow meters in place to measure the flow.

The fluid rate, also called liquid capacity, of a separator is related to the retention time and liquid settling volume. The amount of liquid retention time needed in a vessel governs liquid capacity of a separator. Factors affecting separation are operating pressure, operating temperature, and flow stream composition. Eq. (12.2) is often used to calculate liquid capacity of a separator for separator design during flowback operation. Fig. 12.6 shows three horizontal separators located on a multiwell pad during a flowback operation. Fig. 12.7 illustrates a four-stage horizontal separator from an inside view. Figs. 12.8—12.10 depict the liquid level controller (LLC), back pressure regulator (BPR), and mechanical pop-off on a horizontal separator, respectively.

Figure 12.6 Horizontal separators.

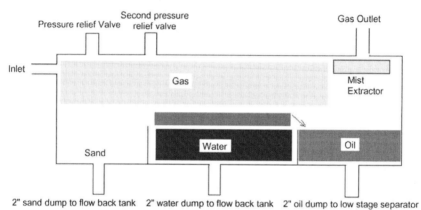

Figure 12.7 Four-phase horizontal separator from inside.

Figure 12.8 Liquid level controller (LLC).

Figure 12.9 Back pressure regulator (BPR).

Figure 12.10 Mechanical pop-off.

$$W = \frac{1440 \times V}{t}$$

Equation 12.2 Liquid capacity

W = Fluid rate (liquid capacity), BBL/day
V = Liquid settling volume, BBL
t = Retention time, min.

Low-Stage Separator

Anytime there is a possibility of producing crude oil, condensate, or wet gas, a low-stage separator is needed. The line coming out of the oil partition of the high-stage separator (horizontal separator) will go into a low-stage separator to give oil more retention time before going to the oil tanks.

Flare Stack

A flare stack is used to burn off flammable gas in certain events. One example is not having the necessary pipeline infrastructures. For instance, in Utica Shale operation (Ohio), many operating companies are still exploring the shale play and some companies do not have the pipeline infrastructure to commercially produce and sell the gas in new exploration areas. As a result, any commercial gas produced from any well is flared. Another example is not having the proper equipment to handle the volume of gas that is being produced. Sometimes the sales line cannot handle the produced volume of gas for whatever reasons and some of the gas is sent to flare. Ideally, any company would like to sell every bit of the gas produced; however, sometimes combinations of various reasons lead to sending the gas to the flare stack. The flare stack is essentially a safety precaution that needs to be used during various shale operations. In the Bakken Shale, gas is occasionally flared since the infrastructure for gas is not there. Bakken Shale is primarily a shale oil formation and this is the primary reason that noncommercial amount of gas produced with oil is flared. Typically there is a line that comes out of the high-stage separator and goes into the flare stack. In addition, there is another line that comes out of the low-stage separator and goes to the flare stack. Some requirements for flare stacks (Fig. 12.12) are as follows:

- Diameter of the flare stack needs to be at least 6".
- Height of flare stack has to be at least 40' for safety reasons depending on the amount of gas that is expected to be flared.
- Flare must be equipped with a check valve and it must also have an autoignition system.

Oil Tanks (Upright Tanks)

Oil tanks are only employed when there is a possibility of producing commercial oil or condensate. The entire purpose of having oil tanks on location is to store the commercial oil or condensate that is being produced. The number of oil tanks depends on the anticipated production volumes. The regular-size oil tanks typically have a capacity of 250 or 450 barrels (Fig. 12.11). There are trucks that are lined up throughout the day to transfer the oil or condensate from the oil tanks. Oil tanks must be

Figure 12.11 Oil tanks (upright tanks).

Figure 12.12 Flare.

closed top for safety reasons. Vapor produced from oil or condensate is heavier than air and can be fatally ignited if open tanks are used. The vapor produced has to be sent to a vapor destructive unit (VDU), which essentially flares any vapor produced from the liquid during flowback. The VDU must be rigged up and used on any location that has a possibility of producing liquid. The VDU is essentially a type of flare.

FLOWBACK EQUIPMENT SPACING GUIDELINES

1. All ignition sources *must* be 100' from the flowback tanks. In addition, flowback tanks must be 100' from the wellhead. The ignition sources should also be upwind of tanks if possible.
2. Choke manifold should be at least 50' from the wellhead.
3. Low- and high-stage separators, along with sand traps, should be placed 75' from the wellhead and 100' from flowback tanks.
4. Flare stack should be 100' from wellhead and flowback tanks.
5. Grounding is very important during flowback as it drains away any unwanted buildup of static electricity.
6. Bonding is also used to connect all metallic equipment to prevent buildup of static electricity.

TUBING ANALYSIS

One of the last steps of completions in unconventional shale reservoirs is running production tubing. Production tubing is used to efficiently remove water from the well until critical rate is reached, at which point the well starts liquid loading and some type of artificial lift will be needed. A well starts liquid loading when the production velocity is unable to carry the liquids from the bottom hole to the surface. Not efficiently and properly removing the liquids from the bottom hole will cause liquid buildup and as a result will lower production, finally killing the well.

The most commonly used production tubing sizes are 2 7/8" and 2 3/8". Some operators do not run production tubing on their well from day one in order to produce as much volume as possible through the casing (typically 5 ½"), especially in very prolific areas and longer lateral length wells. Running production tubing in a well will limit the amount of rate

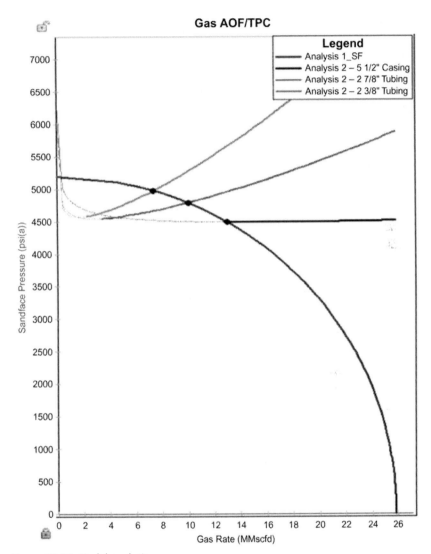

Figure 12.13 Nodal analysis.

that can be produced through the tubing depending on various factors such as reservoir pressure, wellhead flowing pressure, water rate, turbulent factor (*n* value), etc. Therefore, nodal and economic analyses are performed to decide whether production tubing is needed and its size, or simply produce the well through casing until critical velocity of the casing is reached and tubing will have to be run to efficiently unload the water through the tubing. Fig. 12.13 illustrates an inflow performance curve (IPR) versus different tubing performance curves (various pipe sizes) including 5 ½" casing, 2 7/8", and 2 3/8" assuming a reservoir pressure of 5200 psi, n value of 0.5 (fully turbulent), gas rate of 13 MMscfd, wellhead flowing pressure of 3500 psi, and water-to-gas ratio (WGR) of 40 BBL/MMscf. The intersection of the IPR curve with various tubing performance curves (TPCs) indicates the operating point of a particular well at an instantaneous point in time. This analysis is performed at various operating conditions for both IPR and TPCs to determine tubing sizing, drawdown, and compression of a particular well. Other analyses such as erosional and unloading velocity calculations across the lateral are also performed to prevent exceeding the erosional velocity and efficiently unloading fluid from the horizontal section of the wellbore.

CHAPTER THIRTEEN

Rock Mechanical Properties and In Situ Stresses

INTRODUCTION

In general, rock mechanics is a branch of geomechanics where the main focus is on rock deformation and possible failure of rock due to the applied manmade or natural forces. This has been a topic of studies in different earth sciences and engineering programs. In the oil and gas industry and particularly in the field of hydraulic fracturing, the rock and fluid interactions have become a major topic of studies in which fracture initiation, propagation, and geometry due to the applied hydraulic force are investigated. This requires advanced understanding of formation in situ stress conditions and stress behavior around the fracture as it generates and propagates to the formation. Stress, strain, and deformation are essential parameters required for characterization of mechanical properties of the rock. In this section of the book, various concepts of rock mechanics and interactions between induced and in situ stresses, especially during hydraulic fracturing, will be discussed.

YOUNG'S MODULUS (PSI)

Young's modulus is a measurement of stress over strain. Simply put, Young's modulus is the slope of a line on a stress versus strain plot. When hydraulic fracturing occurs, Young's modulus can be referred to as the amount of pressure needed to deform the rock. Young's modulus measures a rock's hardness, and the higher the Young's modulus, the stiffer the rock. A higher Young's modulus will help to keep the fractures open. For a successful hydraulic frac job to occur, higher Young's modulus is required. A higher Young's modulus indicates that the rock is brittle and

Hydraulic Fracturing in Unconventional Reservoirs
DOI: http://dx.doi.org/10.1016/B978-0-12-849871-2.00013-7

© 2017 Elsevier Inc.
All rights reserved.

will help to keep the fractures open for better production after the frac job. Examples of materials with high Young's modulus would be glass, diamond, granite, etc. These materials tend to be very hard but are prone to brittleness. On the other hand, examples of materials with low Young's modulus would be rubber and wax, which are very flexible and resistant to brittleness. The Young's modulus in various unconventional shale plays varies, and the brittleness of the rock will determine the type of frac fluid system to be chosen for the job. Young's modulus can be measured by using sonic log or core data. Core data yields static Young's modulus and sonic log represents dynamic Young's modulus. Eq. (13.1).

$$E = \text{Young}'s \text{ modulus} = \frac{\sigma}{\varepsilon_{xx}}$$

Equation 13.1 Static Young's modulus from core analysis

σ = Stress, psi
ε_{xx} = Strain.

Another method for calculating Young's modulus is using sonic log. The equation listed below can be used to calculate dynamic Young's modulus from a sonic log. Dynamic Young's modulus must then be converted to static Young's modulus.

1. Formation modulus calculation

$$G = 1.34 \times 10^{10} \times \frac{\rho_b}{\Delta t_s^2}$$

Equation 13.2 Formation modulus

G = Formation modulus, psi
ρ_b = Bulk density, g/cc
Δt_s = Shear wave travel time, μs/ft.

2. Dynamic Young's modulus calculation:

$$E = 2G(1 + \nu)$$

Equation 13.3 Dynamic Young's modulus from log analysis

E = Dynamic Young's modulus, psi
G = Formation modulus, psi
ν = Poisson's ratio.

3. Larry Britt came up with a correlation to convert dynamic Young's modulus from log to static Young's modulus as shown in Eq. (13.4).

$$E_{static} = 0.835 \times E_{dynamic} - 0.424$$

Equation 13.4 Static Young's modulus conversion (King, 2010)

Example

A core sample was taken and sent to the lab. After applying 30,000 lbs of force to the core cross-sectional area of 0.3 in.^{2}, the length of the core decreased from 7″ to 6.8″ as shown in Fig. 13.1. Calculate Young's modulus from this core test.

$$\text{Stress} = \sigma = \frac{F}{A} = \frac{30,000}{0.3} = 100,000 \text{ psi}$$

$$\text{Strain} = \varepsilon_{xx} = \frac{\Delta L}{L} = \frac{7 - 6.8}{7} = 0.02857$$

$$E = \text{Young's modulus} = \frac{\sigma}{\varepsilon_{xx}} = \frac{100,000}{0.02857} = 3.5 \text{ MM psi.}$$

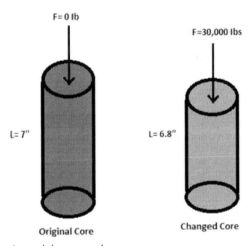

Figure 13.1 Young's modulus example.

POISSON'S RATIO (ν)

Poisson's ratio measures the deformation in material in a direction perpendicular to the direction of the applied force. Essentially Poisson's ratio is one measure of a rock's strength that is another critical rock property related to closure stress. Poisson's ratio is dimensionless and ranges between 0.1 and 0.45. Low Poisson's ratio, such as 0.1−0.25, means rocks fracture easier whereas high Poisson's ratio, such as 0.35−0.45, indicates the rocks are harder to fracture. Please note that Poisson's ratio changes from layer to layer. The best formations to hydraulically fracture have the lowest Poisson's ratios. Poisson's ratio can be measured from a core sample. A core sample is taken to the lab and a compressive force is applied. Afterward, the height and diameter changes are measured (strain in x- and y-directions) and Eq. (13.5) is used to calculate Poisson's ratio.

$$\nu = -\frac{\varepsilon_y}{\varepsilon_x} = \frac{\text{Radial strain}}{\text{Axial strain}}$$

Equation 13.5 Poisson's ratio, core analysis

$\varepsilon_x =$ Strain in the x-direction, which means how much material is deformed when a stress is applied. Compressive strength is positive.
$\varepsilon_y =$ How much material has been deformed after the stress application and is negative because of being tensile strain.

Poisson's ratio can also be calculated by running a sonic log in the depth of an interest. The sonic log provides the shear and compression wavelength travel time, which are used in the calculation of Poisson's ratio using Eqs. 13.6 and 13.7.

$$\nu = \frac{0.5R_\nu^2 - 1}{R_\nu^2 - 1}$$

Equation 13.6 Poisson's ratio, log analysis

where R_ν is:

$$R_\nu = \frac{\Delta t_s}{\Delta t_c}$$

Equation 13.7 R_ν calculation

$\Delta t_s =$ Shear wave travel time, μs/ft
$\Delta t_c =$ Compression wave travel time, μs/ft.

●●●

Example

A core sample is taken from a Marcellus Shale formation. The sample height is 10″ and the diameter is 3″. After applying a compressive force of 150,000 lbs, the height decreases by 0.15″ and the diameter increases by 0.007″. Calculate the Poisson's ratio of the sample.

Strain in the x- and y-directions need to be calculated:

$$\varepsilon_x = \frac{\Delta L}{L} = \frac{0.15}{10} = 0.015$$

$$\varepsilon_y = \frac{\Delta D}{D} = \frac{0.007}{3} = 0.0023.$$

Finally, Poisson's ratio can be calculated.

$$\text{Poisson's ratio} = \nu = -\frac{\varepsilon_y}{\varepsilon_x} = \frac{\text{Radial strain}}{\text{Axial strain}} = \frac{0.0023}{0.015} = 0.16.$$

Example

Calculate Poisson's ratio and Young's modulus with the following data obtained from the sonic log:

Bulk density = 2.6 g/cc, Δt_s = 115 μs/ft, Δt_c = 67 μs/ft

$$R\nu = \frac{\Delta t_s}{\Delta t_c} = \frac{115}{67} = 1.72$$

$$\text{Poisson's ratio} = \nu = \frac{0.5 \, R\nu^2 - 1}{R\nu^2 - 1} = \frac{(0.5 \times 1.72^2) - 1}{1.72^2 - 1} = 0.24$$

$$\text{Formation modulus} = G = 1.34 \times 10^{10} \times \frac{\rho_b}{\Delta t_s^2} = 1.34 \times 10^{10} \times \frac{2.6}{115^2} = 2.63 \times 10^6 \text{ psi}$$

$$\text{Dynamic Young's modulus} = E = 2G(1 + \nu) = (2 \times 2.63 \times 10^6)(1 + 0.24) = 6.5 \times 10^6 \text{ psi}$$

FRACTURE TOUGHNESS (psi/$\sqrt{\text{in}}$)

Fracture toughness modulus is another indicator of a rock's strength in the presence of a preexisting flaw. For example, glass has a high strength, but the presence of a small fracture reduces the strength. Therefore, glass has low fracture toughness. Fracture toughness is an important consideration in hydraulic fracture design. Fracture toughness is an essential parameter in very low fluid viscosity (water) and very low modulus formations. A low fracture toughness value indicates that materials are undergoing brittle fractures, while high values of fracture toughness are a signal of ductility. Fracture toughness ranges from 1000 to

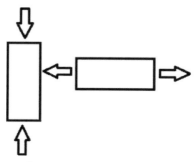

Figure 13.2 Poisson's ratio illustration.

$3500\ psi/\sqrt{in}$. Fracture toughness is measured in the laboratory and is denoted by K_{IC}. Formations with low Poisson's ratio, low fracture toughness, and high Young's modulus are typically the best candidates for slick water hydraulic frac. Fig. 13.2 shows the schematic of a sample under vertical stress and therefore change in length and width of the sample that can be used to calculate the Poisson's ratio as described in Eq. (13.5).

BRITTLENESS AND FRACABILITY RATIOS

Brittleness and fracability ratios are very important to compute and understand in hydraulic fracture design. Calculating Young's modulus and Poisson's ratio separately does not give a clear understanding of the brittleness and fracability of the rock. Therefore, various equations have been developed to combine both parameters into one single variable. The simplest way to find the brittleness of the rock is by taking the ratio of Young's modulus over Poisson's ratio (PR); the higher E/PR, the higher the brittleness. As previously mentioned, various equations were developed for both fracability and brittleness ratios and Eqs. (13.8) and (13.11) are examples of brittleness and fracability ratios that were developed primarily for Barnett shale. The following brittleness ratio was generated after Rickman and Mullen et al. 2008:

$$Brittleness\ ratio = \frac{\left(\left(\frac{E_{static} - 1}{7}\right) \times 100\right) + \left(\left(\frac{\nu - 0.4}{-0.25}\right) \times 100\right)}{2}$$

Equation 13.8 Brittleness ratio

E_{static} = Static Young's modulus
ν = Poisson's ratio.

Fracability ratio was generated after Goodway et al. (2010) using Eq. (13.11) that is a function of incompressibility constant λ Eq. (13.9) and rigidity constant μ Eq. (13.10):

$$\lambda = \frac{E_{static} \times \nu}{(1 + \nu)(1 - 2\nu)}$$

Equation 13.9 Incompressibility constant

$$\mu = \frac{E_{static}}{2(1 + \nu)}$$

Equation 13.10 Rigidity constant

$$\text{Fracability ratio} = \frac{\lambda}{\mu} = \frac{\text{Incompressibility constant}}{\text{Rigidity constant}}$$

Equation 13.11 Fracability ratio

$\lambda = $ Used to relate rock's resistance to fracture dilation
$\mu = $ Describes rock's resistance to shear failure.

For the rock to be brittle and fracable in Barnett shale, the brittleness ratio has to be more than 50 and fracability ratio must be less than 1. These two equations were developed based on the Barnett Shale reservoir, which has the best E and Poisson's ratio as compared to other shale plays across North America. If sonic log is available, obtain dynamic E and Poisson's ratio from the sonic log. Afterward, convert dynamic E to static E. Finally, use brittleness and fracability ratio equations to calculate these two parameters in each section (usually 6") to determine the optimum placement for your lateral well from a completions perspective.

Example

Given the 20 samples below, calculate brittleness, fracability, and E/PR ratios and determine which 10 consecutive zones are the best zones for hydraulic fracturing from a rock brittleness and fracability perspective.

The best zone from a hydraulic fracturing perspective is the zone with the highest E and the lowest PR. This basically means the highest brittleness, the lowest fracability, and the highest E/PR ratios. In Table 13.1, the first 10 consecutive samples must be targeted when drilling the well. This does not take into account other formation properties. In reality, brittleness and fracability along with other formation properties must be taken into account before deciding the landing zone of a well. But for the sake of this example, all the other parameters have been excluded.

Table 13.1 Brittleness and Fracability Ratios Example

	Provided				Results		
Sample	Static Modulus	PR	Brittleness	λ	μ	Fracability	YM/PR
1	4.8	0.33	41.1	3.50	1.80	1.94	14.5
2	5.3	0.35	40.7	4.58	1.96	2.33	15.1
3	4.5	0.27	51.0	2.08	1.77	1.17	16.7
4	3.5	0.22	53.9	1.13	1.43	0.79	15.9
5	3.3	0.25	46.4	1.32	1.32	1.00	13.2
6	5	0.3	48.6	2.88	1.92	1.50	16.7
7	4.5	0.27	51.0	2.08	1.77	1.17	16.7
8	4.1	0.23	56.1	1.42	1.67	0.85	17.8
9	4.3	0.26	51.6	1.85	1.71	1.08	16.5
10	4	0.19	63.4	1.03	1.68	0.61	21.1
11	3.5	0.33	31.9	2.55	1.32	1.94	10.6
12	3.5	0.32	33.9	2.36	1.33	1.78	10.9
13	3.2	0.39	17.7	4.08	1.15	3.55	8.2
14	4.1	0.29	44.1	2.19	1.59	1.38	14.1
15	4.1	0.33	36.1	2.99	1.54	1.94	12.4
16	4.1	0.34	34.1	3.25	1.53	2.13	12.1
17	3.3	0.37	22.4	3.43	1.20	2.85	8.9
18	4.1	0.36	30.1	3.88	1.51	2.57	11.4
19	3.8	0.28	44.0	1.89	1.48	1.27	13.6
20	3.9	0.32	36.7	2.63	1.48	1.78	12.2

PR, Poisson's ratio; *YM*, Young's modulus

VERTICAL, MINIMUM HORIZONTAL, AND MAXIMUM HORIZONTAL STRESSES

The next exciting concept in hydraulic fracture design is the various principal stresses that exist within the rock. There are three principal stresses that exist within a rock.

VERTICAL STRESS

Vertical stress, also referred to as overburden stress, is the sum of all the pressures applied by all of the different rock layers. Every formation contains fluid and rock and each one must be accounted for separately. Porosity correlation can be simply used to define the amount of space that is occupied by fluid versus the amount of space that is occupied by rock. The average density of the rock can simply be calculated using Eq. (13.12).

$$\rho_{avg} = \rho_{rock}(1 - \varnothing) + \rho_{fluid}\varnothing$$

Equation 13.12 Average formation density

ρ_{avg} = Average formation density, ppg
ρ_{rock} = Rock density, ppg
ρ_{fluid} = Fluid density, ppg
$\varnothing$ = Porosity, fraction.

Now that the average formation density is known, the magnitude of the vertical stress in an isotropic, homogeneous, and linearly elastic formation can be calculated using Eq. (13.13).

$$\text{Vertical stress} = \sigma_v = 0.05195 \times \rho_{avg} \times H$$

Equation 13.13 Vertical stress

ρ_{avg} = Average formation density, ppg
H = Height of layer or TVD in ft
0.05195 is conversion from ppg to psi/ft.

Eq. (13.13) can be rewritten as follows if average formation density is reported in lb/ft^3:

$$\text{Vertical stress} = \frac{\rho \times \text{TVD}}{144}$$

Example

The following data is provided from core analysis. Using this data, calculate vertical (overburden stress) at the zone of interest (8000'):

0–4000' → Zone 1 = 9% porosity, rock density = 21.5 ppg, fluid density = 8.35 ppg
4000'–6000' → Zone 2 = 12% porosity, rock density = 23.6 ppg, fluid density = 8.6 ppg
6000–8000 → Zone 3 = 7.5% porosity, rock density = 22.4 ppg, fluid density = 8.4 ppg
Layer 1) $\rho_{avg} = \rho_{rock}(1 - \varnothing) + \rho_{fluid}\varnothing = 21.5(1 - 9\%) + (8.35 \times 9\%) = 20.32$ ppg
Layer 2) $\rho_{avg} = \rho_{rock}(1 - \varnothing) + \rho_{fluid}\varnothing = 23.6(1 - 12\%) + (8.6 \times 12\%) = 21.8$ ppg
Layer 3) $\rho_{avg} = \rho_{rock}(1 - \varnothing) + \rho_{fluid}\varnothing = 22.4(1 - 7.5\%) + (8.4 \times 7.5\%) = 21.35$ ppg.
Now calculate the incremental vertical stress (overburden stress) for each layer:
Layer 1) $\sigma_v = 0.05195 \times \rho_{avg} \times H = 0.05195 \times 20.32 \times 4000 = 4222.5$ psi
Layer 2) $\sigma_v = 0.05195 \times \rho_{avg} \times H = 0.05195 \times 21.8 \times (6000 - 4000) = 2265.0$ psi
Layer 3) $\sigma_v = 0.05195 \times \rho_{avg} \times H = 0.05195 \times 21.35 \times (8000 - 6000) = 2218.3$ psi.

Total vertical stress at 8000':

Total vertical stress $= 4222.5 + 2265.0 + 2218.3 = 8706$ psi.

Total vertical stress gradient at 8000':

$$\text{Vertical stress gradient} = \frac{8706}{8000} = 1.09 \text{ psi/ft.}$$

In the real world, it is very challenging to obtain rock density and fluid density at various depths. Therefore, a density–logging tool can measure the density of the formation every half-foot. A density-logging tool is typically not run all the way to the surface and is only run a few thousand feet around the zone of interest. The vertical stress gradient is typically between 1 psi/ft to 1.1 psi/ft depending on the depth and porosity. In a given formation, the higher the porosity and the shallower the depth, the lower the vertical stress. In contrast, the lower the porosity and the deeper the depth, the higher the vertical stress.

MINIMUM HORIZONTAL STRESS

Minimum horizontal stress is approximated as fracture closure pressure. Units of stress and pressure are both psi. This is not a coincidence because stress and pressure are fundamentally related. The primary difference is that pressure acts in all directions equally, while stress only acts in the direction of the force. Since minimum horizontal stress is a direct result of overburden stress, Poisson's ratio determines the amount of stress that can be transmitted horizontally. Minimum horizontal stress or fracture closure pressure can be obtained from either a diagnostic fracture injection test (DFIT) or by using Eq. (13.14) (if rock properties are available):

$$\sigma_{h,\min} = \frac{\nu}{1 - \nu} \times (\sigma_v - \alpha P_p) + \alpha P_p + P_{\text{Tectonic}}$$

Equation 13.14 Minimum horizontal stress

ν = Poisson's ratio
σ_v = Vertical stress, psi

α = Biot's constant and dimensionless value

P_p = Pore pressure, psi

$P_{Tectonic}$ = Tectonic pressure, psi.

As can be seen in Eq. (13.14), Poisson's ratio, vertical stress, Biot's constant, and pore pressure primarily affect minimum horizontal stress.

BIOT'S CONSTANT (POROELASTIC CONSTANT)

Biot's constant, also known as poroelastic constant, measures how effectively the fluid transmits pore pressure into rock grains. Biot's constant ranges between 0 and 1. In an ideal case where porosity does not change as pore pressure and confining pressure change, Biot's constant can be calculated using Eq. (13.15).

$$\alpha = 1 - \frac{C_{matrix}}{C_{bulk}}$$

Equation 13.15 Biot's constant

C_{matrix} = Compressibility of the matrix

C_{bulk} = Compressibility of the matrix and pore space.

When porosity is high, rock formation (bulk) is very compressible compared to the matrix of the rock. This will cause the ratio of C_{matrix}/C_{bulk} to approach zero resulting in Biot's constant of 1. In contrast, when porosity is low, C_{matrix}/C_{bulk} approaches 1 resulting in Biot's constant of 0.

Biot's Constant Estimation

If a value of porosity is known and there is no information on geomechanical properties of the rock such as bulk modulus and Poisson's ratio, a rough estimate of Biot's constant can be found using Eq. (13.16).

$$\alpha = 0.64 + 0.854 \times \varnothing$$

Equation 13.16 Biot's constant estimation

$\varnothing$ = Porosity, fraction

Example

A formation has a Poisson's ratio of 0.25, overburden pressure of 9000 psi, pore pressure gradient of 0.67 psi/ft, true vertical depth (TVD) of 8500', and porosity of 8.5%. Assuming tectonic pressure of 400 psi, calculate closure pressure.

$\alpha = 0.64 + 0.854 \times \emptyset = 0.64 + (0.854 \times 8.5\%) = 0.713$

Pore pressure = pore pressure gradient × TVD = 0.67 × 8500 = 5695 psi

$$\text{Closure pressure} = \sigma_{h,\min} = \frac{\nu}{1 - \nu} \times (\sigma_v - \alpha P_p) + \alpha P_p + P_{\text{Tectonic}}$$

$$= \frac{0.25}{1 - 0.25} \times (9000 - 0.713 \times 5695) + 0.713 \times 5695 + 400 = 6107 \text{ psi}$$

MAXIMUM HORIZONTAL STRESS

Maximum horizontal stress is more challenging to calculate. Maximum horizontal stress can be determined from the relationship presented by Haimson and Fairhurst (1967). They showed the relationship between the magnitude of near-wellbore stress and the magnitude of horizontal stress through breakdown pressure.

For penetrating fluid (slick water), Eq. (13.17) can be used to calculate maximum horizontal stress.

$$P_b = \frac{3 \times (\sigma_{\min} - P_R) - (\sigma_{\max} - P_R) + T}{\left(2 - \alpha\left(\frac{1 - 2\nu}{1 - \nu}\right)\right)} + P_R$$

Equation 13.17 Breakdown pressure for penetrating fluid

P_b = Breakdown pressure, psi
$\sigma_{\min}$ = Min horizontal stress, psi
α = Biot's constant
P_R = Reservoir pressure, psi
ν = Poisson's ratio
T = Tensile stress.

For nonpenetrating fluid (gelled fluid), Eq. (13.18) can be used to calculate maximum horizontal stress.

$$P_b = 3 \times (\sigma_{\min} - P_R) - (\sigma_{\max} - P_R) + P_R + T$$

Equation 13.18 Breakdown pressure for nonpenetrating fluid

P_b = Breakdown pressure, psi
σ_{min} = Min horizontal stress, psi
P_R = Reservoir pressure, psi
T = Tensile stress.

Example

Calculate vertical and minimum horizontal stresses given the following data and Table 13.2.

Average overburden density = 160 lbs/ft^3, Biot's constant = assume 1, Tectonic stress = 200 psi

Vertical stress for each layer must be calculated first, as follows:

$$\text{Overlaying shale vertical stress(psi)} = \frac{\rho \times TVD}{144} = \frac{160 \times 7350}{144} = 8167 \text{ psi}$$

$$\text{Sandstone vertical stress(psi)} = \frac{160 \times 7400}{144} = 8222 \text{ psi}$$

$$\text{Underlying shale vertical stress(psi)} = \frac{160 \times 7450}{144} = 8278 \text{ psi.}$$

As can be seen from the calculated overburden pressures (above), as the TVD of the rock layer increases the overburden pressure (vertical stress) increases as well.

The minimum horizontal stress of each rock layer is calculated, as follows:

Overlying shale layer:

$$\text{Pore pressure} = 0.64 \times 7350 = 4704 \text{ psi}$$

$$\sigma_{h,min} = \frac{0.28}{1 - 0.28} \times (8167 - 4704) + 4704 + 200 = 6250 \text{ psi.}$$

Sandstone layer:

$$\text{Pore pressure} = 0.64 \times 7400 = 4736 \text{ psi}$$

$$\sigma_{h,min} = \frac{0.22}{1 - 0.22} \times (8222 - 4736) + 4736 + 200 = 5919 \text{ psi.}$$

Table 13.2 True Vertical Depth (TVD), Poisson's Ratio, and Pore Pressure Gradient

Formation	TVD (ft)	Poisson's Ratio	Pore Pressure Gradient (psi/ft)
Overlaying shale	7350	0.28	0.64
Sandstone	7400	0.22	0.64
Underlying shale	7450	0.28	0.65

Underlying shale layer:

$$\text{Pore pressure} = 0.65 \times 7450 = 4843 \text{ psi}$$

$$\sigma_{h,min} = \frac{0.28}{1 - 0.28} \times (8278 - 4843) + 4843 + 200 = 6379 \text{ psi.}$$

Example

Calculate min and max horizontal stresses of a formation with the following properties:

$v = 0.24$, Vertical stress gradient $= 1.1$ psi/ft, TVD $= 11,500'$, Pore pressure gradient $= 0.65$ psi/ft, Tensile stress $= 250$ psi, Breakdown pressure $= 10,500$ psi, Biot's constant of 1, Assume slick water fluid

$$\text{Overburden stress} = \text{vertical stress gradient} \times \text{TVD} = 1.1 \times 11,500 = 12,650 \text{ psi}$$

$$\text{Pore pressure} = \text{pore pressure gradient} \times \text{TVD} = 0.65 \times 11,500 = 7475 \text{ psi}$$

$$\sigma_{h,min} = \frac{0.24}{1 - 0.24} \times (12650 - 7475) + 7475 + 250 = 9359 \text{ psi}$$

$$P_b = \frac{3 \times (\sigma_{min} - P_R) - (\sigma_{max} - P_R) + T}{\left(2 - \alpha\left(\frac{1 - 2v}{1 - v}\right)\right)} + P_R =$$

$$10500 = \frac{3 \times (9359 - 7475) - (\sigma_{max} - 7475) + 250}{2 - \left(\frac{1 - 2(0.24)}{1 - 0.24}\right)} + 7475$$

$$\sigma_{max} = 9398 \text{ psi.}$$

VARIOUS STRESS STATES

There are three different types of geologic environments that can be determined from min, max, and vertical stress magnitudes. These three fault environments are as follows:

1. Normal fault environment:

$$S_V \geq S_{H,max} \geq S_{h,min}$$

2. Strike slip (shear) environment:

$$S_{H,max} \geq S_V \geq S_{h,min}$$

3. Reverse (thrust) fault environment:

$$S_{H,max} \geq S_{h,min} \geq S_V.$$

FRACTURE ORIENTATION

Fracture is always created and propagated (grows) perpendicular to the least principal stress (minimum horizontal stress). Fracture orientation is influenced by various factors such as overburden pressure, pore pressure, tectonic forces, Poisson's ratio, Young's modulus, fracture toughness, and rock compressibility. It is extremely important to understand the principal stresses acting on the rock in the formation of interest for a successful frac job. Engineers, petrophysicists, geologists, and geoscientists are in charge of understanding and calculating the principal stresses. There are two types of fractures that can be achieved via hydraulic fracturing. The first is referred to as a *longitudinal fracture*, which is essentially one big fracture, and the second is called *transverse fractures*, which are combination of long, narrow fractures.

TRANSVERSE FRACTURES

In almost all of the unconventional shale plays across the country, the goal is to create transverse fractures due to the stress directions, magnitudes, production, and economic feasibility. To create transverse fractures, the well needs to be drilled (placed) parallel to the minimum horizontal stress or perpendicular to the maximum horizontal stress. This means the fractures will propagate (grow) perpendicular to the minimum horizontal stress. Stress directions can be typically obtained from a fracture microseismic, formation microimager (FMI) log, or in the worst-case scenario a world stress map, which is free and widely available. The world stress map is a very useful tool that engineers and geologists use to understand various in situ stresses. A world stress map is used to understand the direction of maximum horizontal stress in a particular region of interest. Therefore, after finding the region of interest where fracing needs to take place, the wells must be drilled perpendicular to the stress direction on the world stress map to create transverse fractures (perpendicular to maximum horizontal stress). For example, when looking at development maps in the Marcellus and Utica/Point Pleasant shale plays, almost all of the wells are drilled NW–SE because from the world stress map, the maximum horizontal stress is facing NE–SW in these areas. Therefore, to create transverse fractures, the wells must be drilled perpendicular to the NE–SW direction.

LONGITUDINAL FRACTURES

To create a longitudinal fracture, the well needs to be drilled parallel to the maximum horizontal stress or perpendicular to the minimum horizontal stress. This means the fractures will propagate parallel to the minimum horizontal stress and perpendicular to the maximum horizontal stress, which is exactly the opposite of transverse fractures. Longitudinal fractures are typically created at shallower depths. Fractures created in Bakken, Eagle Ford, Marcellus, Utica, and Barnett Shales, along with in many other shale plays, are confirmed to be transverse fractures from frac microseismic data (Figs. 13.3 and 13.4).

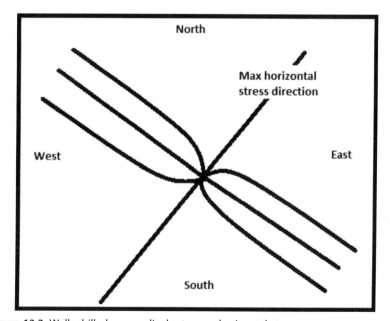

Figure 13.3 Wells drilled perpendicular to max horizontal stress.

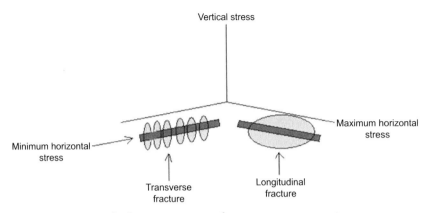

Figure 13.4 Longitudinal versus transverse fractures.

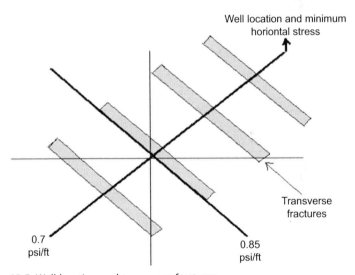

Figure 13.5 Well location and transverse fractures.

●●●

Example

A horizontal well is going to be drilled and hydraulically fractured. The vertical overburden stress gradient is calculated to be 1 psi/ft. One of the principal horizontal stresses has a gradient of 0.7 psi/ft in the direction of N45°E while the other one has a gradient of 0.85 psi/ft in the direction of N45°W. If the goal is to achieve transverse fractures, sketch the direction of the horizontal well and transverse fractures.

The minimum horizontal stress gradient in the direction of N45°E is 0.7 and 0.85 psi/ft is the maximum horizontal stress in the direction of N45°W. To create transverse fractures, the well must be drilled perpendicular to the maximum horizontal stress direction, which means the well must be drilled perpendicular to N45°W as shown in Fig. 13.5. Hydraulic fractures (transverse fractures) on the other hand will grow perpendicular to the minimum horizontal stress.

CHAPTER FOURTEEN

Diagnostic Fracture Injection Test

INTRODUCTION

Diagnostic fracture injection test (DFIT) has become very popular in unconventional shale reservoirs. DFIT is the most commonly used technique in unconventional shale reservoirs to determine various completions and reservoir properties for optimum fracture design. The idea is to create a small fracture by pumping 10–100 BBLs of water at 2–10 bpm and monitor pressure falloff for a specific period of time. DFIT is typically performed a few weeks before the start of a frac job depending on formation permeability. The time of shut-in after pumping the DFIT will be dependent upon the formation permeability and the pump time, which in turn translates into the time it takes to reach pseudoradial flow. After pumping the DFIT test, enough monitoring time should be allowed to reach pseudoradial flow to determine various reservoir properties. Some of the completions properties that can be obtained from DFIT are instantaneous shut-in pressure (ISIP), fracture gradient, net extension pressure, fluid leak-off mechanism, time to closure, closure pressure (minimum horizontal stress), approximation of maximum horizontal stress, anisotropy, fluid efficiency, effective permeability, transmissibility, and pore pressure. It is strongly recommended not to use any volume in excess of 50 BBLs in nanodarcy permeability formations as it might delay the time it takes to reach pseudoradial flow. If permeability is higher, more fluid as high as 100 BBLs can be pumped and still reach pseudoradial flow just in time. The main purpose is to contact the whole net pay to get accurate completions and reservoir properties. The following guideline is an estimation of the shut-in time (post-DFIT shut-in) needed to reach pseudoradial flow and accurately calculate reservoir properties.

1 day	if $k > 0.1$ md
1 week	if $k > 0.01$ md
2 weeks	if $k > 0.001$ md
1 month	if $k > 0.0001$ md

Hydraulic Fracturing in Unconventional Reservoirs
DOI: http://dx.doi.org/10.1016/B978-0-12-849871-2.00014-9
© 2017 Elsevier Inc.
All rights reserved.

Most wells in unconventional reservoirs are shut-in anywhere between 2 weeks and 1 month in order to reach pseudoradial flow.

TYPICAL DFIT PROCEDURE

The following procedure is typically used when performing a DFIT:

- DFIT can be performed through perforations (toe stage) or toe initiation tools.
- If DFIT is performed through perforations, run in the hole (RIH) with TCP (tubing conveyed perforations) guns and perforate the toe stage using 6–10 shots.
- If DFIT is performed through the toe initiation tool, no perforation will be needed.
- Fresh or Potassium Chloride (KCl) water can be used depending on the percentage of clay in the formation. KCl water must be used if the formation is prone to swelling.
- Install the surface self-powered intelligent data retriever (SPIDR) gauge (or any other types of high-resolution gauges) to get accurate pressure measurement (1 psi resolution gauge). If enough money is available in the budget, a bottom-hole pressure gauge is recommended instead for more accurate pressure recording.
- Load the hole with fresh or KCl water.
- Once the hole (casing) is filled, continue pumping at the designed rate until formation breakdown occurs.
- After formation breakdown, continue pumping at 2–10 bpm until the desired DFIT volume is achieved (should not exceed 100 BBLs depending on the permeability).
- It is very important to continuously pump at a constant rate after breakdown because DFIT calculation assumes continuous rate.

DFIT DATA RECORDING AND REPORTING

Record the following data from DFIT:

- type and specific gravity of fluid that was pumped
- pump rate (bpm) during breakdown and while pumping the designed volume

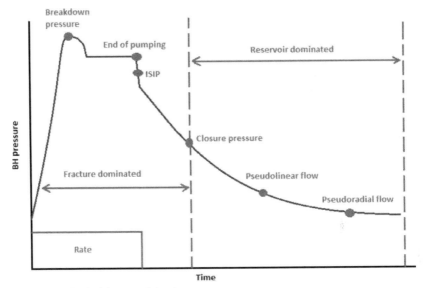

Figure 14.1 Typical fracture injection test.

- ISIP
- formation breakdown pressure
- start and end time
- total pump time
- volume pumped after the breakdown
- any unexpected events, i.e., shutdowns and how the test was restarted, casing and/or surface equipment leaks, pressure spikes while pumping, initial DFIT gauge pressure and time reading, etc.

Fig. 14.1 illustrates a typical fracture injection test that is divided into two sections. The first section is fracture dominated and the second section is reservoir dominated. From the fracture-dominated section, completions properties can be determined, and from the reservoir-dominated section, essential reservoir properties can be obtained.

BEFORE-CLOSURE ANALYSIS

The first analysis in DFIT is called before-closure analysis (BCA), which means the analysis is performed right until closure occurs. The three main plots used for BCA are:

1. Square root plot: plot BHP (*y*-axis) versus square root of time (*x*-axis)

2. Log—log plot: plot log (BH ISIP − BHP) versus log of time
3. BHP versus G-function time

Please note that bottom-hole pressure is typically calculated from surface pressure when a surface pressure gauge is used during DFIT. Time is defined as the time since ISIP. The main idea is to use different types of diagnostic plots since one type of plot may not be suited for every formation. For example, the G-function plot maybe the best for one particular formation and the square root time plot may be the best for another. All plots must be used in conjunction with one another for better estimation of properties. Derivatives in DFIT analysis are used as an aid in the straight line segment of the decline curve.

- First derivative:
  - yields the slope of the curve,
  - a constant slope yields a straight line,
  - yields local minima and maxima.
- Second derivative:
  - yields the curvature of the decline curve.

Square Root Plot

A square root plot is commonly used to determine the closure pressure. When the square root of time (x-axis) versus the bottom-hole pressure (y-axis) is plotted, the linear portion of the plot will lie along a straight line going through the origin. The point at which deviation from the straight line occurs on the superposition plot (second derivative) is referred to as closure pressure. Every square root plot will have three main curves: pressure curve, first derivative, and second derivative (also referred to as superposition). Deviation from the straight line on the pressure curve is used to define minimum closure pressure. In addition, deviation from the smart line going through the origin on the second derivative curve is referred to as fracture closure. In Fig. 14.2, the *blue curve* (dark gray curve in print version) is the pressure curve, the *green curve* (light gray curve in print version) is the first derivative curve, and the *red curve* (gray curve in print version) is the second derivative (superposition curve).

To identify fracture closure, a linear extrapolated line from the origin is drawn on the second derivative curve (*black line*). Fracture closure can be approximated when the second derivative curve deviates from the linear line. After identifying fracture closure on the second derivative curve,

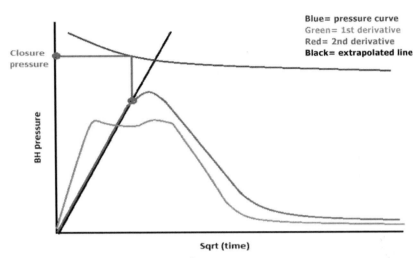

Figure 14.2 BHP versus square root of time.

draw a vertical line from the fracture closure point until the pressure curve is intersected as shown in *red* (gray in print version). After intersecting the pressure curve, closure pressure can be read on the *y*-axis.

In Fig. 14.3 the deviation from the linear extrapolated line going through the origin on the second derivative is referred to as closure pressure, which is around 6845 psi in this example. In addition, closure time is also around 463 minutes.

Log–Log Plot (Log (BH ISIP-BHP) Versus Log (Time))

A log–log plot is derived from a square root plot. This plot should be sufficient to identify closure and various flow regimes before and after closure. Various flow regimes on the second derivative of the log–log plot can be determined:

Before-closure analysis:

Half-slope line (1/2 slope) = Corresponds to linear flow regime

Quarter-slope line (1/4 slope) = Corresponds to bilinear flow regime

After-closure analysis:

Negative half-slope line (−1/2) = Corresponds to linear flow

Negative three-fourth (−3/4) = Corresponds to bilinear flow

Negative unit slope (−1) = Corresponds to pseudoradial flow

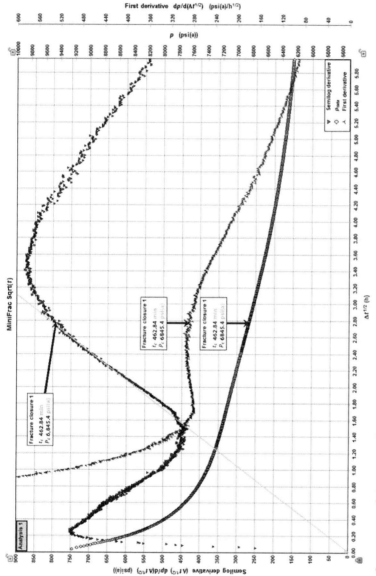

Figure 14.3 Square root of time example.

The log–log plot shows a positive ½ slope on the second derivative curve before closure. In some rare instances, it shows a positive ¼ slope on the second derivative before closure. Closure occurs by the change in slope from positive to negative on the second derivative curve. Pseudolinear flow is indicated when the second derivative curve shows a negative ½ slope in conjunction with a negative 1.5 slope on the first derivative curve. Pseudoradial flow is indicated when the second derivative curve displays a negative unit slope in conjunction with a negative 2 slope on the first derivative curve (Barree et al., 2007).

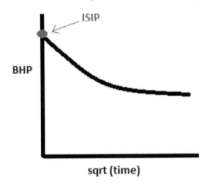

$$y = mx + b$$

$$BHP = m(\sqrt{time}) + ISIP \rightarrow BHP - ISIP = m(\sqrt{time})$$

$$\Delta P = m \times time^{1/2} \rightarrow log\Delta P = log\left(m \times time^{1/2}\right)$$

$$log(\Delta P) = \frac{1}{2}log(time) + log(m)$$

In the log–log plot example shown in Fig. 14.4, the *blue curve* (dark gray curve in print version) represents delta pressure, the *green curve* (light gray curve in print version) represents the first derivative, and the *red curve* (gray curve in print version) represents the second derivative. As can be seen on the second derivative, the slope of the curve changes from being positive to negative. The slope of the open fracture line on the second derivative is ½. Any derivation from this ½ slope line means the fracture would have changed or in this case closed. This represents closure occurrence and that point can be picked as the fracture closure pressure. Negative 1 slope (unit slope) on the second derivative is also an indication of pseudoradial flow. When pseudoradial flow is reached, more

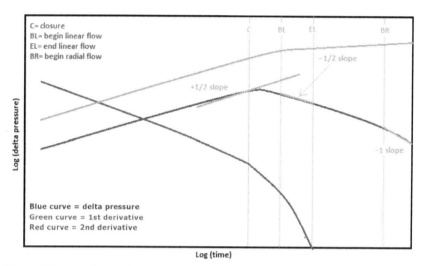

Figure 14.4 Log—log plot.

confidence is obtained when calculating various reservoirs properties, especially pore pressure.

Fig. 14.5 is another log—log plot example where the deviation from positive ½ slope line to negative on the second derivative can be used to determine closure pressure. Closure pressure is identified to be 16,627 psi. In addition, the beginning and the end of linear flow (after closure) can be determined from negative ½ slope line. This well appears to have reached pseudoradial flow but more time will be needed to monitor the pressure falloff to have confidence with the scattered data during pseudor-adial flow, which can be identified by the −1 slope line as shown on the plot. Note that in the log—log plot, derivative slopes before closure are always positive, but after closure slopes are negative. Unit slope indicates storage before closure but −1 slope indicates pseudoradial flow after closure (Barree et al., 2007).

Fig. 14.6 is another log—log plot example where closure is reached by observing the deviation from the positive ½ slope to negative on the sec-ond derivative (9250 psi = closure pressure), but due to other operational and data recording issues, pseudoradial flow cannot be observed from this example. The results from this DFIT must not be used to perform any types of after-closure analyses. Data recording and monitoring is the key to a successful DFIT interpretation.

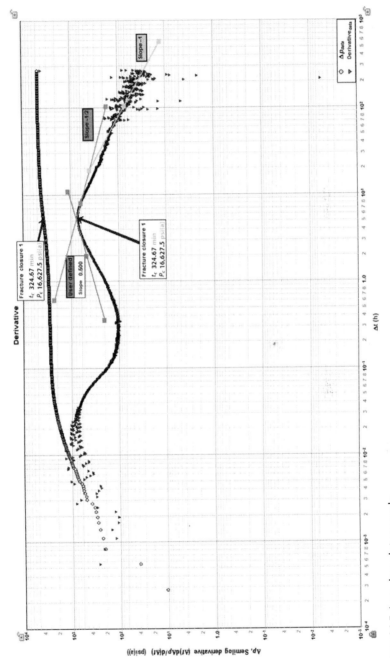

Figure 14.5 Log–log plot example.

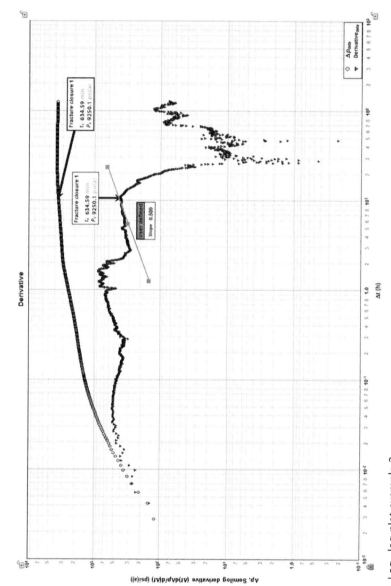

Figure 14.6 Log–log plot example 2.

G-function Analysis

G-function is a variable related to time. G-function (x-axis) versus BHP (y-axis) can be plotted to determine various fracture and formation properties such as fracture closure, fluid efficiency, effective permeability, and leak-off mechanism. G-function assumes constant fracture height, constant pump rate, and stoppage of fracture propagation when pumping stops. Eq. (14.1) can be used to approximate G-function time:

$$G(\Delta t_D) = \frac{4}{\pi}\left[g(\Delta t_D) - g_0\right]$$

$$g(\Delta t_D) = \frac{4}{3}(1+\Delta t_D)^{1.5} - \Delta t_D^{1.5}; \quad \beta = 1.0$$

$$g(\Delta t_D) = (1 + \Delta t_D)\sin^{-1}(1+\Delta t_D)^{-0.5} + \Delta t_D^{0.5}; \quad \beta = 0.5$$

$$\Delta t_D = \frac{t - t_p}{t_p}$$

Equation 14.1 G-function time.

t = Shut-in time, minutes

t_p = Total pump time, minutes

A β value of 1.0 refers to tight formations with low fluid leak-off, while a β value of 0.5 refers to high-permeability formations with high leak-off.

It is important to note that the G-function at shut-in (ISIP) is zero. For example, if total pump time is 5 minutes (t_p = 5 min), t at ISIP will be equal to 5 as well. Therefore, G-function at ISIP is equal to zero. G-function time starts at ISIP. The following steps can be used to find closure pressure on the G-function time:

1. Look for local maximum on the first derivative.
2. Look for deviation from the straight line on the pressure curve.
3. Look for deviation from the straight line going through the origin on the second derivative curve.
4. Closure occurs where the second derivative curve deviates from the straight line.

Fluid Leak-off Regimes on G-function Plot

There are four unique leak-off regimes that can be noted from the G-function plot. The first type of leak-off on a G-function plot is referred to as "normal leak-off," which refers to the occurrence of leak-off through a homogeneous rock matrix (not the typical signature for unconventional reservoirs). Other leak-off types on a G-function plot are pressure-dependent leak-off (PDL), height recession leak-off (transverse storage), and fracture tip extension, which are discussed in detail below.

Pressure-Dependent Leak-Off

Pressure-dependent leak-off (PDL) typically occurs in hard naturally fractured rocks. PDL can indicate complexity during hydraulic fracturing since the rock is naturally fractured. There is a pressure that, when exceeded, will control the opening in natural fractures. Once this pressure is exceeded, the surface area to leak-off will also increase. Since this pressure is driving the leak-off process, it is referred to as PDL. PDL is the most common type of leak-off regime in unconventional shale reservoirs due to the existing natural fractures.

PDL can be easily identified on the G-function plot. The easiest way to identify PDL is when the second derivative curve shows a concave down feature above the extrapolated line going through the origin as shown in Fig. 14.7. In the G-function plot below, *blue* (dark gray in print version) represents the pressure curve, *green* (light gray in print version) represents the first derivative, and *red* (gray in print version) represents the second derivative. When a straight line (*black*) going through the origin is plotted on the second derivative, there is a concave down feature (hump) above the extrapolated line. This represents the existence of natural fractures, which will result in a complex fracture system and indicates that fluid leaks off faster

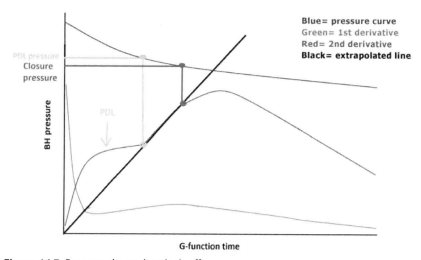

Figure 14.7 Pressure-dependent leak-off.

than expected for a normal bi-wing fracture. As soon as the second derivative curve gets back on the normal linear trend on the extrapolated line, that point is referred to as PDL pressure. This pressure can also be used as an approximation to maximum horizontal stress if and only if natural fractures are believed to be perpendicular to the created hydraulic fractures. Once this point is identified, a vertical line can be drawn from that point to where the pressure curve is intersected. After the intersection of the vertical line with the pressure curve is identified, PDL pressure or maximum horizontal stress can be read on the y-axis as shown in Fig. 14.7.

The next phenomenon that happens on the G-function plot is closure. Fracture closure occurs when the second derivative deviates from the straight line going through the origin. Once that point is identified on the G-function plot, draw a vertical line until the pressure curve is intersected. After the intersection of the pressure curve with the vertical line (as shown in *red* (gray in print version)) is identified, closure pressure can be read on the y-axis. Closure pressure is regarded as the minimum horizontal stress. To summarize this section, PDL pressure represents maximum horizontal stress (if natural fractures are perpendicular to hydraulic fractures) and closure pressure represents minimum horizontal stress. Anisotropy can be calculated using Eq. (14.2).

$$Anisotropy = maximum\ horizontal\ stress - minimum\ horizontal\ stress$$

Equation 14.2 Anisotropy.

When the difference between maximum and minimum horizontal stresses is small (e.g., 200 psi), the created fractures are expected to be complex. In contrast, when the difference is large, the created fractures are expected to be bi-wing. PDL describes the fluid leak-off into fissures that open at a higher pressure than the fracture closure pressure. As a general rule of thumb, when the difference between PDL and closure pressures is less than 5% of the closure, it is an indication that fracture complexity will be created.

Fig. 14.8 shows an example of a G-function plot with a PDL signature. This signature is illustrated by the concave downward feature above the extrapolated line going through the origin on the second derivative curve. Closure pressure can be determined when the second derivative deviates from the linear extrapolated line going through the origin. The closure pressure is around 6845 psi (G-function closure time = 29.4).

Dealing With PDL PDL is very common in naturally fractured formations, especially in unconventional shale reservoirs. Consider pumping

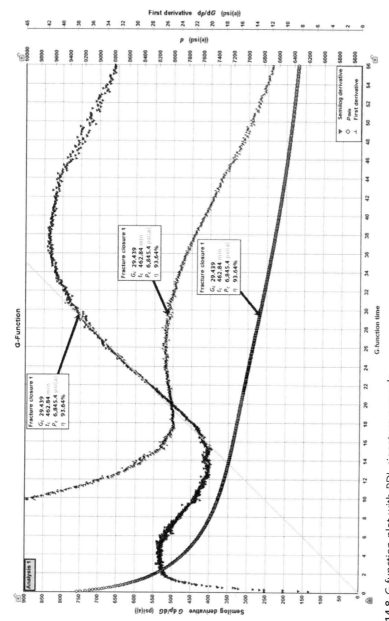

Figure 14.8 G-function plot with PDL signature example.

a smaller sand size and concentration (100 mesh and 40/70) and longer frac stages. Pumping smaller sand size, such as 100 mesh, bridges natural fissures and reduces the chance of screening out. This will increase the fluid efficiency by preventing high fluid leak-off through natural fractures.

Height Recession Leak-Off

The second leak-off regime that can be noted from the G-function plot is height recession. Height recession occurs when the fracture height is decreasing during closure because of contact with an impermeable zone. When these conditions are met, some strange behavior can happen within the fracture. As the fracture closes, the upper and lower zones close more quickly because of the higher stress. However, because permeability is so low in the zones above and below, the fluid is pushed into the main section of the fracture instead of leaking off into the formation. This reduces the apparent leak-off rate in the fracture and, therefore, reduces the leak-off rate from the wellbore into the fracture. In essence, fluid leaks off slower than expected for a normal bi-wing fracture. The resultant fracture may be very narrow and tall (Fig. 14.9).

Identifying Height Recession Leak-Off The second derivative curve on the G-function plot has a concave upward shape below the trend line going through the origin. This concave upward feature below the extrapolated line going through the origin is an indication of height recession.

Dealing With Height Recession Leak-Off If height recession happens, consider reducing rate and proppant amount. Also, lowering sand

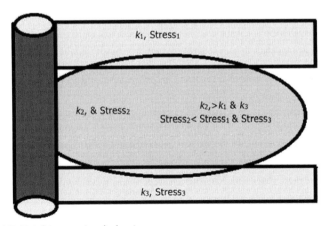

Figure 14.9 Height recession behavior.

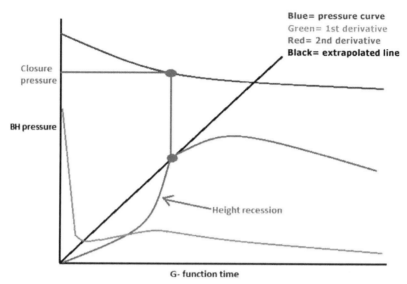

Figure 14.10 Height recession leak-off.

concentration to bridge off impermeable zones can tremendously help when dealing with height recession.

In Fig. 14.10, *blue* (dark gray in print version) is the pressure curve, *green* (light gray in print version) is the first derivative, and *red* (gray in print version) is the second derivative. The concave upward feature on the second derivative below the extrapolated line going through the origin represents height recession.

Fig. 14.11 shows another example of a G-function plot with a height recession signature. This feature can be determined from the concave upward feature below the trend line going through the origin on the second derivative curve. The closure pressure from this example is around 16,743 psi (G-function time = 18.4).

Tip Extension

Fracture-tip extension is a phenomenon that occurs in very low-permeability reservoirs in which a fracture continues to propagate even after injection has been stopped with the well shut-in. The energy that is typically released through leak-off is actually transferred to the tip of the fractures resulting in fracture-tip extension.

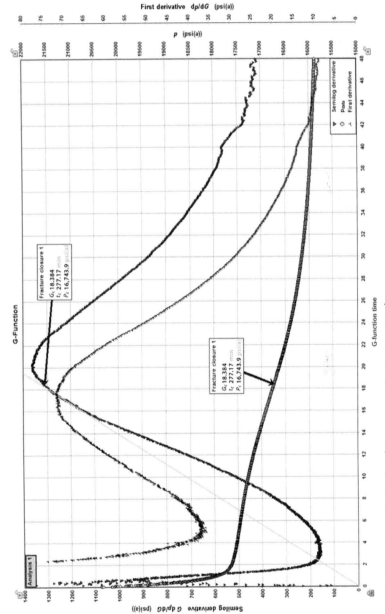

Figure 14.11 G-function plot with height recession signature example.

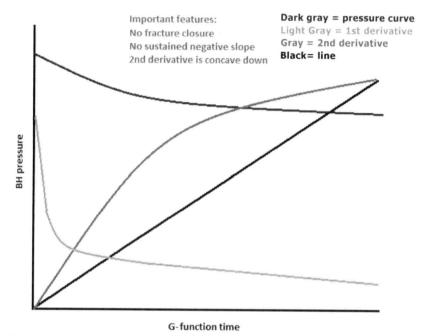

Figure 14.12 Tip extension leak-off.

Identifying and Dealing With Tip Extension Leak-Off When fracture-tip extension exists, the extended straight line on the second derivative curve does not pass through the origin. In addition, no sustained negative slope on the second derivative curve can be seen. Since no negative slope on the second derivative can be realized, fracture closure behavior is not seen when encountering tip extension. The best way to deal with tip extension behavior leak-off is to increase pad volume to assist in creating longer frac lengths.

Fig. 14.12 illustrates a tip extension leak-off behavior. *Blue* (dark gray curve in print version) is the pressure curve, *green* (light gray curve in print version) is the first derivative, and *red* (gray in print version) is the second derivative. As can be seen on the second derivative curve, no sustained negative slope or fracture closure can be seen from this behavior.

Effective Permeability Estimation From G-Function Plot
G-function plot is a powerful tool that can be used to obtain various

fracture and formation properties. In addition to calculating closure pressure and identifying various leak-off regimes, effective permeability can be calculated from the G-function plot using Eq. (14.3).

$$k = \frac{0.0086\mu\sqrt{0.01P_z}}{\emptyset C_t \left(\frac{G_c E r_p}{0.038}\right)^{1.96}}$$

Equation 14.3 Effective permeability using G-function plot (Barree et al., 2007).

k = Effective permeability to reservoir fluid, md
μ = Viscosity of injected fluid used during DFIT, cp
P_z = Net extension pressure or process zone stress, $BHISIP-P_c$
$\emptyset$ = Porosity, fraction
C_t = Total compressibility, 1/psi
G_c = Closure G-function time
E = Young's modulus, MMpsi
r_p = Storage correction factor
For normal and PDL, assume r_p of 1
For height recession and tip extension, assume $r_p < 1$

In addition to the effective permeability calculation, fluid efficiency can also be calculated using the G-function time at fracture closure using Eq. (14.4).

$$\text{Fluid efficiency} = \frac{G_c}{2 + G_c}$$

Equation 14.4 Fluid efficiency using G-function time.

G_c = G-function time at fracture closure

● ● ●

Example

Estimate the effective permeability and fluid efficiency from the following parameters obtained from a G-function plot:

Injected fluid viscosity (μ) = 1 cp, BH ISIP = 7748 psi, P_c = 6338 psi, C_t = 0.0000234 1/psi, G_c = 29.012, E = 3.5 MMpsi, r_p = 1 (since PDL exists), Porosity = 10%

$$P_z = BH\ ISIP - P_c = 7748 - 6338 = 1410\ \text{psi}$$

$$k = \frac{0.0086\mu\sqrt{0.01P_z}}{\varnothing C_t \left(\frac{G_c E f_p}{0.038}\right)^{1.96}} = \frac{0.0086 \times 1 \times \sqrt{0.01 \times 1410}}{0.1 \times 0.0000234 \times \left(\frac{29.012 \times 3.5 \times 1}{0.038}\right)^{1.96}} = 0.00265 \text{ md}$$

$$\text{Fluid efficiency} = \frac{29.012}{29.012 + 2} = 0.9355 \text{ or } 93.55\%$$

A high fluid efficiency of $\sim$94% and a permeability of 0.00265 md indicate high fluid efficiency and low fluid leak-off.

What would the permeability and fluid efficiency have been if the G-function time at fracture closure was 0.554 instead of 29.012?

Replacing the G-function time of 0.554 with 29.012 in the same equation, and keeping all of the other parameters the same, the effective permeability can be solved as follows:

$$k = \frac{0.0086 \times 1 \times \sqrt{0.01 \times 1410}}{0.1 \times 0.0000234 \left(\frac{0.554 \times 3.5 \times 1}{0.038}\right)^{1.96}} = 6.2 \text{ md}$$

$$\text{Fluid efficiency} = \frac{0.554}{0.554 + 2} = 0.2169 \text{ or } 21.69\%$$

If G-function time was 0.554, the formation would have a lower fluid efficiency and higher effective permeability, which is an indication of high fluid leak-off.

AFTER-CLOSURE ANALYSIS (ACA)

ACA refers to various methods used to determine reservoir properties after closure has occurred. The first step in ACA is determining various flow regimes from DFIT analysis. This can be performed using a log–log plot and determining pseudolinear and pseudoradial flows from the log–log plot. Once pseudolinear and pseudoradial flow regimes are identified on the log–log plot, pore pressure, transmissibility, and permeability can then be estimated using various techniques that will be discussed. On certain occasions, if sufficient time is not allowed after DFIT, only pseudolinear flow can be reached and pseudoradial flow will not be reached. In those instances, reservoir pore pressure can be estimated from pseudolinear flow, but this pressure (obtained from pseudolinear flow) is optimistic. Pseudoradial flow occurs after pseudolinear flow. Once this flow regime is reached, pore pressure can be determined using a Horner plot or other available pressure transient methods.

Horner Plot (One Method of ACA)

Horner analysis uses the log of Horner time on the x-axis versus bottom-hole pressure on the y-axis to calculate pore pressure and reservoir permeability. Note that the y-axis is plotted on the Cartesian axis and logarithmic scale is applied to the x-axis. Horner time is defined in Eq. (14.5).

$$\text{Horner time} = \frac{t_p + \Delta t}{\Delta t}$$

Equation 14.5 Horner time.

t_p = Fracture propagation time, minutes

Δt = Elapsed shut-in time, minutes

As shut-in time increases, Horner time decreases. As shut-in time approaches infinity, Horner time approaches 1. A straight line extrapolation to the y-intercept (at Horner time of approximately 1) yields reservoir pressure (pore pressure). One of the biggest limitations with a Horner plot is that pseudoradial flow must be reached or Horner analysis is not recommended to be used. Once pseudoradial flow is identified, the slope of the straight extrapolated line is referred to as m_H. The point at which the extrapolated line reaches the y-intercept (as shown below) is pore pressure. The slope of the Horner plot (m_H) can be used to estimate reservoir transmissibility (kh/μ) and subsequently reservoir permeability using Eq. (14.6).

$$\frac{kh}{\mu} = \frac{162.6(1440)q}{m_H}$$

Equation 14.6 Reservoir transmissibility.

kh/μ = Reservoir transmissibility, md.ft/cp

k = Reservoir permeability, md

h = Net pay height, ft

μ = Far-field fluid viscosity (*not* injected fluid viscosity), cp

m_H = Slope of the Horner plot, psi

q = Average injected fluid rate, bpm

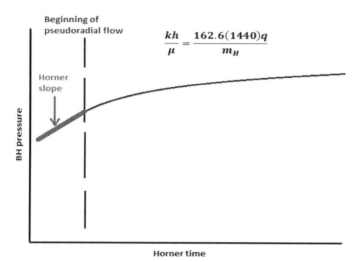

Figure 14.13 Horner analysis.

By assuming a far-field fluid viscosity and net pay height in Eq. (14.6), reservoir effective permeability can be calculated (Fig. 14.13).

Example

Calculate the reservoir transmissibility and reservoir permeability given the following Horner plot and data. Also, estimate the reservoir pressure and reservoir pressure gradient assuming a true vertical depth (TVD) of 8250'.

Avg pump rate $= 7$ bpm, $h = 100'$, far-field fluid viscosity $(\mu) = 0.0452$ cp, $m_H = 568,564$

From Fig. 14.14, Horner slope is 568,564 (m_H), therefore,

$$\text{Transmissibility} = \frac{kh}{\mu} = \frac{162.6(1440)q}{m_H} = \frac{162.6 \times 1440 \times 7}{568,564} = 2.88 \text{ md.} \frac{ft}{cp}$$

$$\text{Effective permeability} = k = \frac{2.88 \times 0.0452}{100} = 0.0013 \text{ md}$$

The extrapolation of the pseudoradial flow line (straight line) to the y-intercept yields reservoir pressure of roughly 4600 psi in this example. Therefore, the reservoir pressure gradient is 4600 psi divided by 8250' which yields 0.56 psi/ft.

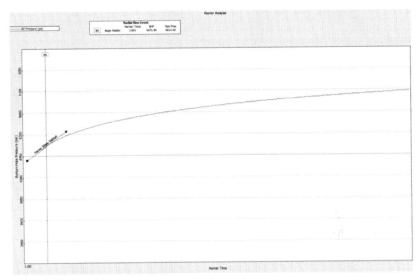

Figure 14.14 Horner analysis example.

Linear Flow-Time Function Versus Bottom-Hole Pressure (Another Method of ACA)

In addition to Horner analysis, reservoir pressure can be determined from the linear flow-time function (x-axis) versus BHP (y-axis). Linear flow-time function is described in Eq. (14.7).

$$F_L(t, t_c) = \frac{2}{\pi} \sin^{-1} \sqrt{\frac{t_c}{t}} \quad \text{for } t \geq t_c$$

Equation 14.7 Linear flow-time function.

t_c = Time to closure, minutes

t = Total pump time, minutes

A straight line extrapolation from the linear flow yields an estimated pore pressure from the linear flow-time function plot. In other words, once after-closure pseudolinear flow is observed during shut-in, the intercept of the extrapolated straight line through the pseudolinear flow data provides an estimate of the pore pressure. Reservoir pore pressure extrapolation is valid and no direct information of transmissibility can be obtained from this analysis. If pseudoradial flow is not obtained from

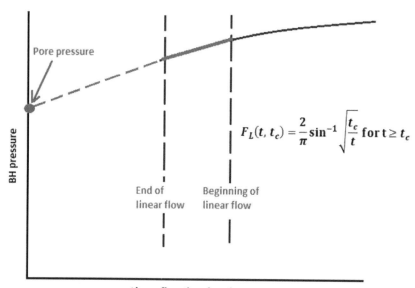

Figure 14.15 Linear flow-time function plot (ACA).

DFIT analysis, this plot can be used to estimate the reservoir pressure (Fig. 14.15).

Radial Flow-Time Function Versus BHP (Another Method of ACA)

Radial flow-time function can also be used to calculate reservoir pressure along with transmissibility when true pseudoradial flow is identified. Radial flow-time function is defined in Eq. (14.8).

$$F_R(t, t_c) = \frac{1}{4} \ln\left(1 + \frac{Xt_c}{t - t_c}\right), \quad X = \frac{16}{\pi^2} \cong 1.6$$

Equation 14.8 Radial flow-time function.

t_c = Time to closure, minutes

t = Total pump time, minutes

In addition to reservoir pressure, when the pseudoradial flow period is properly identified, far-field transmissibility can also be calculated by knowing the slope of the extrapolated line, time to fracture closure, and total volume injected during the test. Transmissibility using a radial flow-time function plot can be obtained using Eq. (14.9) (Fig. 14.16).

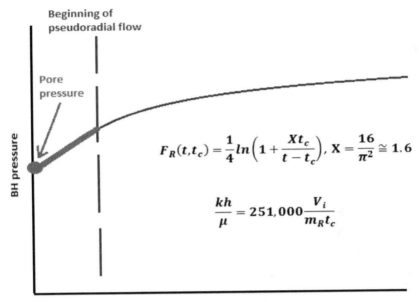

Figure 14.16 Radial flow-time function plot.

$$\frac{kh}{\mu} = 251,000\frac{V_i}{m_R t_c}$$

Equation 14.9 Transmissibility using radial flow-time function.

V_i = Injected fluid during the test, BBLs

m_R = Derived slope, 1/psi

t_c = Time to closure, minutes

h = Net pay, ft

μ = Far–field fluid viscosity, cp

CHAPTER FIFTEEN

Numerical Simulation of Hydraulic Fracturing Propagation

INTRODUCTION

Hydraulic fracturing has been accepted as a technique with a variety of applications. These applications include measurement of in situ stress (Hayashi and Haimson, 1991), underground storage of hazardous materials (Levasseur et al., 2010), heat production from geothermal reservoirs (Legarth et al., 2005), and barrier walls to prevent containment from transportation (Murdoch, 2002). Currently, one of the most important applications of hydraulic fracturing is to improve the recovery of unconventional hydrocarbon reservoirs. Hydraulic fracturing is a coupled process including (1) the deformation of the solid medium, where the fracture width is dependent on the fluid pressure in a global manner and it has the property of non-locality; (2) the fluid flow within the fracture, which is a nonlinear function of fluid pressure and fracture width. These two fundamental properties produce notorious difficulty when investigating hydraulic fracturing.

The conventional methods for the numerical simulation of hydraulic fracturing are the boundary element and finite element methods. The discontinuous displacement (DD) method, which is a variant of the boundary element method, has been greatly used for this purpose. However, it is found to be difficult in the event of complex structures. Compared to the boundary element method, the finite element method has greater flexibility but requires extensive computational power. Recently, advanced techniques such as condensation technique and parallel computing approaches are used to overcome the limitation of well-developed numerical schemes of fracture propagation models (Bao et al., 2014, 2015, 2016).

While significant effort has been put toward the simulation of fracture propagation and fluid flow during injection (Mobbs and Hammond, 2001; Yamamoto et al., 1999; Phani et al., 2004), fracture geometry after

Hydraulic Fracturing in Unconventional Reservoirs
DOI: http://dx.doi.org/10.1016/B978-0-12-849871-2.00015-0

© 2017 Elsevier Inc.
All rights reserved.

flowback has been merely studied. Fracture geometry after flowback is a function of proppant distribution and closure stress, which is significantly different than fracture geometry after injection stops. Proppant transport and distribution in hydraulic fracture is a nonlinear function of injection rate, proppant size, density, and frac fluid properties, i.e., viscosity and density. Therefore, for hydraulic fracturing optimization, a fully coupled numerical simulation coupling governing equations describing fracture opening, fluid flow and leak-off, and solid transport is required. This numerical simulation should also handle different proppant size and density injection with different pumping schedules in order to increase the efficiency of hydraulic fracture stimulation and enhance oil and gas recovery.

STRATIGRAPHIC AND GEOLOGICAL STRUCTURE MODELING

Field stratigraphic and geological structure modeling is necessary to obtain a robust and consistent geometry and petrophysical properties of the formations under study. Three-dimensional (3D) geological models can also be used to provide knowledge of distributions of rock mechanical properties and in situ stress, including maximum horizontal stress, minimum horizontal stress, vertical stress, Young's modulus, Poisson's ratio, and tensile strength. Detailed knowledge of distribution of petrophysical properties is critical to locate the initiation of hydraulic fractures and to evaluate the evolution of fracture-geometry configuration. A number of different commercial software packages are available that provide macromodels for imaging purposes or detailed petrophysical models for reservoir simulation purposes (Aziz and Settari, 1979). They can also be used for stratigraphic studies where the presence of meanders, channels, faults, and discontinuities is important (Mallet, 2002). Conventionally there are two different approaches that have been taken for building geological models and populating the data, namely, statistical approaches including different forms of kriging (Xu et al., 1992) and deterministic geometries obtained from seismic or ground-penetrating radar (GPR) data. Stochastic or deterministic interpolations need to be compared with control horizon geometries obtained from measurements of petrophysical properties using drilling data and well logs, such as spontaneous potential, gamma ray, resistivity, density, and sonic velocity logs. Sufficient well logs and drilling data are necessary to obtain well-defined control horizons providing the required level of accuracy in 3D geological modeling. However, these studies are relatively expensive and provide limited data if abundant wells are not available.

Conventional methods of data analysis obtained from well logs and seismic are knowledge-driven and ignore underlying physical relationships between correlated petrophysical parameters. Recently, advanced techniques such as decision trees (DTs), support vector machines (SVMs), data mining, and artificial neural network (ANNs) have been used for the identification of sedimentary facies and lithology variation using limited log and core data. However, the applications of these techniques also need special attention since they might not consider important geological phenomena embedded in geological formations. Therefore, a combination of advanced mathematical and knowledge-driven techniques is required to obtain sound geological models.

DEVELOPMENT OF HYDRAULIC FRACTURING SIMULATORS

Multistage hydraulic fracturing of horizontal wells enabled the oil and gas industry to economically enhance production from unconventional resources, especially organic-rich shale reservoirs. The process involved the creation of multiple fractures in any single stage of hydraulic fracturing by injecting significant amounts of fracturing fluid and proppant at high pressures. This process is expected to generate highly conductive flow paths for oil and gas to flow from the reservoir to the production well. After reaching the predesigned fracture length, the injection will stop and fracturing fluid will be produced during the flowback process. However, injected proppants will remain in the fracture to prevent fracture closure due to overburden pressure. There have been extensive studies on hydraulic fracturing optimization to generate maximum oil and gas production from unconventional resources. These studies mainly focus on the impact of different reservoir and operational parameters on the efficiency of hydraulic fracturing, multiple hydraulic fracturing interactions, and hydraulic and natural fracture interactions (Ozkan et al., 2009; Olson and Dahi 2009; Cheng, 2012). They showed that the change in local stresses due to earlier hydraulic fracturing stages or preexisting natural fractures can significantly impact the dimensions and orientation of subsequent fractures.

Numerical schemes used to model the hydraulic fracture propagation and optimization are mostly based on the theory of linear elastic fracture mechanics (LEFM), which was developed in the 1920s and introduced fundamental equations governing the process of hydraulic fracturing. The major assumption of LEFM is an isotropic and linear elastic formation. This neglects deformation at the fracture tip or it assumes the deformation at the fracture tip is negligible as compared to the fracture dimensions. This assumption is not valid for fracture tip behavior in soft formations with significant plastic deformation. In this case, the crack-tip plasticity (CTP) method might be more applicable. As the plastic properties of the formation increase, hydraulic fracturing and fracture propagation in the formation becomes difficult to perform since most of the energy that otherwise would be used for fracture propagation will be absorbed by the formation. Even though the CTP technique is more promising for modeling fracture-tip behavior, it has not been used due to its complexity. Modifications have been applied to the theory of LEFM to capture some nonlinear fracture-tip behaviors.

In modeling fracture propagation using LEFM, the stress field near the fracture tip will be calculated and compared with fracture toughness, which is the formation property that needs to be obtained through experimental studies. When the stress field exceeds fracture toughness, the fracture propagates inside the material. In addition to fracture toughness, one also needs to have a good understanding of variables such as normal and shear stresses, strain, Young's modulus, Poisson's ratio, tensile strength, and yield strength to understand the basics of the theory of elasticity.

Stress "σ" is defined as force or load per area and can be shown using Eq. (15.1):

$$\sigma = \frac{F}{A}$$

Equation 15.1 Stress

F is the force applied to the cross-section area A. The dimension of stress is the same as pressure and can be measured in pascals (newtons per square meter) in SI units or psi (pound per square inch) in field units. The component of stress applied perpendicular to the surface area is called normal stress, usually shown by σ, and the component of stress parallel to the surface area is called shear stress, or τ. Three-dimensional space stress encompasses nine components that can be shown using a 3×3 matrix as follows:

$$\begin{bmatrix} \sigma_{xx} & \tau_{xy} & \tau_{xz} \\ \tau_{yx} & \sigma_{yy} & \tau_{yz} \\ \tau_{zx} & \tau_{zy} & \sigma_{zz} \end{bmatrix}$$

By studying the behavior of shear stresses acting on an infinitesimally small volume, one can show that $\tau_{xy} = \tau_{yx}$, $\tau_{xz} = \tau_{zx}$, and $\tau_{yz} = \tau_{zy}$, which is the basis of the theory of shear stress reciprocity. Based on the theory of shear stress reciprocity, changing the shear stress indicates only changes in the direction of shear stress and does not change the magnitude of the shear stress. A coordinate system in which stresses are calculated can be transformed to any coordinate system given that the shear stress components of the total stress becomes zero and only the diagonal component of stress remains through coordinate transformation. In this case, the normal stresses in the x-, y-, and z-directions are called principal stresses, where σ_1 is the maximum and σ_3 is the minimum principal stress, shown as follows:

$$\begin{bmatrix} \sigma_1 & 0 & 0 \\ 0 & \sigma_2 & 0 \\ 0 & 0 & \sigma_3 \end{bmatrix}$$

Principal stresses can be obtained from components of the general stress matrix. In a two-dimensional system, the maximum $"\sigma_1"$ and minimum $"\sigma_2"$ principal stresses can be obtained as follows:

$$\sigma_1 = \left(\frac{\sigma_x + \sigma_y}{2}\right) + \sqrt{\left(\frac{\sigma_x - \sigma_y}{2}\right)^2 + \tau_{xy}^2}$$

$$\sigma_2 = \left(\frac{\sigma_x + \sigma_y}{2}\right) - \sqrt{\left(\frac{\sigma_x - \sigma_y}{2}\right)^2 + \tau_{xy}^2}$$

Strain "ϵ" is used to quantify the deformation of solid material, which is defined as a relative change in displacement in the x-, y-, and z-directions as follows:

$$\epsilon = dL/L$$

In this equation, dL is defined as a change in displacement and L is the initial length. The stress and strain relationship is defined using constitutive equations such as Hooke's law, which assumes a linear relationship between applied load and displacement in the range of the elastic behavior of the material. To quantify this relationship, Young's modulus "E" is used, which is a ratio of stress to strain and is an indication of formation stiffness.

$$E = \frac{\sigma}{\epsilon}$$

Tensile strength and yield strength are defined as the maximum pressure a stress formation can bear before it breaks. This is the point of stress at which the formation is permanently damaged. Even though linear elastic fracture mechanics have been used extensively in hydraulic fracture simulation, they suffer from a high computational cost and decreased accuracy when predicting the fracture-tip behavior. LEFM especially cannot predict the formation failure ahead of the fracture tip. This is due to the fact that LEFM only considers the local stress criteria at the fracture tip (i.e., where the fracture propagates when stress intensity factor K_I overcomes the fracture toughness K_{IC}).

On the other hand, cohesive zone models (CZM) are more suitable to model fracture-tip behavior. CZM extends the fracture tip area to a "cohesive zone" ahead of the fracture tip within which the fracture propagation processes occur gradually. Cohesive zone modeling is based on the determination of two important parameters: cohesive strength and separation energy. This introduces both strength and energy criteria for fracture propagation, and enables CZM to predict formation failure ahead of fracture tip. These parameters can be measured experimentally or obtained using numerical simulations developed for interface behavior predictions. Different methods other than LEFM and CZM are also used for hydraulic fracturing simulation purposes such as crack-tip open displacement (CTOD), but these are not as common as the first two techniques. Gao et al. (2015) applied the DD technique using a boundary element model to investigate the changes that occur in multiple hydraulic fracture pressures on local stress changes and any geological discontinuities such as faults. However, their model assumed a predefined/fixed fracture length and pressure at the fracture surface and neglected the poroelastic effect of the formation. Morrill and Miskimins (2012) applied the finite element technique to optimize the fracture spacing, neglecting the fracture interactions.

Different numerical simulators have been developed in spite of LEFM or CZM to simulate fracture propagation, fracture geometry, and magnitude and the direction of stress change around hydraulic fractures. These are either two-dimensional, pseudo—three-dimensional (pseudo-3D), or three-dimensional hydraulic fracturing models depending on the complexity of the problem and the amount of information available. These models are useful when studying the general behavior and physics of the simplified hydraulic fracturing process. The following is a brief discussion of the different models available that enable the development of more accurate hydraulic fracturing simulators.

TWO-DIMENSIONAL HYDRAULIC FRACTURING MODELS

Hydraulic fracture geometry is a complex function of initial reservoir stress conditions (global and local), reservoir rock properties such as heterogeneous and anisotropic rock mechanical properties (Young's modulus and Poisson's ratio), permeability, porosity, natural fracture system, and operational conditions such as injection rate, volume, and pressure. To model this complicated process, specific assumptions have been made to simplify the problem while capturing the major characteristics of hydraulic fracture geometry. For this, scientists had first assumed the hydraulic fracturing process would occur in a homogeneous and isotropic formation that would lead to a symmetric, bi-wing fracture from the point or line source of the injecting fluid. There are three common fracture modeling methods introduced based on these assumptions: (1) the Khristianovic—Geertsma de Klerk (KGD) model, (2) the Perkins and Kern (PKN) model, and (3) the radial fracture geometry or penny-shaped model (Abe et al., 1976).

The KGD model assumes a two-dimensional plane—strain model in a horizontal plane with a constant fracture height that is larger than the fracture length. In the KGD model, an elliptical horizontal cross-section and rectangular vertical cross-section are assumed where the fracture width is independent of the fracture height and is constant in the vertical direction. The rock stiffness is also only considered in the horizontal plane. Fig. 15.1 shows the schematic of fracture geometry in the KGD model.

The Perkins and Kern (PKN) model assumes constant fracture height independent of fracture length. In the PKN model a two-dimensional plane—strain model is assumed in the vertical plane where the fracture has an elliptical cross-section both in the horizontal and vertical directions.

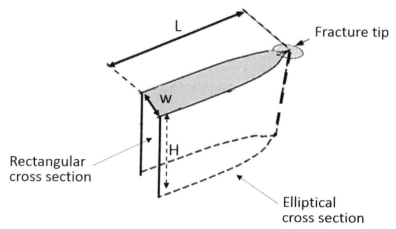

Figure 15.1 Khristianovic—Geertsma de Klerk (KGD) fracture geometry, schematic diagram.

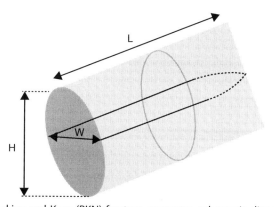

Figure 15.2 Perkins and Kern (PKN) fracture geometry, schematic diagram.

Unlike the KGD model, the PKN model assumes a fracture height much smaller than the fracture length. The PKN model also assumes the hydraulic fracturing energy applied by the fluid injection would only be consumed by an energy loss from fluid flow (viscosity-dominated regime) and ignores fracture toughness. Fig. 15.2 shows the schematic of fracture geometry in a PKN model.

The PKN and KGD models assume fluid flow in a fracture as a one-dimensional problem in the direction of the fracture propagation or fracture length governed by the lubrication theory and Poiseuille's law. They

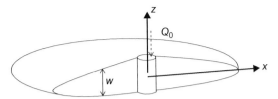

Figure 15.3 Radial fracture geometry.

assume that the fracture is confined and there is no change in horizontal stress, reservoir pressure, and temperature.

The third model used to simulate hydraulic fracture propagation in a two-dimensional plane is called the penny-shaped or radial fracture model. This model has found application in shallow formations where overburden stress became equal to minimum horizontal stress. In this case, the symmetric geometry was assumed to be at the point of line-injection source. In this model, the injection rate and fluid pressure within the fracture are assumed to be constant. Fig. 15.3 shows the schematic for fracture geometry assuming the penny-shaped model.

In all hydraulic fracturing models, the hydraulic fracture propagation is a function of injection of fracturing fluid "Q_0" from the injection point or injection line representing the well perforations. As a result, a bi-wing symmetric fracture is assumed to propagate in the formation perpendicular to minimum principal stress "σ_0" of the formation. The fracture width generated is therefore a function of effective stress, which is the difference between the pore pressure and minimum principal stress "σ_0" ($P_e = P_f - \sigma_0$). Effective pressure is a good indicator of fracture width and is a likely indicator of well performance after hydraulic fracturing. The higher this effective pressure measured during the hydraulic fracturing, the better well productivity is expected.

FLUID FLOW IN HYDRAULIC FRACTURES

Fluid flow in hydraulic fracture is governed by one- or two-dimensional Poiseuille's law and lubrication theory. In fluid dynamics, lubrication theory is used for cases where fluid flows through a media where one dimension is significantly smaller than another is considered. In hydraulic fracturing, this translates into having a fracture width much smaller than the fracture height and length. Given a two-dimensional model, one-dimensional fluid flow along the fracture length is assumed, and can be shown as follows:

$$q = -\frac{w^3}{12\mu}\nabla p_f$$

In this equation, q is the flow rate, μ is the fluid viscosity, w is the fracture width, and (∇p_f) is the gradient of fracture pressure defined in a direction of fracture length. Assuming incompressible fracturing fluid and fluid leak-off governed by Carter's leak-off model, Eq. (15.2) describes the conservation of fluid in the fracture.

$$\frac{\partial w}{\partial t} + \nabla \cdot q + \xi = 0 \qquad (15.2)$$

Equation 15.2 Mass conservation in fracture

It is very common in the oil and gas industry to attribute the fluid leak-off to the surrounding formations using Carter's model as described by the following equation:

$$\xi(x, t) = \frac{2C}{\sqrt{t - t_0(x)}}$$

Equation 15.3 Carter's leak-off model

C is the leak-off coefficient, t is the time, and t_0 is the fracture-tip arrival time.

The boundary conditions for the fluid flow equation can be obtained by assuming a constant flow rate of $Q_0/2$ at the injection point of the symmetric bi-wing fracture and a zero flow rate at the fracture tip (assuming no fluid lag or zero flow rate at waterfront having fluid lag). Fluid lag refers to a zone between the fluid front and the fracture tip. Depending on formation permeability and mechanical properties, fluid lag may not be present, which is due to a fracture tip and fluid front moving with the same velocity.

SOLID ELASTIC RESPONSE

The solid elastic response of the medium is governed by three equations: equilibrium condition, constitutive law, and geometry, which can be defined using Eqs. (15.4)−(15.6), respectively. The equilibrium condition is defined as follows:

$$\nabla.\sigma + g = 0$$

Equation 15.4 Equilibrium condition

The constitutive law of linear elasticity is governed by:

$$\sigma(x) = \kappa: \varepsilon(x)$$

Equation 15.5 Constitutive law

The geometry, which is a function of solid displacement, is expressed as:

$$\varepsilon = \left[\nabla D + (\nabla D)'\right]/2$$

Equation 15.6 Displacement

By definition, σ is the stress tensor, g is the gravitational acceleration, κ is the elastic stiffness, ε is the strain tensor, and D is the displacement. Superscript "'" here denotes the transpose of the matrix. The stress boundary conditions also need to be defined based on specific upper, lower, and fracture surface conditions.

PSEUDO−THREE-DIMENSIONAL (PSEUDO-3D) HYDRAULIC FRACTURING MODELS

Even though using 2D models is useful to understand the fundamentals of hydraulic fracturing, they cannot be used for practical purposes. Therefore, pseudo-3D models are developed with the assumption of a constant fracture height as described in the PKN model. Two different models are introduced to consider variations in fracture height, namely, the equilibrium and dynamic height pseudomodels. At equilibrium height, pseudo-3D models with a uniform pressure distribution in vertical cross sections are assumed. In these models, toughness criteria for fracture propagation (i.e., the fracture propagates when stress intensity factor K_I overcomes fracture toughness K_{IC}), are also considered. Given a

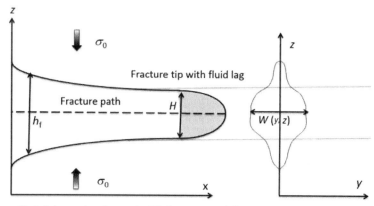

Figure 15.4 Schematic of pseudo-3D fracture model.

dynamic height, pseudo-3D models with two-dimensional fluid flows (parallel and perpendicular to fracture path) are assumed (note in 2D models, fluid flow was in one dimension and along the fracture path), and fracture height calculation followed KGD model solutions (Dontsov and Peirce, 2015). Pseudo-3D models are more practical than 2D models since they consider fracture height variation as a function of location in the direction of fracture propagation and time. However, they are still restricted to certain geometries and follow the plane—strain conditions in each cross-section perpendicular to the fracture path. They also have different accuracies in different hydraulic fracturing regimes. For example, these models are inaccurate in toughness-dominated hydraulic fracturing regimes due to local elasticity assumption. These models are also inaccurate in viscosity-dominated hydraulic fracturing regimes due to viscous losses perpendicular to fracture path. Fig. 15.4 shows the schematic of pseudo-3D hydraulic fracture.

Different hydraulic fracturing regimes are defined based on energy dissipation through the process of hydraulic fracturing due to fracture toughness or viscous flow of fluid in the fracture. A parameter called K_m independent of time is used to distinguish between these two energy dissipation regimes. High K_m value represents a toughness-dominated regime ($K_m > 4$) and low K_m refers to a viscosity-dominated regime, i.e., $K_m < 1.0$. A K_m value between 1 and 4 is referred to as an intermediate case. Eq. (15.7) shows the definition of K_m.

$$K_m = \frac{4K_{IC}\sqrt{2/\pi}}{(E/1-\nu^2)}\left[\frac{(E/1-\nu^2)}{12\mu Q_0}\right]^{1/4}$$

Equation 15.7 K_m definition

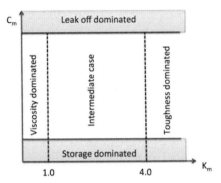

Figure 15.5 Different hydraulic fracturing regimes.

In Eq. (15.7), K_{IC} is the fracture toughness and rock property, E is the Young's modulus, ν is the Poisson's ratio, Q_0 is the injection rate, and μ stands for the dynamic viscosity of the fluid. Hydraulic fracturing regimes divided into leak-off-dominated or storage-dominated processes are quantified using the parameter "C_m." Given nonzero fluid leak-off to the formation, C_m value will vary between zero and infinity. A higher C_m value denotes higher fluid leak-off in the formation and therefore lower fracture efficiency. C_m is a function of time and is defined in Eq. (15.8).

$$C_m = 2C\left[\frac{\left(E/1-\nu^2\right)t}{12\mu Q_0^3}\right]^{1/6}$$

Equation 15.8 C_m defenition

Fig. 15.5 shows the schematic of different hydraulic fracturing regimes (Bunger et al., 2005).

THREE-DIMENSIONAL HYDRAULIC FRACTURING MODELS

Extending the pseudo-3D models to full 3D models, in addition to the high computational cost associated with full 3D models, would be difficult considering the nonlocal dependency of fracture geometry to fluid pressure and confining stresses. The exact model should be able to fully combine multiphase fluid flow in the fracture, leak-off to the formation, and rock deformation. This can be achieved by solving sets of coupled partial differential equations governing this multiphysics process simultaneously. Finding the solution for fully coupled problems is an extremely difficult task, and therefore simplified models have been introduced where one-way coupling or weak coupling was previously used. In one way coupling, partial differential equations governing the fluid flow in the fracture are solved at each time step and pressure distributions are obtained. Pressure distributions will then be used as an initial condition for differential equations governing the rock deformation and fracture propagation. In this technique, the fracture pressure distributions at each time step are assumed to be independent of rock deformations; therefore, they will not be updated due to a change in fracture geometry. There is also an intermediate technique between fully coupled and one-way coupling called weak coupling. In this technique, similar to one-way coupling, fracture pressure distribution at each time step is calculated independent of change in the fracture geometry at that time step. However, after certain time intervals, the fracture pressure distribution is updated based on fracture geometry.

Fracture-tip behavior and dynamics of fluid lag at the fracture tip create additional difficulties in hydraulic fracturing simulation due to the dynamic boundary conditions imposed at the fluid front following the fracture tip. In the literature, there has been a great effort to take these effects into consideration by different research groups (Garagash 2006, 2007; Adachi and Detournay, 2008; Shen, 2014; Dontsov and Peirce, 2015). The conventional method implemented in these studies is the DD method, which is a modified version of the boundary element method and can be applied to simulate models with arbitrary fracture geometry. In this technique, displacement along the fracture propagation path will be discretized into a series of elements where displacement is assumed to be constant for each element. Having an analytical solution based on Green's function that describes the displacement and stress tensor relationship for a single element, the total displacement will be calculated by the summation of all displacements in each element. The advantage of this method is that the key equations for the coupled process are built upon the fracture surface rather than on the whole model. This significantly reduces the computational cost of numerical simulation. The disadvantage of the technique is to find its nonlocal kernel function when the model has a complex structure (Siebrits and Peirce, 2002).

Recently, different finite element methods have been used for hydraulic fracturing simulation. These techniques have more flexibility when compared to the DD method, in that they do not require explicit calculation of the kernel function. In these techniques, two coupled nonlinear finite element equations are defined. One is describing the elastic response of the elastic medium, and the other describes the fluid and solid transport within the fracture. The first system will be used to find the relationship between the net pressure in the fracture and fracture width, and the second system to simulate the fluid and solid transport in the fracture. The investigation will be accomplished by solving the coupled equations from the two systems using the Newton—Raphson iteration algorithm.

HYDRAULIC AND NATURAL FRACTURE INTERACTIONS

Hydraulic fracturing in naturally fractured formations is completely different than hydraulic fracturing in homogeneous and isotropic formations, which is assumed in most of the numerical simulators. This is due to the interactions between hydraulic and natural fractures. In the presence of natural fractures, and depending on their density and major direction with respect to local minimum and maximum in situ stresses, hydraulic fracture might cross the natural fracture, locally merge with natural fractures, and break out in a short distance, or completely follow natural fracture directions. Different experimental studies on hydraulic fracture and natural fracture interactions showed that a hydraulic fracture tend to cross the natural fractures, if they approach the natural fractures at a high angle (close to perpendicular) and also where there is a significant difference between fracture pressure and natural fracture stresses. If a hydraulic fracture reaches the natural fractures with a low angle and similar stress conditions, the natural fractures will open up and the hydraulic fracture merges with natural fractures (Lamont and Jessen, 1963; Daneshy, 1974; Blanton, 1982). In the oil and gas industry, microseismic data and core characterization and imaging have been used to map the hydraulic fractures and investigate the natural fracture and hydraulic fracture interactions. These studies in shale gas reservoirs show that it is not uncommon to have complex fractures generated instead of the expected conventional symmetric and biwing hydraulic fractures. However, microseismic studies are expensive and not available for every frac job. There are also significant concerns regarding the upscaling of the experimental studies from laboratory scale to actual field applications. Therefore, different numerical and analytical techniques have been used to investigate these effects. Potluri et al. (2005) studied the effect of natural fractures on hydraulic fracture propagation using Warpinski and Teufel's criteria (1987) and concluded that the hydraulic fracture will pass the natural fractures, if the normal stress on the natural fractures is higher than the rock fracture toughness. They also defined different criteria for a hydraulic fracture when it merges with a natural fracture and extends from the natural fracture tip and when it merges and breaks out after a short distance based on the angle of the hydraulic and natural fracture, fracture toughness, and hydraulic and natural fracture pressures. Recent studies also investigated the major parameters impacting the hydraulic fracturing behavior in the presence of natural fractures using numerical simulations based on the extended finite element method (XFEM). Dahi and Olson (2011) investigated the interactions between hydraulic fractures and cemented and uncemented natural fractures using XFEM. They showed that the anisotropy in a stress field can significantly enhance the hydraulic fracture and natural fracture interactions and suggested that further detailed studies are required to quantify these impacts.

HYDRAULIC FRACTURE STAGE MERGING AND STRESS SHADOW EFFECTS

Developing unconventional resources such as organic–rich shale reservoirs using horizontal well technology and multistage hydraulic fracturing introduced a whole new area of research in both academia and industry to optimize these activities. One of the major concerns in designing and optimizing the multistage hydraulic fracturing jobs is the merging hydraulic fracturing stages. This has not been seen using commercial hydraulic fracturing numerical simulators due to oversimplified assumptions made during their developments. Therefore, detailed studies on quantifying the magnitude of induced stresses and reorientation of the stress fields during multistage hydraulic fracturing jobs are required.

Recently, different numerical and analytical studies have been published concerning the magnitude of stress change and stress reorientation in single pressurized fractures or multiple fractures. Cheng (2012) studied the change in fracture geometry due to the change in stress field around three pressurized fractures. In her model, she assumed fixed fracture lengths and used the DD method to quantify the change in the fracture width as a function of the change in the stress field around the pressurized fractures. Soliman et al. (2004) used an analytical technique to calculate the magnitude of the stress change around multistage hydraulic fractures.

In general, the magnitude of the stress change around a propagating fracture is a function of fracture dimensions, target, and over- and underlying formation characteristics such as fracture length, width, height, the formation's Poisson's ratio, relative magnitude of the target formation's Young's modulus, and under- and overlaying formations, magnitude, and direction of in situ stresses.

Fisher et al. (2004) introduced the stress shadowing effect in multistage hydraulic fracturing where the local maximum and minimum horizontal stresses are changed due to hydraulic fracture propagation. These changes in local state of stress will highly impact the subsequent hydraulic fracture paths and will result in hydraulic fracture stage merging or deviation depending on the magnitude of the change. If the pressure applied during hydraulic fracturing at the fracture surface falls between local minimum and maximum horizontal stresses, it will not be expected to have a huge change in local stresses and subsequent fracture paths. However, if the pressure exceeds the maximum horizontal stress, a phenomenon called principal stress reversal will occur, leading to a significant change in subsequent hydraulic fracture paths. Taghichian (2013) studied the stress shadowing around single and multiple pressurized fractures in confined and unconfined environments. He demonstrated that there is a nonlinear and direct relationship between fracture pressure, formation's Poisson's ratio, and stress shadow size. Increasing the fracture pressure leads to an increase in the stress shadow zone but a decrease in gradient. Waters et al. (2009) showed that the shadow effect around a single hydraulic fracture leads to locally increased compressive stresses perpendicular to the fracture propagation plane. This leads to reorientation of local maximum stresses and thus unintended change in the direction of the subsequent fractures, if they happen to fall in the stress shadow zones. Therefore, optimized hydraulic fracture spacing is required to increase the efficiency of hydraulic fracturing stimulation. The effect of stress shadowing is investigated not only in single-well multistage hydraulic fracturing but also in multihorizontal well fracturing. Recently, new publications have been focused on multihorizontal well stimulations in which simultaneous hydraulic fracturing of parallel horizontal wells is studied. Mutalik and Gibson (2008) showed that this technique could increase the efficiency of the stimulation between 21% and 100%. Rafiee et al. (2012) applied a similar concept to zipper frac to increase the efficiency of stimulation by using a staggered pattern. Other studies in this line tried to precondition the stress field using outer hydraulic fractures to prevent deviation in the middle stages, such as the work published by Roussel and Sharma (2011).

CHAPTER SIXTEEN

Operations and Execution

INTRODUCTION

A frac job in general is a massive operation because it takes a lot of manpower and equipment to accomplish the job. Slick water frac requires high pump rates. As a result, lots of high-pressure pumps are required for each pad. The design and fracture modeling of a hydraulic frac job for optimum production enhancement is important; however, being able to operate and execute the design treatment schedule is more important. Therefore, exploration and production (E&P) companies typically develop the best operations and execution practices for each field in order to minimize nonproductive time (NPT), reduce unnecessary capital expenditures, eliminate safety accidents and compliance issues, and increase operations efficiency in an attempt to optimize the economics of the project. A frac job is logistically a large operation that will need a lot of discipline and coordination. There are lots of opportunities for improvement throughout the completions operation. When historically going back in time and reviewing the first 10 wells drilled and completed in each shale play, it can be easily seen that with time, efficiency and savings were obtained. Some of the first wells in each field cost two to four times more than the current capital expenditure in the same field for the same completions design. This can be tied back to the learning curve associated with comprehending the formation and coming up with ideas to minimize encountering any issues related to the operations of the hydraulic frac job. Drilling and completing one expensive well in a new exploration area typically does not scare a true E&P company because they have enough experience, expertise, and knowledge to know that with time, efficiency and savings will be obtained.

Hydraulic Fracturing in Unconventional Reservoirs
DOI: http://dx.doi.org/10.1016/B978-0-12-849871-2.00016-2
© 2017 Elsevier Inc.
All rights reserved.

WATER SOURCES

There are a few sources for frac water. The most common water sources are as follows:

- freshwater from rivers, lakes, etc.
- municipal water
- reused flowback and produced water (100% produced water)
- treated water from a treatment facility
- mixing fresh and reused water from flowback and producing wells.

Water Storage

As discussed earlier, water is one of the most important aspects of hydraulic fracturing. Water storage is used to make up for lack of supply, or to hold water for reuse for upcoming frac jobs. There are multiple ways to store water used during the frac job and the most common methods are as follows:

1. **Centralized impoundment (in-ground pits)** can be built on a side of the pad and water can be stored. The capacity of in-ground pits can vary from pit to pit. With stricter environmental rules and regulations, some states do not give permits for building centralized impoundments. In addition, in-ground pits can be very expensive to build, monitor, and reclaim. More regulations have been applied to in-ground pits since 2012 restricting their use due to possible detectable leakage concerns into the ground. Centralized impoundments come in various sizes and can typically hold 5 + million gallons of water ($\sim$ 120,000 BBLs). Freshwater centralized impoundment is typically single-lined but the impaired water impoundment is usually double-lined with leak detection.

2. **Above-ground storage tanks (ASTs)** have become more common since 2011 because of ease of building and monitoring associated with this type of water storage. Instead of going through the hassle of in-ground pits that must be reclaimed once done, above-ground storage tanks can be built in 2—3 days and function exactly the same as in-ground pits. One of the biggest disadvantages of AST is the high cost as compared to in-ground pits but many E&P companies have been pushed to get away from using in-grounds pits in some states. Some states that do not have stricter environmental rules still use in-ground pits as they are considerably cheaper. Water storage regulations associated with in-ground pits highly depend on each state. Above-ground storage tanks can be rigged down in a few days depending on the size of the tank. Since regulations are becoming harder and a lot

of operators are getting away from using centralized impoundments, AST has become more common in the past few years. The primary reason from an environmental perspective for using an AST is the ease and ability to detect any leaks much easier than in-ground pits. Typical sizes for ASTs are 10K, 20K, 40K, and 60K BBLs. In Pennsylvania, an OG71 permit is needed to use an AST for impaired water. Temporarily storage of freshwater in an AST requires a single liner while impaired water requires a double liner and secondary containment.

3. **Centralized tanks (tank batteries)** are essentially combinations of many frac tanks that are connected to each other through a manifold. Tank batteries can be anywhere from 5 to 60 or more frac tanks (depending on location) that are connected through a manifold. The capacity of each frac tank is typically 500 BBLs and enough frac tanks must be located onsite for continuous operation. For example, if each stage requires 8000 BBLs of water, 16 frac tanks (assuming a 500-BBL frac tank) are required for hydraulically fracturing only one stage. In slick water frac using conventional plug and perf, typically three to eight stages are done per day (depending on the amount of sand designed per stage). Therefore, depending on the pump rate (bpm) coming into the frac tanks, the number of frac tanks for the job can be determined. Since water is being continuously pumped from a surrounding pit, AST via buried or above-ground temporary or permanent water lines, there are only five to six frac tanks on frac location. Occasionally where permanent or temporary water infrastructures are not in place, water must be trucked to the location, which can get expensive; there are also environmental or location impacts associated with trucking water to the location. Figs. 16.1–16.3 show examples of an in-ground pit, AST, and frac tank batteries.

Water delivery:

Water is delivered in two ways:

1. Pipeline. In developed fields with lots of frac activities, there are either buried or above-ground water pipelines that are used for water delivery and transfer. Some commonly used water pipelines are 8" to 16" HDPE, PE4710. This pipeline comes in various ratings including DR7 (315 psi), DR9 (250 psi), and DR11 (200 psi).

2. Trucking. In undeveloped fields or areas where water infrastructure does not exist, water is trucked to location. For instance, if a well in a new exploration area with no water infrastructure is being completed and 8 stages are expected to be done per day using 9000 BBLs of

Figure 16.1 In-ground pit.

Figure 16.2 Above-ground storage tank (AST).

water per stage, 720 trucks per day will be needed to deliver water on location (assuming truck capacity of 100 BBLs). This can get very costly and have some local impacts.

Pipeline and pump system design for water delivery uses Bernoulli's principle. Bernoulli's principle for incompressible flow is shown in

Figure 16.3 Tank batteries.

Eq. (16.1): h_L or head loss is the energy losses in the system from pipe, friction (material), bends, pipeline size changes, etc. Due to pipeline distance, pipeline friction is the main energy loss.

$$\frac{V_1^2}{2g} + Z_1 + \frac{P_1}{\rho g} - h_L = \frac{V_2^2}{2g} + Z_2 + \frac{P_2}{\rho g}$$

Equation 16.1 Bernoulli's principle for incompressible flow

In Eq. (16.1), V is the fluid velocity at a point, Z is the elevation of the point from baseline, P is the pressure at the point of interest, and ρ is the density of the fluid. The methods discussed in this section are commonly used to store reused water or freshwater. In the winter, it is very important to have sufficient brine water onsite to prevent the frac iron from freezing. In addition, sufficient heaters have to be used for certain equipment to avoid the possibility of freezing, and as a result, this slows down the operation.

HYDRATION UNIT ("HYDRO")

The hydration unit, also referred to as "hydro" in the field, is a big tank used to provide sufficient time for hydration of linear gel. If gel is used in some stages to overcome tortuosity along with other benefits associated with using gel, the hydration unit provides the gel enough time to hydrate. Without a hydration unit pumping gel is not possible. The importance of the hydration unit becomes more evident in the winter months when gel will need a longer time to hydrate due to the cold weather. If gel is not part of the job design, a hydration unit will not be needed. Some operators do not believe in pumping gel in slick water frac jobs and do not use hydration units in their equipment rig up. A hydration unit is located right after the frac tanks where water is stored. There are typically five to seven frac tanks that are connected through a manifold right before the hydration unit. The hydration unit has a suction side that sucks the water from the frac tanks and a discharge side that releases the water to the next equipment (blender, will be discussed). Some hydration units have a discharge pump but some do not. It is not a necessity to have a discharge pump on the hydro since the suction pump of the following equipment located after the hydro (blender) is used to suck the water out of the hydro. There are injection ports located on the suction side of the hydro in the event gelling agents need to be started. In addition, chemicals are stored in the special containers called "totes." There are small liquid additive (LA) pumps (such as stator or positive displacement pumps) located on the chemical totes to pump the chemicals to the hydro through the injection ports. Therefore, when gel is started, linear gel and buffer are pumped via the LA pumps through the injection ports to the suction side of the hydro. A hydration unit's capacity is typically between 170 and 220 BBLs depending on the type and size of the hydro. Therefore, it takes 170−220 BBLs of fluid (to give sufficient time for gel hydration) from the time gel is started until it leaves the hydro. Therefore, it takes some time from the time gel is started until gel reaches the perforations.

●●●

Example: Calculate the volume it takes for the gel to hit the perforations assuming the following parameters:

Bottom perf measured depth (MD) = 15,500′ with a casing capacity of 0.0222 BBL/ft, Hydration unit capacity = 180 BBLs, Surface line volume capacity = 50 BBLs

Casing capacity = $0.0222 \times 15,500 = 344$ BBLs $\rightarrow 344 + 50 + 180 = 574$ BBLs

Therefore, it takes 574 BBLs for the gel to reach the perforations as soon as it starts.

This example shows the importance of starting gel at the right time since gel will not yield an instantaneous relief. If surface-treating pressure increases rapidly during a slick water frac job, starting gel to increase fracture width and reducing the pressure will not help because it takes some time for the gel to reach the perforations. Typically, it is recommended to cut sand and start over again to prevent a costly screen-out. This example shows the importance of starting gel early enough to see the impact.

BLENDER

The blender is the heart of the frac operation. The blender is used to mix water, proppant, and some chemicals in the blender tub before sending the slurry fluid downhole. The blender is typically located right after the hydro in a frac setup. There is a tub on every blender and there is an agitator at the bottom of the tub. The tub agitator consists of two sets of blades on a shaft. The main function of the agitator is to keep the proppant suspended in the fluid without carrying air. If the agitator speed is too low, there is a high possibility of proppant building up and settling at the bottom of the tub and suddenly getting picked up as a sand slug. If the agitator speed is too high, it can entrain air in the fluid, causing the booster pump to pick up air, and as a result, causing a decrease in boost pressure.

The heart of a blender is the centrifugal pump. The main reason centrifugal pumps are utilized is because they are very tolerant to abrasive frac fluids which will result in an increase in the life of the pump. As previously discussed, millions of pounds of proppant are pumped when hydraulically fracturing a well and it is very important to use quality pumps, such as centrifugal pumps, on a blender. A centrifugal pump consists of one or more impellers equipped with vanes. The impeller is located on a rotating shaft and fluid enters the pump at the center of the impeller. Fig. 16.4 illustrates a centrifugal pump from inside.

There are two centrifugal pumps located on the blender. The first one is called a suction pump, which sucks the water from the hydro and sends it to the blender tub. The second pump is called a discharge pump (also referred to as a "boost pump"), which sends the mixed slurry from the tub to high-pressure pumps. In other words, a blender has two sides:

1. **Suction side (clean side)**. A suction pump is located on this side of the blender. A centrifugal suction pump sucks the frac water from the hydro and sends it to the tub. This side is also referred to as the "clean

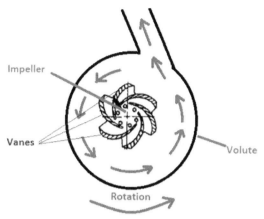

Figure 16.4 Centrifugal pump from inside.

side" because proppant has not yet been mixed up on this side and only frac water and chemicals enter the blender.

2. **Discharge side (dirty side)**. Combinations of water, chemicals, and proppant exit this side and this is why it is referred to as the "dirty side."

The boost pump (or discharge pump) is actually the means of providing the rate (boost) for all of the high-pressure pumps located after the blender.

SAND MASTER (SAND MOVER, SAND KING, OR SAND CASTLE)

Since millions of pounds of proppant are required for each pad (depending on the number of frac stages), proppant has to be stored onsite using sand masters. Sand masters have different bins used for placing various types of proppant and mesh sizes (note: some sand masters do not have any bins). Proppant is delivered on location using sand cans or sand trucks. Depending on the state regulations and guidelines, each sand truck on average can hold 40,000 to 50,000 lbs of proppant because of the weight limitations. For example, if a frac design of a pad consists of 100 stages and each stage is designed to use 400,000 lbs of proppant, 800 sand trucks (assuming each sand truck can haul 50,000 lbs) must travel to the pad to blow the proppant into the sand masters throughout the job. Placing the proppant into the sand master is called blowing sand. There

Figure 16.5 Sand masters.

are new technologies that do not require blowing proppant anymore; instead, proppant can be placed into sand masters through gravity. During a large-sized hydraulic frac job, sand trucks are entering and exiting the pad all day long to provide the proppant needed for the job. The sand master is also referred to as the sand mover, sand king, or sand castle. There are other patent industry names for sand masters but the concept of all of this equipment is the same regardless of the names and shapes of the movers. They are all used to store proppant on location. Fig. 16.5 illustrates three sand masters side by side during a hydraulic frac job.

T-BELT

Proppant falls out of the sand bins onto the T-belt. Once on the T-belt, proppant is carried through the T-belt and falls into the blender hopper. Once at the blender hopper, proppant gets picked up via the blender's sand screws and is dumped into the blender tub. The sand screws are another important part of the blender: they are the means of picking up proppant and dumping the proppant into the blender tub. Therefore, sand screws are essential in conveying proppant to the blender tub. Fig. 16.6 illustrates the point at which proppant gets dumped on the T-belt and carried via the T-belt until it reaches the blender hopper. Fig. 16.7 illustrates the point at which proppant is dumped into the blender hopper, the point where proppant is picked up through the sand

Figure 16.6 Sand masters and T-belt.

Figure 16.7 Hopper and blender screws.

Figure 16.8 Sand screws with proppant in the hopper.

screws from the blender hopper, and finally the point at which proppant is dumped into the blender tub where proppant, water, and chemicals will be mixed. Fig. 16.8 demonstrates a blender hopper full of proppant along with sand screws.

Sand Screws

As previously mentioned in a hydraulic frac job, proppant concentration is gradually increased each time proppant hits the perforations (more aggressive design schedules do not wait for the proppant to hit the perforations and proppant is staged up faster) depending on the design schedule and pressure response throughout the stage. There are typically two to three sand screws on a blender depending on the blender manufacturer and type. In slick water frac, two screws are typically used and the third one is a backup. The third screw is normally used with very high proppant concentration frac jobs such as cross-linked jobs. Every blender has a maximum rpm per screw that can be obtained from the blender manufacturer. The reason more than one rpm is needed for the job is to be able to pump higher sand concentrations at higher rates. The most commonly used sand screws have 12" and 14" diameters. Normally the maximum output for a 12" screw is approximately 100 sacks per min (one sack of proppant is equal to 100 lbs), and for a 14" screw is 130 sacks per min with a maximum rpm of 350–360 (different depending on the blender manufacturer and type). Note that maximum rpm of 350–360 is only for one screw and since typically two screws are used in slick water frac jobs, up to 700 rpm can be obtained to fulfill the client's needs and design schedule. Different types of proppant yield different pounds per revolution (PPR). PPR decreases at higher sand concentrations and as the screws wear out. For example, if proppant delivery of Ottawa sand is about 36 PPR with a brand new screw, as sand concentration increases, PPR decreases. In addition, if lower sand concentration is run (e.g., 0.25 ppg) and PPR is run at 29 (if typical is 36), there could be a high possibility that the screws are worn out.

$$rpm = \frac{Q \times SC \times 42}{PPR}$$

Equation 16.2 Round per minute (rpm) calculation

rpm = Round per minute
Q = Slurry rate, bpm
SC = Sand concentration, ppg
PPR = Pounds per revolution.

Eq. (16.2) is constantly used in the field to calculate the amount of rpm needed on the screws to achieve the designed sand concentration. For example, after pumping the designed volume of pad, and once the sand stage is ready to start, the person responsible for adjusting the sand screws on the blender is notified by radio to bring his/her rpm to a certain value to achieve the required proppant concentration requested by the operating company's designed proppant schedule. This is referred to as running the blender on "manual." On the other hand, the majority of service companies run their blender screws on "auto" for simplicity. Running the blender on auto means entering the proppant concentration needed on the blender, which will automatically calculate the required rpm. This is the preferred method since every time the slurry rate is changed throughout the stage for any reason, the auto system calculates the new rpm needed and adjusts the screws. For example, if a pump is dropped for any reason (e.g., mechanical issues), a new rpm is automatically calculated. If a manual system was being used, the new rpm would need to be manually calculated and changed on the blender, which might take some time. It is strongly recommended that any service company knows how to run the blender in both auto and manual. This way, if there are any issues throughout the stage while running the sand system in auto, a manual system can be substituted and the frac stage can continue instead of coming offline while the problem is being fixed.

●●●──

Example: Calculate the rpm needed at 100 bpm slurry rate if 0.25 ppg proppant concentration is to be achieved with 36 PPR.

$$rpm = \frac{Q \times SC \times 42}{PPR} = \frac{100 \times 0.25 \times 42}{36} = 29 \text{ rpm}$$

The rpm needed to achieve 0.25 ppg proppant concentration is 29 rpm. At this stage, typically the company representative waits until proppant hits the perforations to see the reaction of the formation. If everything looks promising on the surface-treating pressure chart, proppant concentration is increased by increasing rpm. Let's assume the next designed proppant concentration is 0.5 ppg and the rate had to be dropped to 94 bpm. Calculate the new rpm needed to achieve this concentration.

$$pm = \frac{Q \times SC \times 42}{PPR} = \frac{94 \times 0.5 \times 42}{36} = 55 \text{ rpm}$$

As can be seen in this example, as proppant concentration increases, rpm increases as well. Please note that if, throughout the stage, rate is increased or decreased, rpm needs to be adjusted as well if the system is not set up for auto. Rate is directly proportional to rpm and as rate increases or decreases, rpm increases or decreases as well. There is a person called a treater who is responsible for calculating the new rpm every time rate, proppant concentration, or PPR is altered if and only if the manual rpm system is used. If the auto system is used the only parameter that needs to be entered is proppant concentration and everything else will be automatically calculated.

PPR is typically adjusted throughout the job as well to stay at the required proppant concentration and volume. Throughout the job, there is a person on the sand master who takes proppant straps (the amount of proppant that is left in the bin or sand master). Newer sand master systems can actually measure the amount of proppant pumped out of each bin or sand master via a scale. There are two ways to measure proppant. The first one is the calculated amount located on the frac monitors inside the frac van from the blender screws. The second one is through the gentleman/lady on the sand master who measures the amount of proppant that is left in the bin (or in the newer system via a scale). For example, let's assume that after taking a bin strap, the person responsible for keeping track of the proppant announces 30,000 lbs of proppant has been pumped. On the other hand, the monitor located in the frac van shows the total proppant pumped is 40,000 lbs (calculated from sand screws). This means less proppant has been pumped compared to the proppant volume that must have been pumped. This condition is referred to as sand light. In this situation, PPR needs to be decreased in order to increase the rpm and catch up with the required proppant.

In contrast, if the monitor shows 20,000 lbs instead, proppant is referred to as sand heavy because more proppant (30,000 lbs) was actually pumped. In this case, PPR is increased in order to decrease rpm and slow down the actual amount of proppant. The difference in proppant can be easily caught up if and only if the difference is 5000–15,000 lbs at lower proppant concentrations. If the difference is drastic, such as 30,000 lbs, it is strongly recommended not to catch up. If proppant is running 30,000 lbs light and we have 50,000 lbs of proppant left to go in a frac stage, it means more proppant needs to be pumped in order to compensate for the lack of proppant concentration accuracy throughout the stage. This can be devastating because proppant is now run at very high concentrations in an attempt to catch up, which can cost the operating company a screen-out. Therefore, it is very important to remind the person in charge that if proppant is light or heavy by a drastic amount, do not try to catch up and run it as it is all the way to the end of the frac stage. The most important part about blender screws is the difference between rpm and PPR. As can be seen from Eq. (16.2), rpm and PPR are inversely proportional. When PPR decreases, rpm increases and more proppant will be pumped. PPR is typically decreased by a small amount when proppant is running light. When PPR increases, rpm decreases and proppant will be pumped slower. PPR is typically increased by a small amount when proppant is running heavy. Fig. 16.9 is another example of sand screws and a blender tub where sand, water, and some of the chemicals get mixed.

Chemical Injection Ports

There are multiple chemical injection ports located on the blender. Some chemicals such as friction reducer (FR), biocide, scale inhibitor, etc. can be directly pumped from LA pumps to the suction side or tub of the blender. Other chemicals such as cross-linker have to be pumped by the LA pump to the injection ports located on the discharge side of the blender. This is because the tub agitator is not able to mix such a viscous fluid in the tub. In addition, surfactant has to be injected in the discharge side as well because special types of surfactants can foam up and block the

Figure 16.9 Blender tub and blender screws.

Figure 16.10 Chemical totes.

view in the blender tub. Therefore, depending on the chemical and frac setup, some chemicals are pumped into the suction side while others are pumped into the discharge side. Fig. 16.10 shows chemical totes where various chemicals used during the hydraulic frac job are stored on location. As can be seen from Fig. 16.10, frac chemical totes are located on containment, which prevents any type of spill from reaching the ground. Any type of spill on the containment can be easily cleaned up with no environmental damage, however, the majority of companies take many precautions to avoid any type of spill regardless of its amount on the containment. Many companies even tie incentives back to any type of spill in an attempt to make sure that all of the employees place 100% of their effort in having zero environmental incidents and staying in compliance with all of the environmental regulations and laws. The containments are inspected constantly throughout the frac job to make sure that there are no holes in the containment. Any holes in the containment are reported and fixed immediately.

Densometer ("Denso")

The densometer, also referred to as the "denso," measures the density of the frac fluid going downhole. There is always one densometer on any blender located on the discharge side of the blender to read the proppant concentration of the fluid coming out of the blender tub. This is the most accurate way of measuring the proppant concentration that is being pumped downhole. Some service companies try to hide this value from the client (operating company) and instead show a corrected proppant concentration value since this value fluctuates throughout the stage due to the level of proppant in the hopper dropping or gaining. Therefore, to make the service look better, service companies usually hide this value and instead show a corrected value on the screen inside the frac van. There is one more densometer located at the end of the main line. This densometer is usually used to make sure all of the proppant has cleared the surface lines. Once all of the proppant is cleared from the surface lines, the flush stage starts. In the flush stage, all of the proppant must be placed into the formation to make sure the wireline can go downhole to set the plug and perforate the next stage in a conventional plug–and–perf setup.

MISSILE

The missile is located right after the blender. Water, proppant, and chemicals are sent to the missile with the boost rate obtained from the boost pump on the blender. The missile is a big manifold that allows multiple hoses and frac irons to be connected. This reduces the amount of frac lines and hoses used during the frac job. The missile has two sides. The low–pressure side is where frac fluid is transferred from the missile to frac pumps. The

Figure 16.11 High-pressured iron and low-pressured hose on the missile.

high-pressure side is where the slurry comes out of the frac pumps and goes back to the missile. The low-pressure side of the missile has a low pressure of normally 60–120 psi. The discharge hoses can be used to transfer the water from the blender to the missile and the missile to the frac pumps. The high-pressure side of the missile is approximately the frac pressure obtained from the surface-treating pressure chart. The main reason this side is called the high-pressure side is to denote the difference (Fig. 16.11).

Below are the simplified steps of transferring slurry frac fluid from the boost pump (located on the blender) to the wellhead:

1. The boost pump located on the blender provides the rate needed to transfer the slurry to the missile and all the pumps on location.
2. Slurry fluid (water + proppant + chemical) enters the missile and subsequently the frac pumps via discharge hoses.
3. High-pressure frac pumps are used to shoot the slurry fluid back to the high-pressure side of the missile via frac irons.
4. The missile is a big manifold that takes several frac lines (depending on the size of the missile) and turns them into two to six lines depending on the rate needed for the job.

FRAC MANIFOLD (ISOLATION MANIFOLD)

Frac manifold is sometimes used in zipper frac operation to isolate one well from another. While a frac job is performed on one well, the

Figure 16.12 Three-leg frac manifold.

other well can be safely perforated by having two barriers. The frac manifold isolates the frac well from the wireline well with two barriers. The first barrier is a hydraulic valve located on the frac manifold and is operated via the accumulator. The second barrier is a manual valve operated manually. Two-leg or three-leg frac manifolds are the most commonly used manifolds in the industry. Each leg has hydraulic and manual valves as shown in Fig. 16.12. A two-leg frac manifold is used to zipper frac two wells while a three-leg frac manifold is used to zipper frac three wells at a time. A frac manifold is not a necessity in frac operations since zipper frac can be done without it. However, the use of a frac manifold eliminates rigging up and rigging down frac irons between stages. As a result, using a frac manifold saves time and in the oil and gas industry, time is money. Therefore, the decision on whether to use a frac manifold or not is an economic decision based on economic analysis. Fig. 16.13 illustrates an overview of the frac equipment from a frac site.

FRAC VAN (CONTROL ROOM)

The frac van is where the company representative ("company man") along with the frac supervisor oversees the entire operation via various charts. There are various charts that are monitored during hydraulic frac jobs and crucial decisions are taken based on those

Figure 16.13 An overview of frac site.

charts. The most important charts monitored during the frac job are as follows:

1. **Surface-treating pressure chart**

    This chart typically has surface-treating pressure, bottom-hole pressure, slurry rate, blender concentration (sand concentration at the blender), and formation concentration (calculated sand concentration at formation). Surface-treating pressure is read directly from the transducer on the main line. Bottom-hole pressure is a calculated pressure using the bottom-hole surface-treating pressure equation. Some companies display the actual blender concentration from the densometer on the blender. Slurry rate shown on the chart is obtained from the flow meter located on the blender. Fig. 16.14 shows an example of a typical slick water stage along with treating pressure, bottom-hole pressure, slurry rate, and sand concentration.

2. **Net bottom-hole pressure chart**

    The net bottom-hole pressure (NBHP) or Nolty chart is another main diagram illustrated in the frac van. In conventional reservoirs, the Nolty chart is used to make critical decisions based on pressure trends. Even though this chart was mainly developed for conventional reservoirs, it is still widely used and followed during unconventional plays, such as various shale plays across the United States, as part of tradition.

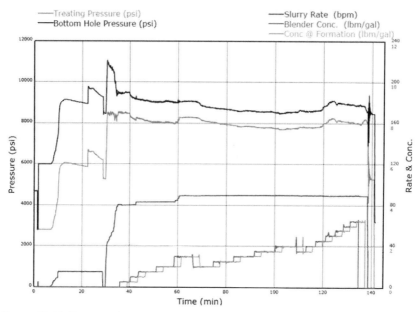

Figure 16.14 Pressure chart.

3. Chemical (chem.) chart

This chart illustrates each particular chemical and the amount of chemical used throughout the stage. It is very important to monitor all of the chemicals that are being pumped downhole throughout the stage to make sure the right type and amount of chemical concentrations are being used. In addition, as soon as the personnel in charge of the chemical starts having issues at any point during the job, it can be easily seen on the chemical chart and immediate actions must be taken to correct the problem. For example, FR is one of the most important chemicals that must be run throughout the stage during slick water frac to reduce pipe friction. If at any point throughout the stage FR is not pumped downhole due to equipment malfunctioning or any other reason, it is very important to cut sand and flush the well right away since pumping the job without FR is impossible. Also a few buckets of FR must be placed right by the blender tub to survive the flush stage in the event FR is lost. If FR is not being pumped during the flush because there are no FR buckets by the blender tub, rate must be significantly dropped to stay below the maximum allowable surface-treating

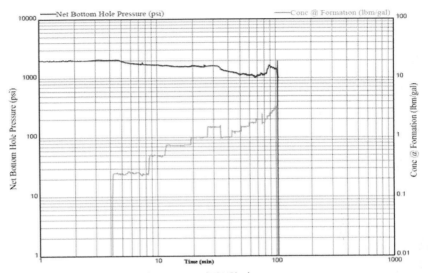

Figure 16.15 Net bottom-hole pressure (NBHP) chart.

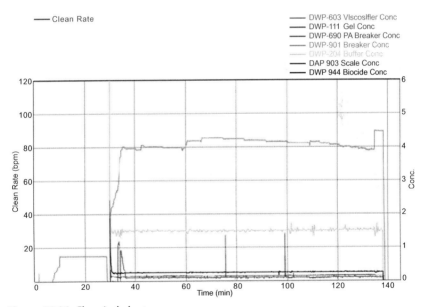

Figure 16.16 Chemical chart.

pressure. Significantly dropping the rate will eventually cause the formations to give up and might result in a costly screen-out. Figs. 16.15 and 16.16 illustrate NBHP behavior and chemical chart in a slick water frac stage, respectively.

OVERPRESSURING SAFETY DEVICES

Frac operations can be very complicated and unpredictable. In frac operations, precautionary actions must be taken to prevent overpressuring the iron, casing, and equipment during the frac job. As a result, the following two minimum precautionary actions are taken by service companies to prevent any unpleasant consequences such as parting iron, bursting the casing, blowing up the wellhead, and other well-control issues during the frac operation:

1. **Pump trips**

   Pump trips can be easily placed on all of the pumps in the event of an emergency and to prevent overpressuring the iron, wellhead, and casing. Pump trips are determined by the operator and vary from operating company to company. For example, if maximum allowable surface-treating pressure is calculated to be 10,000 psi, pump trips are staggered between 9500 and 9900 psi depending on the operator's preferences and guidelines. Pump trips need to be staggered in a pressure range. If the trips on all of the pumps are set at 9400 psi, all of the frac pumps will trip at the same time in the event the pressure exceeds 9400 psi for a short period of time. The main reason for not wanting all of the pumps to trip at the same time is because a lot of the time the pressure is still under control and can be controlled by dropping one pump at a time instead of giving up and bringing all of the pumps offline. Fig. 16.17 shows a typical frac equipment set up plan. Fig. 16.18 illustrates a frac stage that was screened out due to a 1000 psi pressure spike. All of the pumps tripped out even with the pump trips being staggered. In this situation, pump trips avoided overpressuring the iron, casing, and equipment by catching the pressure spike and bringing all of the pumps offline. In these occasions where the formation completely gives up, the pump operator does not have sufficient time to take any action no matter how fast the reaction occurs. The pressure spike shown in Fig. 16.18 happened in 1 second, and this shows the importance of having mechanical and automatic pressure control equipment during a frac job to ensure the safety of the frac operation.

2. **Pressure–relief valves (PRVs)**

   Pressure–relief valves (PRVs), also known as pop-offs, are another safety precautionary action taken to prevent overpressuring the iron,

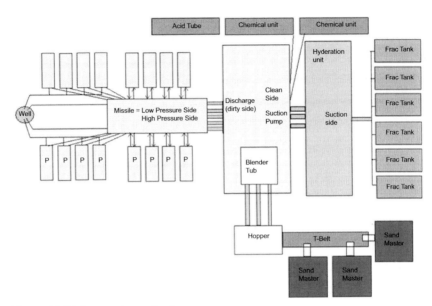

Figure 16.17 Frac equipment setup.

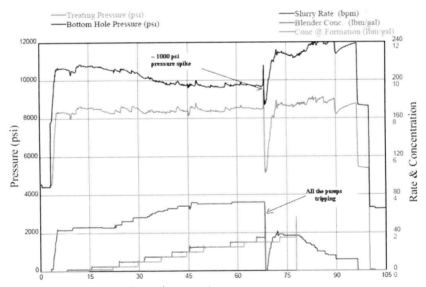

Figure 16.18 Pressure spike and pump trips.

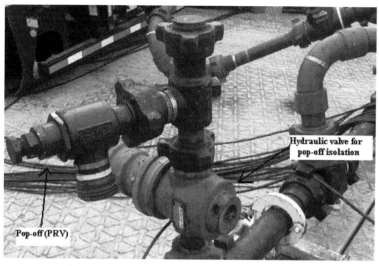

Figure 16.19 Mechanical pop-off.

casing, and equipment in the event pump trips fail or as a secondary safety preventative. PRVs can be easily set to any specific pressure and will go off as soon as that particular set pressure is reached. For example, if PRV is set at 9900 psi during the frac job, PRVs will go off and release the pressure as soon as 9900 psi is reached. The first type of PRV used during a hydraulic frac job is referred to as mechanical pop-off. Mechanical pop-offs have been known to malfunction. Therefore, there are new patent technologies that have been introduced that guarantee the activation of pop-offs at a certain pressure, and have been replacing conventional mechanical pop-offs. It is crucial to stay out of the pop-off area (danger zone) when pops are going off due to the release of pressure into the atmosphere. It is recommended to rig up the frac iron from the pop-offs to the flowback tank to prevent releasing any kind of pressure to the atmosphere as a safety precaution. Fig. 16.19 shows a mechanical pop-off.

Pressure Transducer

Pressure transducers are used to measure the surface-treating pressure during frac jobs. Each frac pump has a pressure transducer on the discharge side. In addition, real-time surface-treating pressure is obtained via the transducer located on the main line. Pressure transducers are usually covered with a plastic cover to prevent them from getting wet during summer or winter.

Figure 16.20 Pressure transducer.

Figure 16.21 Pressure transducer on a pump.

Some pressure transducers are known to yield inaccurate readings when wet. Figs. 16.20 and 16.21 are two examples of pressure transducers.

Check Valves and Manual Valves

Flapper or dart check valves are very commonly used in frac operations. Every frac pump used during the frac job needs to have a check valve and a manual valve on the discharge side. Check and manual valves located on the discharge side of the pump provide isolation between the pump and the rest of the equipment and iron. For example, if a frac iron located before the check valve on a frac pump parts (comes apart) during the live frac stage treatment, fluid will take the path of least resistance. Without check and manual valves, pumps cannot be isolated and all of the pumps have to come offline to fix the problem. When all of the pumps have to come offline to fix the problem without being able to flush the well, a costly screen-out could be the consequence. Check valves located on the discharge side of the pump isolate the pumps from the rest of the equipment so the frac operation can continue in the event of any leakage on the iron. In some occasions, flapper or dart-type check valves fail due to pumping millions of pounds of abrasive fluid. In those particular events, the manual valve located after the check valve can be closed to continue the operation without having to come offline. Note that some companies do not have a manual valve (wheel valve) after the check valve. Having a manual valve right after the check valve on each pump is highly recommended in the event the first check valve fails. Check valves fail and leak all the time. Therefore, it is very important to have a manual valve right after the check valve on each pump.

WATER COORDINATION

A frac operation is an enormous operation that needs lots of organization and coordination from both service and operating companies. As previously mentioned, each frac stage uses lots of water, proppant, and chemicals. To give some perspective on the amount of water used during slick water frac jobs, it is important to note that every two stages that the industry pumps on average is equivalent to one Olympic-sized swimming pool, which has a capacity of approximately 15,724 BBLs. One of the most challenging parts of the frac operation is obtaining sufficient water needed for the frac job. Since millions of gallons of water will be pumped per stage, each operating company usually has a water group to make sure a water transfer plan is scheduled, known, and determined before the actual operation begins to minimize any downtime. Water is essentially pumped into the pit, above-ground storage tank, or tank battery via PVC or poly lines. From that point water is transferred to onsite working tanks via polyvinyl chloride (PVC) or poly lines. The rate at which water is transferred into the in-ground pit, above-ground storage tank, or tank battery depends on the amount of water used per stage and the number of stages to be completed per day. For example, if six frac stages are estimated to be pumped per day and each frac stage uses approximately 8000 BBLs of water,

48,000 BBLs of water must be transferred and pumped to maintain the pit level at all times. This means water needs to be pumped at approximately 33.3 bpm to keep up with the frac job and avoid any nonproductive time (NPT). Therefore, water transfer and coordination are not as easy as they sound and require 24 hour supervision during the frac job to make sure a sufficient amount of water is available to be pumped every day.

SAND COORDINATION

Every frac stage can use anywhere between 100,000 and 700,000 lbs of proppant and it is very important to have an excellent sand coordinator in charge to make sure an adequate amount of proppant is delivered on time for a continuous frac operation to take place. Proppant is stored in sand masters and each sand master has a limited capacity. As a result, sand trucks are constantly delivering proppant during the frac job. It is the sand coordinator's responsibility to make sure proper size and type of proppant are placed into each sand master. No frac stage must be started until enough proppant is present on location to pump a stage. For example, if the designed proppant stage is 300,000 lbs and there are only 230,000 lbs on location and two sand trucks are estimated to be on location at any moment, it is not a good practice to start a stage without having the total amount of designed proppant available. There is always the possibility of sand trucks breaking down or delaying for various reasons. Therefore, until the total amount of proppant is present in the sand masters, no stage must be started.

CHEMICAL COORDINATION

Having a sufficient amount of chemicals per stage is another important aspect of the hydraulic frac operation. Eq. (16.3) is used to find the amount of each chemical used during each stage. Based on the chemical usage per stage, more chemicals can be ordered and coordinated throughout the day.

$$\text{Chemical needed per stage} = 0.042 \times \text{chemical concentration}$$

$$\times \text{clean volume}$$

Equation 16.3 Amount of chemical needed per stage

Chemical needed/stage = gallons

Chemical concentration = Chemical concentration that will be pumped during the stage, gpt

Clean volume = Total estimated clean volume per stage.

Example:

How many gallons of FR are needed if 8000 BBLs of water are estimated to be pumped in a stage at 1.5 gpt FR concentration?

$$FR \text{ (gallons)} = 0.042 \times 1.5 \times 8000 = 504 \text{ gallons of FR}$$

The person in charge of chemical coordination needs to have a sufficient amount of chemicals onsite for at least twice the volume of the calculated number. In addition, the number of hours of downtime can be completely prevented by performing a simple calculation. Time is money in the oil and gas industry and whenever there is downtime due to lack of water, sand, or chemical coordination, money is lost. NPT caused by the service company has to be reported and recorded for the end-of-the-year evaluation and continuous improvement.

STAGE TREATMENT

As previously mentioned, a frac operation is big, stressful, exciting, and live. One of the main aspects of hydraulic fracturing is that typically each stage treats differently. This makes hydraulic fracturing an interesting operation. Formal education and understanding the theories definitely help as far as visualizing and estimating the formation treatment. However, the most important aspect of the hydraulic fracturing job is experience. This is the main reason the majority of companies across the United States hire very experienced people to be in charge of the operation. Some companies hire two people to be in charge of a slick water frac operation due to the extent of the job.

TIPS FOR FLOWBACK AFTER SCREENING OUT

An important tip known in the field is to avoid a screen-out for the first few stages because there is not enough energy downhole to have a successful flowback after screening out. Typically when a well screens out on the first few stages, the possibility of a successful flowback is very low. The energy downhole needed to clear the wellbore from proppant when flowing back is not available during very early screen-outs. Without this energy, it is not possible to successfully flow a well back. Not being able to flow a well back is very costly and time consuming. It usually takes at least a day to rig up (R/U) coiled tubing, perform a clean out run, and rig down (R/D) coiled tubing. In some instances, coiled tubing cannot reach all the way to the bottom depth where the screen-out occurred (from drag model analysis). Therefore, a snubbing unit has to be used to perform a clean-out run. Thus, due to time and expense, special care must be taken to avoid the possibility of screening out on the first few stages or stages (depths) where reaching coiled tubing is not possible in long lateral wells. The industry has been moving toward drilling longer lateral wells (lateral lengths in excess of 8000 ft) since drilling longer lateral wells is significantly better from an economic perspective, as long as insignificant or no detrimental damage is observed in production results.

When screened out, it is very important to flow the well back within minutes to prevent the proppant settling in the heel. For a safe and efficient flowback operation to take place, it is important to have a safety and operational meeting with the flowback crew on a daily basis. This way the crew knows their responsibility and the company representative's expectations when screening out. It is not recommended to have a meeting after screening out because flowback needs to start within minutes of a screen-out for a successful flowback operation.

The idea behind flowing a well back after screening out is to have a balance of enough flowback rate while not pulling more proppant from the formation and previous zones into the wellbore. Flowing a well back in different formations varies. However, it is recommended to flow back at 8−10 bpm when the flowback is taking place through a 5 ½" casing and 5−7 bpm through a 4 ½" casing. Flowing the well hard is the best way to clean the wellbore from sand but as previously mentioned getting sand from the formation and previous stages must be minimized and avoided. When slick water is used, a minimum of two hole volumes (plug depth) needs to be flowed back before attempting an injection test. In cross-linked gel frac, 1 ½ hole volume is probably sufficient before attempting to perform an injection test.

Flowback tanks and lines must be rigged up in a manner that is ready to accept high fluid rates and large sand volumes without having to shut-in during flowback for any adjustments. Flow back after screening-out is like the cementing operation. Once the operation starts, it is to be continued without any stoppage unless there is an absolute emergency. The reason being is that having to shut-in during flowback truly jeopardizes the chances of success. A sufficient number of flowback tanks must be available to flow the well back 2–4 hole volumes at a high rate. If only one gas-buster tank is available, there needs to be a transfer line (such as poly line) accessible to pump out the flowback fluid to an existing pit. Essentially it is very important to have enough room available to flow a well back without having to shut-in. If more than one flowback tank is used, equalizing hoses must be high enough on the tank to avoid getting plugged up with high volumes of sands.

After flowing the well back for a minimum of two hole volumes, returns must be monitored to make sure sand is not being recovered anymore. For accurate volume measurements, periodic straps must be taken from the tanks to make sure proper flowback rate is obtained.

Post–Screen-out Injection Test

An injection test after screening out requires patience and experience. The key to success in injection testing is taking sufficient time and fluid to slowly increase rate. Increasing the rate rapidly as pressure drops dramatically has proven to be unsuccessful in many different shale plays. Increasing rate quickly causes too much sand to be picked up at a time and as a result sending sand slug at lower rate and plugging off the perforations.

Below is the recommended post–screen-out injection test procedure:

1. Roll over a pump truck at the lowest possible rate (usually 1.5–2 bpm depending on the pump).

2. Once pressure is stabilized, increase the rate to 3–4 bpm by pumping 5 lbs linear gel and 1–2 gpt FR. After about 100–150 BBLs, increase the gel to a 10-lb system for two full wellbore volumes. Having the 10-lb gel will help land the ball softly (if not retrieved during flowback) and prevent a dramatic pressure spike, which can cause all of the pumps to trip out.

3. Afterward, walk up the rate 1 bpm at a time for ½ to 1 full wellbore volume depending on the pressure reaction. For example, it is important not to exceed 8000 psi (if max pressure is 9500 psi) to have plenty of room for pressure to increase and roll over. The name of the game is patience.

4. Do not increase rate more than 2 bpm at a time even if pressure decreases dramatically. As previously mentioned, grabbing too much rate at a time causes too much sand to be picked up at one time and may cause failure.

5. After reaching about 10–12 bpm, cut gel and keep the FR running. Hold rate until all of the pumped gel clears the perforations. Continue working the rate up very slowly as pressure allows until 30–35 bpm is reached; 30–35 bpm is the rule of thumb and the desired rate among operating companies in conventional plug-and-perf operations to ensure clean wellbores and pumping wireline down for the next stage.

6. Once 30–35 bpm is reached, pump 100–200 BBL gel sweep until the sweep clears the perforations.

A typical post–screen-out injection test should take at least 5 hours and can be as long as 10–15 hours depending on the measured depth of the screened-out stage.

FRAC WELLHEAD
Tubing Head (B Section)

The tubing head is one of the main components of a wellhead, which is placed after the drilling process is over and before frac operations start. Tubing head is used to land the production tubing, and the back-side pressure (casing pressure) can be monitored throughout the life of the well via the wing valves located on the tubing head. The tubing head also provides the means of attaching the Christmas tree to the wellhead. The casing head is referred to as the "A section," the tubing head is called the "B section," and the Christmas tree is referred to as the "C section." Fig. 16.22 shows a tubing head with production tubing hung inside the tubing hanger.

Lower Master Valve (Last Resort Valve)

The lower master valve is used to control the flow of fluid from the wellbore and is located directly above the tubing head. This valve is the last valve that needs to be operated as the last chance in the event that all of the primary well-control barriers fail. It is very crucial to perform bolt check and other visual inspections on this and all of the other valves

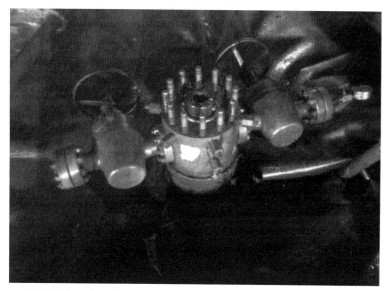

Figure 16.22 Tubing head with production tubing hung inside the tubing hanger.

frequently during the frac job. During the frac job operation, a lot of proppant is being pumped downhole at a high pressure and rate, which can cause the bolts or threads to gradually come loose. To prevent this issue, regular visual inspections on all of the valves are recommended and a must. In addition, every valve has a certain number of turns to open or close provided by the manufacturer. The most important valve in the frac operation is the lower master valve. If this valve washes out (by pumping abrasive fluid) during the frac job, there is a major well-control issue. If this valve starts leaking, wireline needs to be stabbed on the wellhead as soon as possible to set a kill plug inside the casing to control the flow of fluid before it gets worse. If the valve washout is uncontrollable, the site needs to be evacuated as soon as possible and a well-control company should be called to control the live well.

Hydraulic Valve

The hydraulic valve is another important valve during frac jobs because in the event of an emergency during the frac operation, this valve can be hydraulically closed from the accumulator, which has to be located at least 100' from the wellhead for safety purposes. Some operators do not use a hydraulic valve to save cost. The use of a hydraulic valve is recommended in the event of an emergency situation where a frac line close to the

Figure 16.23 Hydraulic valve.

wellhead comes apart (parts). Please note that the hydraulic valve is only designed to close on open wells with no tubing or wireline in the hole. The hydraulic valve will easily shear wireline in the hole even though it is not designed for this type of application. Therefore, it is important to label this valve accurately to prevent cutting wireline by accident during a plug-and-perf operation. Fig. 16.23 shows the hydraulic valve.

Flow Cross

The flow cross is located above the hydraulic valve and in the event of a screen-out, the well can be flowed back through either side of the flow cross. One of the main applications of a flow cross is to flow a well back after screening out during frac jobs. Flow cross can also be used to pump down wireline in zipper frac operations. In a zipper frac operation, since one well is being fracked while the other one is being perforated, one side of the flow cross could be used to pump the wireline down to the desired depth. Any line that comes off of the flow cross must have an ESD, which stands for emergency shut-down valve. The ESD needs to be function tested or cycled before the operation begins. It is absolutely necessary to rig up the ESD in the event of an emergency or parting iron at surface. In those emergency situations, ESD is automatically or hydraulically closed without approaching the wellhead. There are two types of ESD valves used in the industry. The first one is referred to as pneumatic ESD, which will automatically close once a substantial pressure drop has occurred during the flowback. The second type of ESD, which is not as commonly used as a pneumatic one, is referred to as hydraulic ESD. Hydraulic ESD will not automatically shut itself down in the event of an

Figure 16.24 Flow cross, 2" and 4" sides.

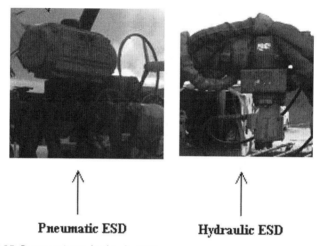

Pneumatic ESD   **Hydraulic ESD**

Figure 16.25 Pneumatic vs. hydraulic ESD.

emergency and the small accumulator located away from the well has to be used. The ESD accumulators are recommended to be placed 100' from the wellhead. Fig. 16.24 shows a flow cross used as part of the frac wellhead during frac jobs. Fig. 16.25 shows pneumatic vs. hydraulic ESD.

Manual Valve (Upper Master Valve, Frac Valve, Swab Valve, Top Valve)

The manual valve is also referred to as the upper master valve, frac valve, swab valve, or top valve, and is located above the flow cross. This valve is typically the main valve used for opening and closing the well during a frac job. For example, after each frac stage is completed, the manual valve is closed and the pressure above the manual valve is bled off. If this valve fails or leaks, a hydraulic valve or manual valve can be used to close the well and pressure can be bled off above the hydraulic valve while the manual valve can get greased, fixed, or replaced. Finally, if the hydraulic

Figure 16.26 Manual valve.

Figure 16.27 Four-way entry frac head (goat head).

valve fails as well, the last resort (lower master valve) will be used to close the well (Fig. 16.26).

Frac Head (Goat Head)

The frac head, also referred to as the goat head, is located above the manual valve and typically 2–6 frac irons (referred to as candy canes) are connected to the goat head for the frac job. The goat head is the head of the frac job and is the means of injecting water, sand, and chemicals at high pressure and rate into the well. During slick water frac jobs, high rate is

Figure 16.28 Typical frac wellhead.

required to pump each stage; therefore, to get to the desired rate, two to six 4" or 3" lines are rigged up to the goat head to obtain the desired rate. The main application of frac head (goat head) in frac and perforation operation is being able to R/U 2−6 lines to achieve the designed rate. The rule of thumb for obtaining the maximum rate through each line is $OD^2 \times 2$. For instance, one 4" line yields a maximum rate of 32 bpm; therefore, four 4" lines yield a maximum rate of 128 bpm. In cross-linked fluid system jobs, not much rate is required. Therefore, there is no need to rig up four 4" lines. Essentially, the rig-up for each frac type and formation is different. If the designed rate for a cross-linked job is 50 bpm, only two 4" lines are required to perform the job. Fig. 16.27 is a four-way entry frac wellhead used during frac jobs. Fig. 16.28 illustrates a typical wellhead configuration during the frac job.

CHAPTER SEVENTEEN

Decline Curve Analysis

INTRODUCTION

Economic analysis is one of the most important aspects of any sales, acquisition, drilling, and completions design in any of the shale plays across North America. In fact, it is so important that the decision to complete a well using new technology will solely depend on the economics of the well. The first step in performing any type of economic analysis is to forecast the expected production volumes with time. Forecasting the production volumes and well behavior with time can be quite challenging especially in new exploration areas or areas with limited production data. In this chapter of the book, primary methods for determining production volumes with time will be discussed.

DECLINE CURVE ANALYSIS (DCA)

Decline curve analysis (DCA) is used to predict the future production of oil and gas, and it has been widely used since 1945. Arnold and Anderson (1908) presented the first mathematical model of DCA. Cutler (1924) also used the log—log paper to obtain a straight line for hyperbolic decline, so the curve shifted horizontally. Larkey (1925) proposed the least square method to extrapolate the decline curves. Pirson (1935) proposed the loss ratio method and concluded that the production decline curve rate/time has a constant loss ratio. Furthermore, Arps (1944) categorized the decline curve using the loss ratio method, and he then defined the rate/time and rate/cumulative production. He defined three types of decline curve models: exponential, harmonic, and hyperbolic decline curves. The hyperbolic decline curve can be considered as a general model, and exponential and harmonic decline curves can be derived from it. The decline curve consists of three parameters [q_i, D_i, and b] that could

Hydraulic Fracturing in Unconventional Reservoirs
DOI: http://dx.doi.org/10.1016/B978-0-12-849871-2.00017-4
© 2017 Elsevier Inc.
All rights reserved.

be found from production data. Furthermore, the following differential equation was used to define the three decline curve models:

$$d = -\frac{1}{q}\frac{dq}{dt} = Kq^b$$

Equation 17.1 Decline curve differential equation

$b = $ Hyperbolic decline exponent

$K = $ Proportionality constant.

In Eq. (17.1) "d" is called decline factor that is a slope of the natural log of production rate versus time. The decline curve equations assume that production decline is proportional to reservoir pressure decline. In addition, conventional DCA assumes constant flowing bottom-hole pressure, drainage area, permeability, skin, and existence of boundary-dominated flow. Most of these assumptions are not valid in unconventional shale reservoirs. The reason DCA is still widely used is because it is an easy and quick tool to estimate production decline (rate) with time on producing and nonproducing wells. In today's business model, DCA drives the business by providing near- and long-term production forecasts and booking economic reserves. In fact, various forms of DCA are taught in short courses for reserve booking and estimation. Other tools such as rate transient analysis (RTA) and numerical simulation can also be used to forecast the future behavior of wells.

ANATOMY OF DECLINE CURVE ANALYSIS

There are a few crucial parameters used in DCA that are as follows:

Instantaneous Production (IP)

Instantaneous production (IP) is measured in MSCF/D or BBL/D. IP is often mistaken for 24-hour initial production. However, IP in DCA refers to the instantaneous production rate at a point in time that the well has been able to reach.

Nominal Decline (D_i)

Nominal decline is the instantaneous slope of the decline.

Effective Decline (D_e)

Effective decline is the percentage change in flow rate over a time interval. Effective decline is usually calculated 1 year from time zero. For

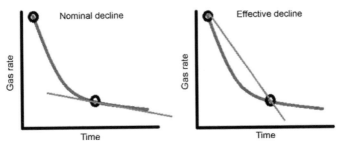

Figure 17.1 Nominal versus effective decline.

example, if the IP of a well is 15 MMscf/D and after 1 year, the flow rate is 4 MMscf/D, the effective decline is 73.3%. Effective decline is the percentage reduction in production volumes over a 1-year period. The lower the nominal decline, the less it varies from an effective decline. Effective decline is defined in Eq. (17.2). Fig 17.1 illustrates the difference between nominal and effective decline.

$$\text{Effective decline } (D_e) = \frac{q_i - q}{q_i}$$

Equation 17.2 Effective decline

Hyperbolic Exponent (*b*)

The *b* value is referred to as the hyperbolic exponent and reduces effective decline over time. Hyperbolic exponent is the rate of change of the decline rate with respect to time. In other words, hyperbolic exponent is the second derivative of the production rate with respect to time. Fig. 17.2 illustrates that as *b* value increases, the rate of deceleration of effective decline increases as well. In addition, as effective decline decreases, *b* value will have less of an impact.

Shape of the Decline Curve

The most important parameter in DCA that drives the shape of the decline curve is the *b* value. Table 17.1 shows the approximate range of *b* values for various reservoir drive mechanisms. As can be seen, unconventional shale and tight gas reservoirs typically have a *b* value in excess of 1 due to the long transient period caused by low permeability. In conventional reservoirs that have hyperbolic decline, *b* is typically between 0 and 1 depending on the reservoir drive mechanism.

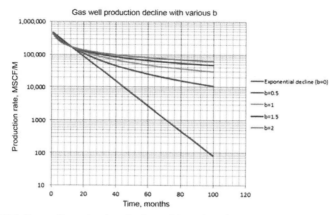

Figure 17.2 Gas well production decline with various *b*.

Table 17.1 Reservoir Drive Mechanism Versus *b* Values

b Value	Reservoir Drive Mechanism
0	Single-phase liquid expansion (oil above bubble point)
0.1−0.4	Solution gas drive
0.4−0.5	Single-phase gas expansion
0.5	Effective edge water drive
0.5−1	Layered reservoirs
>1	Transient (tight gas, shales)

Before discussing different types of decline curves, it is very important to understand different well-flow behaviors in multistage horizontal fracs.

Unsteady (Transient Flow) State Period

Transient flow is observed in unconventional shale reservoirs with low permeability (<0.1 md) and it is the time period when reservoir boundaries have no effect on pressure behavior. The reservoir acts as if it is infinite in size. Wellbore storage effect does take place during this period. In general, transient flow is defined as the pressure pulse traveling through the reservoir without any interference by the reservoir boundaries.

Late Transient Period

This is the period of time that separates the transient state from the steady or pseudosteady state. It is when the well drainage radius has reached some parts of the reservoir boundaries.

Pseudosteady State (Boundary-Dominated Flow)

Pseudosteady state occurs when there is boundary-dominated flow and the transient period ends. The boundary-dominated flow is a flow regime that starts when the drainage radius of the well reaches the reservoir boundaries. Boundary-dominated flow is a late-time flow behavior when the reservoir is in a state of pseudoequilibrium. One of the aspects that makes the unconventional shale production analysis quite challenging is that the flow stays in transient mode for a very long period of time. As a result, determination of fracture geometry is difficult from modern production analysis such as rate transient analysis.

PRIMARY TYPES OF DECLINE CURVES

There are three primary types of decline curves and they are as follows:

Exponential Decline

When production rate (y-axis) versus time (x-axis) is plotted on a semi-log plot, the plot will be a straight line or exponential. In exponential decline, b is equal to 0. Exponential decline is also known as "constant-rate" decline. Exponential decline has two terms. The first term is the initial production rate (IP) and the second one is the decline rate. Decline rate in exponential decline refers to the rate of change of production with time, which stays constant.

Hyperbolic Decline

When production rate (y-axis) versus time (x-axis) is plotted on a semi-log plot, the plot will be a curved line. Hyperbolic decline has three terms. The first term is referred to as the initial production rate (IP), the second term is initial decline rate @ initial production rate, and finally the third term is hyperbolic exponent or b value. In hyperbolic decline as opposed to exponential decline (where the decline rate stays constant with time), the decline rate decreases as a function of the hyperbolic exponent with time. This is because the data shows a hyperbolic behavior on a semi-log plot. Hyperbolic decline rate varies and is typically 40–80% depending on many factors such as reservoir pressure, reservoir characteristics, completions characteristics, pressure drawdown strategy, etc. The decline rate depends on the way a well is produced. The way in which a well is produced is just as important as the way in which a well is completed. The harder that a well is produced (higher pressure drawdown), the sharper the decline rate. The slower that a well is produced (minimizing pressure drawdown), the shallower the decline rate. For example, two wells side by side with the exact same completions design and formation properties can have varying decline rates depending on the way in which each well was produced. The well with higher pressure drawdown will have a higher decline rate (e.g., 85%) and the well with lower pressure drawdown will have a lower decline rate (e.g., 55%). For comparison reasons between wells, it is very important to produce all wells in the same manner operationally. Figs. 17.3 and 17.4 are examples of exponential and hyperbolic declines.

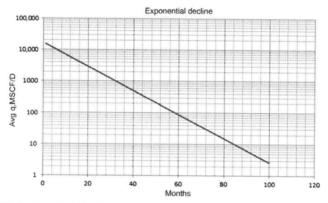

Figure 17.3 Exponential decline.

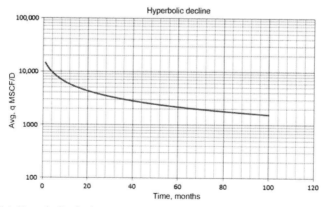

Figure 17.4 Hyperbolic decline.

Harmonic Decline

Harmonic decline occurs when the b value is equal to 1 and the decline rate of change is constant.

Modified Hyperbolic Decline Curve (Hybrid Decline)

With the development of unconventional shale reservoirs, choosing only hyperbolic decline could cause an overestimation of estimated ultimate recovery (EUR). This is because hyperbolic decline without limit tends to overestimate cumulative production during the life of a well. In an attempt to account for this, modified hyperbolic decline is typically used in unconventional shale reservoirs and reserve booking. Reserve engineers will typically transition a decline curve to an exponential decline to compensate for this overestimation. The transition to an exponential decline in later stages of production is called the terminal decline. Terminal decline is the rate at which the hyperbolic decline switches from hyperbolic to exponential decline. For example, if the initial D_e (annual effective decline) for a Haynesville Shale well is 65%, once D_e reaches around 4–11%, the hyperbolic decline switches over to exponential decline. Determination of terminal decline for reservoirs that have not produced long enough is very challenging. Companies typically assume the terminal decline to be anywhere between 4% and 11%. The higher the terminal decline, the faster the transition from hyperbolic to exponential and as a result, lower EUR. In areas with limited production data, higher terminal decline is assumed to be conservative. Fig. 17.5 shows the difference between a hyperbolic decline and a modified hyperbolic decline in which terminal decline is assumed to be 5%. As can be seen, as soon as annual effective decline (D_e) reaches 5%, hyperbolic decline switches to exponential for the remaining life of the well (50 years in this example).

Other DCA Techniques

There are other types of DCA techniques that were developed recently. Some of those techniques are as follows:
1. **Power law exponential decline model (PLE):**
    Power law exponential (PLE) decline was developed by Ilk et al. (2008) by modifying Arps' exponential decline (Seshadri and Mattar, 2010). This methodology was developed specifically for tight gas wells

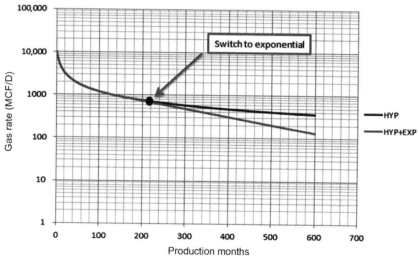

Figure 17.5 Hyperbolic versus modified hyperbolic decline.

to model the decline in a transient period of production data. The PLE decline model is defined in Eq. (17.3) (McNeil et al., 2009):

$$q = q_i e^{\left[-D_\infty t - \frac{D_1}{n} t^n\right]}$$

Equation 17.3 Power law exponential decline model

The equation above can be reduced to power law loss ratio as defined below:

$$q = q_i e^{\left[-D_\infty t - D_i t^n\right]}$$

In Eq. (17.3) "D_1" is the decline constant at specific time such as 1 day, D_∞ is the decline constant at infinite time, D_i is the initial decline rate % per year, and n is the time exponent. The PLE method does not consider the b value as a constant value but as a declining function in contrast to the Arps method. Moreover, by using the PLE model it is easier to match the production data in transient and boundary-dominated regions without overestimating the reserve (McNeil et al., 2009).

2. **Stretched exponential:**

A newer variation of the Arps model adding a bounding component to provide a limit on EUR was developed by Valkó in 2009. Stretched exponential rate time relationship is defined as follows:

$$q = q_i \exp\left[-\left(\frac{t}{\tau}\right)^n\right]$$

Equation 17.4 Stretched exponential decline

τ is a characteristic time for stretched exponential and n is the time exponent. This technique is similar to the PLE model, however, it ignores the behavior at late times. Stretched exponential has an advantage over PLE by providing the cumulative time relation as follows:

$$Q = \frac{q_i \tau}{n} \left\{ \Gamma\left[\frac{1}{n}\right] - \Gamma\left[\frac{1}{n}, \left(\frac{t}{\tau}\right)^n\right] \right\}$$

Equation 17.5 Cumulative time relationship

3. Duong decline:

Duong (2011) developed the rate decline analysis for fractured shale reservoirs. In this model, the long-term linear flow was taken into consideration. This model is defined based on Eq. (17.6).

$$q(t) = q_i t(a, m) + q_\infty$$

Equation 17.6 Duong decline model

Parameters "a" and "m" are determined by using Eq. (17.7):

$$\frac{q}{G_p} = at^{-m}$$

Equation 17.7 Determination of parameters a and m in Duong decline model

q = Flow rate, volume/time
a = Intercept (log−log plot of $\frac{q}{G_p}$ vs t)
G_p = Cumulative gas production.

Furthermore, a plot of "q" versus "t (a, m)" should provide a straight line with a slope of q_1 and intercept of q_∞:

$$t(a, m) = t^{-m} \exp\left[\frac{a}{1 - m}(t^{1-m} - 1)\right]$$

Note that q_∞ can be positive, zero, or negative depending on the operating conditions. A cumulative gas production can be determined using "q_∞" is equal to zero as:

$$G_p = \frac{q_1 t(a, m)}{at^{-m}}$$

Duong examined different types of wells such as tight, dry, and wet gas to prove the accuracy of his model. He also found that most of the shale models have "a" values ranging from 0 to 3 and m values ranging from 0.9 to 1.3. His model yields reasonable estimation of cumulative production compared to the power law and Arps models.

 ## ARPS DECLINE CURVE EQUATIONS FOR ESTIMATING FUTURE VOLUMES

As previously discussed, nominal decline is simply the conversion of the effective decline.

Exponential Decline Equations

Nominal decline as a function of effective decline is written in Eq. (17.8).

$$D = -\ln\left[(1-D_e)^{\frac{1}{12}}\right]$$

Equation 17.8 Monthly nominal exponential

D = Monthly nominal exponential, 1/time
D_e = Annual effective decline, 1/time.
Exponential decline rate equation can also be written in Eq. (17.9).

$$q_{\text{exponential}} = \text{IP} \times e^{-D \times t}$$

Equation 17.9 Exponential decline rate

$q_{\text{exponential}}$ = Exponential decline rate, MSCF/D
IP = Initial production (instantaneous rate), MSCF/D
D = Monthly nominal exponential, 1/time
t = Time, months.

Exponential example: Calculate the production rate of an exponential decline at the end of two years if IP is 800 MSCF/D and annual exponential effective decline (D) is 6%.

$$D = \text{Monthly nominal exponential} = -\ln\left[(1-D)^{\frac{1}{12}}\right] = -\ln(1-6\%)^{\frac{1}{12}} = 0.515\%$$

$$\text{Exponential } q @ \text{ the end of 2 years} = \text{IP} \times e^{-D \times t} = 800 \times e^{-0.515\% \times 24}$$

$$= 707 \text{ MSCF/D}$$

Hyperbolic Decline Equations

Initial decline rate can be defined in three ways for hyperbolic decline. Nominal, tangent effective, and secant effective decline equations can be used in defining initial decline rate. Secant effective decline is the preferred methodology used in unconventional shale reservoirs. Fig. 17.6 illustrates the difference between secant and tangent effective decline rates.

Nominal decline as a function of tangent effective decline is written in Eq. (17.10).

$$D_{i,\text{tangent}} = -\ln\left[(1 - D_{ei})^{\frac{1}{12}}\right]$$

Equation 17.10 Monthly nominal tangent hyperbolic

$D_{i,\text{ tangent}}$ = Monthly nominal tangent hyperbolic, 1/time
D_{ei} = Initial annual effective decline rate from tangent line, 1/time.

Nominal decline as a function of secant effective decline can be written as shown in Eq. (17.11).

$$D_{i,\text{secant}} = \left[\frac{1}{12b}\right] \times \left[(1 - D_{eis})^{-b} - 1\right]$$

Equation 17.11 Monthly nominal secant hyperbolic

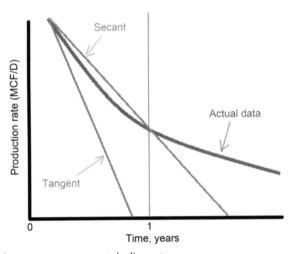

Figure 17.6 Secant versus tangent decline rates.

$D_{i,secant}$ = Monthly nominal secant hyperbolic, 1/time
D_{eis} = Initial annual effective decline rate from secant line, 1/time
b = Hyperbolic exponent.
Secant effective decline rate is calculated from two rates. The first rate is at time 0 and the second rate is exactly after 1 year.
The hyperbolic decline rate equation is shown in Eq. (17.12).

$$q_{hyperbolic} = IP \times (1 + b \times D_i \times t)^{-\frac{1}{b}}$$

Equation 17.12 Hyperbolic decline rate

$q_{hyperbolic}$ = Hyperbolic decline rate, MSCF/D
IP = Initial production (instantaneous rate), MSCF/D
D_i = Monthly nominal hyperbolic, 1/time
b = Hyperbolic exponent
t = Time, months.

Hyperbolic example: Calculate the production rate of a hyperbolic decline at the end of 2 years with an IP of 8000 MSCF/D, initial annual secant effective hyperbolic decline (D_{eis}) of 66%, and b value of 1.3.

$$D_i = \left[\frac{1}{12b}\right] \times \left[(1 - D_{eis})^{-b} - 1\right] = \left[\frac{1}{12 \times 1.3}\right] \times \left[(1 - 66\%)^{-1.3} - 1\right] = 19.65\%$$

$$q = IP \times (1 + b \times D_i \times t)^{-\frac{1}{b}} = 8000 \times (1 + 1.3 \times 19.65\% \times 24)^{-\frac{1}{1.3}}$$

$$= 1765 \, MSCF/D$$

Monthly hyperbolic cumulative volume:
Monthly hyperbolic cumulative volume can be calculated using Eq. (17.13).

$$N_p = \left\{ \left[\frac{IP}{(1 - b) \times \text{Monthly Nominal Hyp}}\right] \right.$$

$$\left. \times \left[1 - (1 + b \times \text{Monthly Nominal Hyp} \times time)^{1 - \frac{1}{b}}\right] \right\} \times \frac{365}{12}$$

Equation 17.13 Monthly hyperbolic cumulative volume

N_p = Monthly hyperbolic cum volume, MSCF
IP = Initial production, MSCF/D
Time = months.

Example

Calculate monthly hyperbolic cumulative volume and monthly hyperbolic volume for the first 24 months assuming the following parameters:

IP = 10,500 MSCF/D, $b = 1.5$, Initial annual secant effective decline (D_{eis}) = 61%

Step 1) Calculate monthly nominal secant hyperbolic:

$$D_i = \left[\frac{1}{12b}\right] \times \left[(1-D_{eis})^{-b} - 1\right] = \left[\frac{1}{12 \times 1.5}\right] \times \left[(1-61\%)^{-1.5} - 1\right] = 17.25\%$$

Step 2) Calculate hyperbolic cumulative volume for each month starting with month 1:

$$N_p = \left\{\left[\frac{IP}{(1-b) \times \text{Monthly Nominal Hyp}}\right] \times [1-(1+b \times \text{Monthly Nominal Hyp} \times \text{time}]^{1-\frac{1}{b}}\right\}$$

$$\times \frac{365}{12}$$

$$N_{p,\text{month 1}} = \left\{\left[\frac{10,500}{(1-1.5) \times 17.25\%}\right] \times [1-(1+1.5 \times 17.25\% \times 1]^{1-\frac{1}{1.5}}\right\}$$

$$\times \frac{365}{12} = 295,208 \frac{MCF}{1 \text{ month}}$$

$$N_{p,\text{month 2}} = \left\{\left[\frac{10,500}{(1-1.5) \times 17.25\%}\right] \times [1-(1+1.5 \times 17.25\% \times 2]^{1-\frac{1}{1.5}}\right\}$$

$$\times \frac{365}{12} = 552,264 \frac{MCF}{2 \text{ month}}$$

$$N_{p,\text{month 3}} = \left\{\left[\frac{10,500}{(1-1.5) \times 17.25\%}\right] \times [1-(1+1.5 \times 17.25\% \times 3]^{1-\frac{1}{1.5}}\right\}$$

$$\times \frac{365}{12} = 781,523 \frac{MCF}{3 \text{ month}}$$

$$N_{p,\text{month 4}} = \left\{\left[\frac{10,500}{(1-1.5) \times 17.25\%}\right] \times [1-(1+1.5 \times 17.25\% \times 4]^{1-\frac{1}{1.5}}\right\}$$

$$\times \frac{365}{12} = 989,466 \frac{MCF}{4 \text{ month}}$$

$$N_{p,\text{month 5}} = \left\{\left[\frac{10,500}{(1-1.5) \times 17.25\%}\right] \times [1-(1+1.5 \times 17.25\% \times 5]^{1-\frac{1}{1.5}}\right\}$$

$$\times \frac{365}{12} = 1,180,447 \frac{MCF}{5 \text{ month}}$$

Step 3) Calculate monthly production volumes simply by subtracting each month's cumulative volume from the previous month:

$$\text{Hyperbolic rate for month 1} = 295,208 \frac{MCF}{month}$$

$$\text{Hyperbolic rate for month 2} = 552,264 - 295,208 = 257,056 \frac{MCF}{month}$$

$$\text{Hyperbolic rate for month } 3 = 781,523 - 552,264 = 229,259 \frac{MCF}{month}$$

$$\text{Hyperbolic rate for month } 4 = 989,466 - 781,523 = 207,943 \frac{MCF}{month}$$

$$\text{Hyperbolic rate for month } 5 = 1,180,447 - 989,466 = 190,981 \frac{MCF}{month}$$

Cumulative and monthly production volumes for 24 months are summarized in Table 17.2.

Table 17.2 Cumulative and Monthly Production Volumes Example

Time	CUM Volumes	Monthly Rate
Months	MCF	MCF/M
1	295,208	295,208
2	552,264	257,056
3	781,523	229,259
4	989,466	207,942
5	1,180,447	190,981
6	1,357,553	177,106
7	1,523,059	165,506
8	1,678,695	155,639
9	1,825,814	147,119
10	1,965,493	139,679
11	2,098,606	133,113
12	2,225,875	127,269
13	2,347,902	122,027
14	2,465,196	117,293
15	2,578,189	112,994
16	2,687,257	109,068
17	2,792,723	105,466
18	2,894,871	102,148
19	2,993,949	99,079
20	3,090,179	96,230
21	3,183,757	93,578
22	3,274,859	91,101
23	3,363,641	88,782
24	3,450,246	86,605

As previously discussed, in modified hyperbolic decline once annual effective decline reaches terminal decline, the decline curve is switched from hyperbolic to exponential. Therefore, annual effective decline must be calculated monthly in an attempt to find the transition point from hyperbolic to exponential decline. Before being able to calculate the annual effective decline, monthly nominal decline must be calculated using Eq. (17.14).

$$\text{Monthly nominal decline}\,(D) = \frac{\text{Monthly nominal hyperbolic}}{1 + b \times \text{monthly nominal hyperbolic} \times \text{time}}$$

Equation 17.14 Monthly nominal decline

Monthly nominal decline $= 1/\text{time}$

Time $=$ months.

Afterward, annual hyperbolic effective decline can be calculated using Eq. (17.15).

$$\text{Annual effective decline} = D_e = 1 - (1 + 12 \times b \times D)^{-\frac{1}{b}}$$

Equation 17.15 Annual effective decline

$D =$ Monthly nominal decline, $1/\text{time}$.

Example

Calculate annual effective decline after 1, 5, 24, and 50 months subsequently if the initial annual secant effective decline is 85% with a b value of 1.3.

Step 1) Calculate monthly nominal hyperbolic:

$$D_i = \left[\frac{1}{12b}\right] \times \left[(1 - D_{eis})^{-b} - 1\right] = \left[\frac{1}{12 \times 1.3}\right] \times \left[(1 - 85\%)^{-1.3} - 1\right] = 69.09\%$$

Step 2) Calculate monthly nominal decline after 1, 5, 24, and 50 months:

$$D_{\text{month }1} = \frac{\text{Monthly nominal hyperbolic}}{1 + b \times \text{monthly nominal hyperbolic} \times \text{time}} = \frac{69.09\%}{1 + 1.3 \times 69.09\% \times 1} = 36.40\%$$

$$D_{\text{month }5} = \frac{69.09\%}{1 + 1.3 \times 69.09\% \times 5} = 12.6\%$$

$$D_{\text{month }24} = \frac{69.09\%}{1 + 1.3 \times 69.09\% \times 24} = 3.1\%$$

$$D_{\text{month }50} = \frac{69.09\%}{1 + 1.3 \times 69.09\% \times 50} = 1.5\%$$

Step 3) Calculate annual effective decline after 1, 5, 24, and 50 months:

$$D_{e,\text{month }1} = 1 - (1 + 12 \times b \times D)^{-\frac{1}{b}} = 1 - (1 + 12 \times 1.3 \times 36.40\%)^{-\frac{1}{1.3}} = 76.8\%$$

$$D_{e,\text{month } 5} = 1 - (1 + 12 \times 1.3 \times 12.6\%)^{-\frac{1}{1.3}} = 76.8\% = 56.6\%$$

$$D_{e,\text{month } 24} = 1 - (1 + 12 \times 1.3 \times 3.1\%)^{-\frac{1}{1.3}} = 76.8\% = 26.0\%$$

$$D_{e,\text{month } 50} = 1 - (1 + 12 \times 1.3 \times 1.5\%)^{-\frac{1}{1.3}} = 76.8\% = 15.0\%$$

●●●

Example

A well drilled in the Barnett Shale has a hyperbolic shape with an IP of 6500 MSCF/D, initial annual secant effective decline rate of 55%, and b value of 1.4. Calculate the monthly production rates along with annual effective decline for the first 12 months.

Step 1) Calculate monthly nominal secant hyperbolic:

$$D_i = \left[\frac{1}{12b}\right] \times \left[(1 - D_{eis})^{-b} - 1\right] = \left[\frac{1}{12 \times 1.4}\right] \times \left[(1 - 55\%)^{-1.4} - 1\right] = 12.25\%$$

Step 2) Calculate hyperbolic cumulative volume for each month starting with month 1:

$$N_p = \left\{\left[\frac{IP}{(1 - b) \times \text{Monthly Nominal Hyp}}\right] \times [1 - (1 + b \times \text{Monthly Nominal Hyp} \times \text{time}]^{1 - \frac{1}{b}}\right\}$$

$$\times \frac{365}{12}$$

$$N_{p,\text{month } 1} = \left\{\left[\frac{6500}{(1 - 1.4) \times 12.25\%}\right] \times [1 - (1 + 1.4 \times 12.25\% \times 1]^{1 - \frac{1}{1.4}}\right\}$$

$$\times \frac{365}{12} = 186,661 \frac{\text{MCF}}{1 \text{ month}}$$

$$N_{p,\text{month } 2} = \left\{\left[\frac{6500}{(1 - 1.4) \times 12.25\%}\right] \times [1 - (1 + 1.4 \times 12.25\% \times 2]^{1 - \frac{1}{1.4}}\right\}$$

$$\times \frac{365}{12} = 354,699 \frac{\text{MCF}}{2 \text{ month}}$$

$$N_{p,\text{month } 3} = \left\{\left[\frac{6500}{(1 - 1.4) \times 12.25\%}\right] \times \left[1 - (1 + 1.4 \times 12.25\% \times 3]^{1 - \frac{1}{1.4}}\right\}$$

$$\times \frac{365}{12} = 508,034 \frac{\text{MCF}}{3 \text{ month}}$$

$$N_{p,\text{month } 4} = \left\{\left[\frac{6500}{(1 - 1.4) \times 12.25\%}\right] \times [1 - (1 + 1.4 \times 12.25\% \times 4]^{1 - \frac{1}{1.4}}\right\}$$

$$\times \frac{365}{12} = 649,420 \frac{\text{MCF}}{4 \text{ month}}$$

$$N_{p,\text{month }5} = \left\{ \left[\frac{6500}{(1-1.4) \times 12.25\%} \right] \times \left[1 - (1+1.4 \times 12.25\% \times 5)^{1-\frac{1}{1.4}} \right] \right.$$

$$\times \frac{365}{12} = 780,873 \frac{\text{MCF}}{5\,\text{month}}$$

$$N_{p,\text{month }6} = \left\{ \left[\frac{6500}{(1-1.4) \times 12.25\%} \right] \times \left[1 - (1+1.4 \times 12.25\% \times 6)^{1-\frac{1}{1.4}} \right] \right.$$

$$\times \frac{365}{12} = 903,921 \frac{\text{MCF}}{6\,\text{month}}$$

$$N_{p,\text{month }7} = \left\{ \left[\frac{6500}{(1-1.4) \times 12.25\%} \right] \times \left[1 - (1+1.4 \times 12.25\% \times 7)^{1-\frac{1}{1.4}} \right] \right.$$

$$\times \frac{365}{12} = 1,019,747 \frac{\text{MCF}}{7\,\text{month}}$$

$$N_{p,\text{month }8} = \left\{ \left[\frac{6500}{(1-1.4) \times 12.25\%} \right] \times \left[1 - (1+1.4 \times 12.25\% \times 8)^{1-\frac{1}{1.4}} \right] \right.$$

$$\times \frac{365}{12} = 1,129,292 \frac{\text{MCF}}{8\,\text{month}}$$

$$N_{p,\text{month }9} = \left\{ \left[\frac{6500}{(1-1.4) \times 12.25\%} \right] \times \left[1 - (1+1.4 \times 12.25\% \times 9)^{1-\frac{1}{1.4}} \right] \right.$$

$$\times \frac{365}{12} = 1,233,317 \frac{\text{MCF}}{9\,\text{month}}$$

$$N_{p,\text{month }10} = \left\{ \left[\frac{6500}{(1-1.4) \times 12.25\%} \right] \times [1 - (1+1.4 \times 12.25\% \times 10)^{1-\frac{1}{1.4}} \right.$$

$$\times \frac{365}{12} = 1,332,445 \frac{\text{MCF}}{10\,\text{month}}$$

$$N_{p,\text{month }11} = \left\{ \left[\frac{6500}{(1-1.4) \times 12.25\%} \right] \times [1 - (1+1.4 \times 12.25\% \times 11)^{1-\frac{1}{1.4}} \right.$$

$$\times \frac{365}{12} = 1,427,196 \frac{\text{MCF}}{11\,\text{month}}$$

$$N_{p,\text{month }12} = \left\{ \left[\frac{6500}{(1-1.4) \times 12.25\%} \right] \times [1 - (1+1.4 \times 12.25\% \times 12)^{1-\frac{1}{1.4}} \right.$$

$$\times \frac{365}{12} = 1,518,006 \frac{\text{MCF}}{12\,\text{month}}$$

Step 3) Calculate monthly rates by simply subtracting cumulative volumes from the previous month as shown in Table 17.3:

Step 4) Calculate monthly nominal decline for each month:

$$D_{\text{month }1} = \frac{\text{Monthly nominal hyperbolic}}{1 + b \times \text{monthly nominal hyperbolic} \times \text{time}} = \frac{12.25\%}{1 + 1.4 \times 12.25\% \times 1} = 10.5\%$$

$$D_{\text{month }2} = \frac{12.25\%}{1 + 1.4 \times 12.25\% \times 2} = 9.1\%$$

Table 17.3 Monthly Production Rate Example

Time	CUM Volumes	Monthly Rate
Months	MCF	MCF/Month
1	186,661	186,661
2	354,699	168,038
3	508,034	153,335
4	649,420	141,386
5	780,873	131,454
6	903,921	123,047
7	1,019,747	115,826
8	1,129,292	109,545
9	1,233,317	104,025
10	1,332,445	99,128
11	1,427,196	94,751
12	1,518,006	90,810

$$D_{\text{month } 3} = \frac{12.25\%}{1 + 1.4 \times 12.25\% \times 3} = 8.1\%$$

$$D_{\text{month } 4} = \frac{12.25\%}{1 + 1.4 \times 12.25\% \times 4} = 7.3\%$$

$$D_{\text{month } 5} = \frac{12.25\%}{1 + 1.4 \times 12.25\% \times 5} = 6.6\%$$

$$D_{\text{month } 6} = \frac{12.25\%}{1 + 1.4 \times 12.25\% \times 6} = 6.0\%$$

$$D_{\text{month } 7} = \frac{12.25\%}{1 + 1.4 \times 12.25\% \times 7} = 5.6\%$$

$$D_{\text{month } 8} = \frac{12.25\%}{1 + 1.4 \times 12.25\% \times 8} = 5.2\%$$

$$D_{\text{month } 9} = \frac{12.25\%}{1 + 1.4 \times 12.25\% \times 9} = 4.8\%$$

$$D_{\text{month } 10} = \frac{12.25\%}{1 + 1.4 \times 12.25\% \times 10} = 4.5\%$$

$$D_{\text{month } 11} = \frac{12.25\%}{1 + 1.4 \times 12.25\% \times 11} = 4.2\%$$

$$D_{\text{month } 12} = \frac{12.25\%}{1 + 1.4 \times 12.25\% \times 12} = 4.0\%$$

Step 5) Calculate annual effective decline for each month:

$$D_{e,\text{month } 1} = 1 - (1 + 12 \times b \times D)^{-\frac{1}{b}} = 1 - (1 + 12 \times 1.4 \times 10.5\%)^{-\frac{1}{1.4}} = 51.5\%$$

$$D_{e,\text{month } 2} = 1 - (1 + 12 \times 1.4 \times 9.1\%)^{-\frac{1}{1.4}} = 48.5\%$$

Table 17.4 Hyperbolic Example Summary

Time Months	CUM Volumes MCF	Monthly Rate MCF/Month	Monthly Nominal %	Annual Effective %
1	186,661	186,661	10.5	51.5
2	354,699	168,038	9.1	48.5
3	508,034	153,335	8.1	45.8
4	649,420	141,386	7.3	43.4
5	780,873	131,454	6.6	41.3
6	903,921	123,047	6.0	39.4
7	1,019,747	115,826	5.6	37.6
8	1,129,292	109,545	5.2	36.0
9	1,233,317	104,025	4.8	34.5
10	1,332,445	99,128	4.5	33.2
11	1,427,196	94,751	4.2	31.9
12	1,518,006	90,810	4.0	30.8

$$D_{e,\text{month } 3} = 1 - (1 + 12 \times 1.4 \times 8.1\%)^{-\frac{1}{1.4}} = 45.8\%$$

$$D_{e,\text{month } 4} = 1 - (1 + 12 \times 1.4 \times 7.3\%)^{-\frac{1}{1.4}} = 43.4\%$$

$$D_{e,\text{month } 5} = 1 - (1 + 12 \times 1.4 \times 6.6\%)^{-\frac{1}{1.4}} = 41.3\%$$

$$D_{e,\text{month } 6} = 1 - (1 + 12 \times 1.4 \times 6.0\%)^{-\frac{1}{1.4}} = 39.4\%$$

$$D_{e,\text{month } 7} = 1 - (1 + 12 \times 1.4 \times 5.6\%)^{-\frac{1}{1.4}} = 37.6\%$$

$$D_{e,\text{month } 8} = 1 - (1 + 12 \times 1.4 \times 5.2\%)^{-\frac{1}{1.4}} = 36.0\%$$

$$D_{e,\text{month } 9} = 1 - (1 + 12 \times 1.4 \times 4.8\%)^{-\frac{1}{1.4}} = 34.5\%$$

$$D_{e,\text{month } 10} = 1 - (1 + 12 \times 1.4 \times 4.5\%)^{-\frac{1}{1.4}} = 33.2\%$$

$$D_{e,\text{month } 11} = 1 - (1 + 12 \times 1.4 \times 4.2\%)^{-\frac{1}{1.4}} = 31.9\%$$

$$D_{e,\text{month } 12} = 1 - (1 + 12 \times 1.4 \times 4.0\%)^{-\frac{1}{1.4}} = 30.8\%$$

A summary of this example is located in Table 17.4.

CHAPTER EIGHTEEN

Economic Evaluation

INTRODUCTION

There are three commonly used models to determine profit. Each model has a unique way of defining cost. These models are as follows:

- net cash-flow (NCF) model
- financial model
- tax model.

NET CASH-FLOW MODEL (NCF)

From the above three models, the net cash-flow (NCF) model is the most commonly used model in the oil and gas industry due to the fact that it accounts for time value of money, which will be discussed later in this chapter. This model is typically used in any oil and gas property evaluation to calculate profit and net present value (NPV) and other important capital budgeting and financial parameters as needed. One unique feature of the NCF model is the time zero. Time zero refers to the day when the first investment is made. For example, if Company ABC decides to invest roughly $10 million for exploration and development of one well in the Bakken Shale (located in North Dakota), $10 million will be inputted in time zero. Time zero is the point at which the future profits are discounted. If long-term economic analysis is being performed in which wells will be turned in line (TIL) many years from today (e.g., four years), all of the future cash flows are typically discounted back to today's dollar to get a comprehensive understanding on the value of an asset today. There are two important concepts in the NCF model. The first one is referred to as *cash outflow,* which is essentially the cash spent (i.e., money coming out of the business) on a project. Examples of cash outflows are investment, operating costs, and income taxes. Companies invest in a particular project to recover the original investment and

Hydraulic Fracturing in Unconventional Reservoirs
DOI: http://dx.doi.org/10.1016/B978-0-12-849871-2.00018-6
© 2017 Elsevier Inc.
All rights reserved.

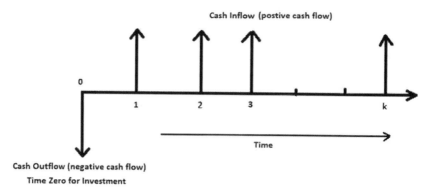

Figure 18.1 Net cash-flow (NCF) model.

additionally make a profit on top of what was originally invested. The second important concept in the NCF model is *cash inflow*, which is basically the amount of cash that the company generates from the project. An example of cash inflow is revenue. Fig. 18.1 shows the flow chart of the NCF concept.

ROYALTY

Royalty is the amount of money paid to the landowners who own the mineral rights. In the oil and gas industry, the first payment is typically the amount of money paid to the landowner per acre to lease the acreage. For example, if a landowner owns 5000 acres in which an oil or gas company is interested (due to the potential or proven reserves), the oil and gas company will end up paying a certain amount of money per acre to the landowner to be able to drill and complete on a particular property. This amount varies from state to state and can be as low as $500/acre to as high as $15,000/acre depending on the formation potential and rate of return of the project. For example, if Mr. Hoss Belyadi owns 5000 acres in Pennsylvania and an operator is interested in leasing his acreage for $2500/acre, Mr. Belyadi will get a big check for $12,500,000 for allowing the company to drill and complete on his property since he owns the mineral rights. Many people became millionaires overnight after signing an agreement with an oil and gas company at the beginning of the shale boom. Aside from getting a big lump sum of money initially, Mr. Belyadi will also get a percentage of the produced hydrocarbon's profit from each well called a *royalty* as soon as the well starts producing hydrocarbon. This percentage varies from contract to contract again and can be anywhere between 12.5% and 20%.

This can add up to a lot of money every year, and many landowners have become rich by earning money after the shale boom. Royalty percentage and other conditions and circumstances of the lease contract can be discussed and agreed upon between the landowner and the operating company when signing the lease and the contract. Some companies' strategy is to buy the land instead of going through the hassle of leasing various properties and renewing leases once expired. Leasing property from mineral-rights owners is more common. One of the biggest disadvantages of leasing a property is that operating companies have a limited period of time to begin operations, which is usually 5 years (can be extended). If the oil and gas market stays healthy by sustaining a profitable oil and gas price, companies would be able to make decisions more easily on the number of wells to be drilled and completed each year. However, when the price of oil or natural gas fluctuates to uneconomic prices, companies that do not have their gas hedged at a certain price will have a hard time staying focused on drilling and completing wells in undeveloped areas where leasing has been signed. Therefore, it is very important for companies to be ready and have strategic plans in place before leasing a property. Companies need to consider all kinds of development plans. If a particular strategic plan does not exist within a particular operating company, renewing expiring leases or losing leases will cost the company millions of dollars each year. Some companies do not have to go through the hassle of leasing and paying royalties to the landowners to a big extent due to the fact that the land and mineral rights have been previously owned from prior activities (coal mining, etc.) in a particular area.

WORKING INTEREST (WI)

Working interest (WI) is fundamentally a percentage of ownership in an oil and gas lease or property that gives the owner of the interest the right to drill, complete, and produce oil and gas on the leased acreage. For example, if an operating company XYZ has an 80% WI in a particular property, this means that this particular company is obligated to pay

80% of any investment and costs incurred. These investments and costs are included but not limited to cost of acquisition, exploration, drilling, completion, operating costs, and so forth. Generally, there are two main considerations in which an operating company elects to obtain the WI percentage. The first consideration is the amount of capital that a company has. If a company has a small amount of capital (i.e., private owner or sometimes family owner), it is very important to only own a small percentage. This is due to the fact that large amounts of capital are required to drill and complete wells. Owning a high percentage of WI means a lot of capital will be needed to pay for all of the previously discussed investments and costs. There are small family business owners in Marcellus and other shale plays that own as low as 1% or less WI. Large and medium developed exploration and production (E&P) companies typically own larger WI percentages but can have small WI in various non-core acreage position. Companies who have 50% or more WI are normally in charge of the operation.

The second important consideration is the risk and confidence level associated with a project that will determine the WI percentage. For example, let's assume Mr. Hoss Belyadi owns an $8 billion company and he is trying to invest in a particular prospect for the drilling program. The only issue and drawback that he is facing is not being confident in the outcome of that particular prospect upon completion. To split the risk associated with that project, Mr. Belyadi has the option of finding a business partner who is willing to buy 50% of the WI of that prospect under a joint operating agreement (JOA). By doing this, the risk associated with that project will be 50% in the event that particular project does not meet the expectation. This is called a risk mitigation practice and is common amongst some operators that do not have the confidence level to invest big lump sums of money into a particular project or simply do not have the capital and would like to develop the acreage position faster than their capital budget would allow. There are various methods to obtain third party acreage and the most common ones are as follows:

— **Exchange**:

Exchange is performed by trading leases of equal values, which are typically located in a similar area with similar number of acres. Exchange is recommended when disposing leases that a company does not have any plans to develop. Section 1031 of the IRS tax code of 1986 and treasury regulations permits investors (e.g., companies) to postpone capital gains taxes on any exchange of like-kind properties, which are properties of the same nature, class, or character.

— **Assignment**:

Assignment is an agreement assigning leasehold for a certain period of time. In an assignment, the owner leases to operator ABC. Operator ABC assigns all or a portion of those rights to another operator. Assignment is common when a small chunk of acreage in which an operator is trying to develop (within the operator's development plan) is owned or leased by another operator. In this scenario, the operator who has the lease or ownership in that chunk of acreage will have to run economic analysis to determine whether to participate in the well (through a joint operating agreement, or JOA) or simply not participate and assign the acreage in return for a sign-on bonus, overriding royalty interest, well data, etc. (depending on the JOA). Assignments typically include upfront bonus per acre, overriding royalty interest (ORRI), well data, and site visits. Overriding royalty interest is a royalty in excess of the royalty provided in an oil and gas lease and is usually added through an assignment of the lease.

— **Participation/Joint operating agreement (JOA)**:

In a JOA, a third party retains ownership of their leasehold and becomes part working interest owner in wells. In addition, the third party will have to pay its proportionate share of well costs (depending on the WI%). Economic analysis and other risk factors associated with participating in a well will determine whether to participate in a well or not.

NET REVENUE INTEREST (NRI)

Net revenue interest (NRI) is the percentage of production that is actually received after all the burdens, such as royalty and overriding royalty, have been deducted from the WI. For example, if a company's WI is 100% but it has agreed to pay 18.5% royalty interest to the landowner, NRI will be a smaller percentage than 100% (in this example 81.5%) since this is the money that the company actually receives after paying off the royalty to the landowner. NRI percentage can be calculated using Eq. (18.1).

$$NRI = WI - (WI \times RI)$$

Equation 18.1 Net revenue interest (NRI).

WI = Working interest, %
NRI = Net revenue interest, %.

Example

Calculate the NRI for the following prospects given the WI and royalty percentages located in Table 18.1.

Prospect A NRI% = [80% − (80% × 18.5%)] × 100 = 65.2%
Prospect B NRI% = [100% − (100% × 12.5%)] × 100 = 87.5%
Prospect C NRI% = [76% − (76% × 15%)] × 100 = 64.6%

Table 18.1 Net Revenue Interest (NRI) Example

Prospect	WI (%)	Royalty (%)
A	80	18.5
B	100	12.5
C	76	15.0

Every company's NRI will vary depending on the royalty percentage agreement, and this has to be taken into account for accurate economic analysis calculations.

BRITISH THERMAL UNIT (BTU) CONTENT

Another important concept to consider in the oil and gas economic evaluation is BTU content. As previously discussed, BTU stands for British thermal unit and is defined in every textbook as the amount of energy needed to cool or heat one pound of water by one degree Fahrenheit. The higher the BTU content, the hotter it will burn. Since gas is sold per MMBTU, it needs to be converted into proper units when performing economic analysis calculation. BTU content can be obtained from the gas composition analysis as shown in Chapter 1. BTU factor is simply BTU divided by 1000. For example, if the BTU of a dry gas well is 1040 (from gas composition analysis), the BTU factor is 1.04. The current and forecasted gas pricing must be adjusted for BTU content by using Eq. (18.2).

$$\text{Adjusted Gas Price} = \text{Gas Price} \times \text{BTU factor}$$

Equation 18.2 Adjusted gas pricing.

Adjusted gas price = $/MSCF
Gas price = $/MMBTU.

SHRINKAGE FACTOR

Shrinkage factor in dry gas areas is typically low (0.5–3%). Shrinkage in dry gas areas refers to the volumes lost due to possible line losses or field usage (e.g., fueling the compressor station). Since compressor stations are sometimes fueled by natural gas in some areas, a small percentage of produced volume from each well will be used to fuel the compressor station. As a result, a very small percentage of shrinkage must be considered when performing economic analysis calculations in dry gas areas.

Shrinkage factor becomes more important in wet gas areas with much higher BTU. Shrinkage factor is used to convert the produced wet gas into dry sales gas. Hydrocarbons exit the wellhead and hit the separator. Condensate/oil and water exit the bottom of the separator and are metered. Wet gas/dry gas exits the top of the separator and is metered, and that is usually the wellhead volume reported to the state and used for reserve forecasting. At this point, shrinkage can come into play, but it depends on the situation. For example, if wet gas is sold to the market (assuming no processing required), shrinkage factor will only be line losses and field usages. Let's assume the line losses and field usages add up to be roughly 5% of the total volume. The wet gas in this case is reduced by 5% due to the losses. Therefore, a shrinkage factor of 95% must be taken into account when performing the calculations. On the other hand, if gas is processed at a plant and both residue dry gas and natural gas liquid (NGL) revenue are obtained (based on what comes out of the exit of the plant), a much larger shrinkage factor is applied, which will contain field usage, processing shrink, liquid shrinkage, line losses, and so forth. Therefore, it is very important to consider the shrinkage factor when performing economic analysis on any well specifically in wet gas regions where the gas will be processed. Total shrinkage factor for wet gas areas can be calculated using Eq. (18.3). Liquid shrinkage can be calculated based on the inlet gas composition along with the plant removal percentage (varies from plant to plant) for each gas component.

$$S_T = [(1 - \text{field usages})(1 - \text{liquid shrinkage})$$

$$\times (1 - \text{processing plant shrinkage})] \times 100$$

Equation 18.3 Total shrinkage factor.

S_T = Total shrinkage factor, %
Liquid shrinkage = %

Field usages $=$ %

Processing plant shrinkage $=$ %.

Example

Calculate the liquid shrinkage assuming the following gas composition and plant removal percentage.

As can be seen from the table below, liquid shrinkage can be calculated as follows:

$$100\% - 88.47\% = 11.53\%$$

Gas Component	Known		Calculated
	Inlet Gas Composition (%)	Plant Removal (%)	(1-Plant Removal %) *Inlet Gas Composition (%)
Methane (C_1)	77.9731	0.00	77.973
Ethane (C_2)	14.6177	35.00	9.502
Propane (C_3)	4.7239	90.00	0.472
i-Butane (i-C_4)	0.4634	98.00	0.009
Butane (n-C_4)	1.0839	99.00	0.011
i-Pentane (i-C_5)	0.2671	99.90	0.000
Pentane (n-C_5)	0.1496	99.90	0.000
Hexane$^+$	0.2225	99.90	0.000
Nitrogen (N_2)	0.4379	0.00	0.438
Carbon dioxide (CO_2)	0.0609	0.00	0.061
Sum	**100.00**		**88.47**

OPERATING EXPENSE (OPEX)

Operating expense (Opex) or cost is the ongoing cost for running a business. Unfortunately, it is a common mistake among the general public, who thinks once the oil or gas well is produced, there are no more operating costs associated with it. This is a wrong assumption because there are many different operating costs associated with producing a BBL of oil or an MSCF of gas. Some of the most important operating costs associated with operating a well are as follows:

— **Lifting cost**:

Lifting cost is the cost of lifting oil or gas out of the ground and bringing it to the surface. Lifting cost typically includes labor cost, cost of supervision, supplies, cost of operating the pumps, electricity, and general maintenance/repairs on the wellhead and surface production equipment. One major part of the lifting cost is the cost of labor and supervision or well-tending cost. Well tenders are the contractors that the operating company hires to go on different well sites, often on a daily basis to perform routine maintenance and make sure the well and surface equipment installed on the well are functioning properly. Well tenders could also be hired on a full time basis by the operator. The lifting cost is often divided into two categories. The first lifting cost is referred to as the variable lifting cost, which is a function of producing one BBL or MSCF of oil, NGL, condensate, or gas. The second category of the lifting cost is called the fixed lifting cost, which is not a function of the amount of hydrocarbon produced but is a fixed monthly cost associated with the well. It is up to the operating company to classify which costs fall under fixed or variable lifting costs. Every company's categorization can be different.

For example, fixed and variable lifting costs on a producing dry gas well can be $650 per month per well and $0.26/MSCF in sequence. If the entire cost is assigned as only fixed lifting cost per month per well, the economics of the well will end prematurely (which is considered to be very conservative). If the entire cost is assigned as only variable lifting cost, the Opex will be too high initially. Reserve auditors do not typically like all costs to be variable lifting since it would be very optimistic from a reserve perspective as there are fixed lifting costs associated with low-producing wells (e.g., 30 MSCF/D) and the reserve life could be prolonged. Therefore, it is crucial to have a combination of both fixed and variable to create a balance in the Opex. The categories that fall under the fixed lifting cost do not depend on the production volume over time. For example, snow removal and vegetation control are considered as fixed lifting costs because no matter how much a well produces, snow must be removed from the access road and site to perform routine maintenance on the well. On the other hand, well tending is considered a variable cost because this cost is typically a function of the amount of hydrocarbon produced from a well.

- **Gathering and compression cost (G&C)**:

  Gathering and compression (G&C) is the cost of gathering gas from the sales line located on every well site and sending it to the compressor station to compress the gas before sending it to the market. In almost every gas field in this country and other countries across the world, compression is an essential operation. Gathering is typically the cost of gathering the gas from the well site to the compressor station, and compression is the cost of compressing the gas per MSCF at the compressor station. Compression is used to increase natural gas pressure to successfully meet various markets' pipeline pressure before injecting gas into the transmission pipeline. For example, if the pipeline in which the gas is being sent has a pressure of 1000 psi, the compressed gas has to be over 1000 psi to send the gas to the transmission pipeline and consequently consumers. Gas always moves from high pressure to low pressure. For the gas to be sent into transmission pipelines, there are minimum requirements such as pressure and vapor percentage that must to be met. Examples of gathering and compression costs are leased compression equipment, dehydration, repairs and maintenance of electrical flow meters, etc. The unit for gathering and compression cost is a function of the amount of gas produced and it is in $/MMBTU or $/MSCF. G&C costs also have a small fixed portion associated with each that is considered a fixed cost regardless of the amount of gas compressed.

- **Processing cost**:

  Processing cost is the cost of processing oil, condensate, wet gas, etc. into more useful products. The petroleum that comes out of the ground has to be processed to obtain products such as gasoline, diesel, heating oil, kerosene, and so forth. Just like all of the other costs discussed, there is a fee associated with processing petroleum. This cost is typically considered in $/BBL for oil fields and $/MMBTU for wet gas and retrograde condensate fields. Processing cost does not apply to fields that only produce dry gas unless the gas has a high percentage of H_2S (hydrogen sulfide).

- **Firm transportation (FT) cost**:

  Firm transportation (FT) cost is the cost of transporting natural gas from the compressor station to the consumers. FT cost depends on many factors such as the pipeline that the gas flows into and the contract associated with the FT purchased. This cost is typically in $/MMBTU or $/MSCF.

— **General and administrative (G&A) cost**:

General and administrative (G&A) cost is basically the cost of running the company such as office expenses, employee salaries, professional fees, personal costs, etc. This cost is typically in $/MSCF or $/BBL. This cost is typically not included when performing economic analysis because it is considered to be a sunk cost.

— **Water disposal cost**:

Another important cost is the water disposal cost per BBL of produced water. Once a well is turned in line (TIL), it will most likely produce water for the rest of the life of the well. The wells that have been hydraulically fractured produce more water initially and the water production decreases with time afterward. For example, on average unconventional shale reservoirs are known to typically produce 10—30% of the total injected fluid throughout the life of the well, depending on many factors such as the amount of water injected, water saturation, target depth, etc. This produced water can be reused on a different frac job or must be disposed of. Water disposal costs the operating companies a lot of money. Water disposal cost is typically in $/BBL. Many operating companies that have continuous operation in a particular basin mix the produced water with freshwater and pump the mix back downhole on the next hydraulic frac job instead of spending lots of money for disposal. This technique works when there are continuous frac operations in a particular area; otherwise, lots of money must be spent on water disposal. Another alternative when not having a continuous frac operation is to sell or give away the water to a nearby operating company not because of charity, but because it is sometimes cheaper to give away the water than spend more money on the disposal of the water.

TOTAL OPEX PER MONTH

Total operating costs (Opex) for a dry gas or wet gas well can be calculated. The equation below assumes that the operating company will be responsible for paying all of the operating costs since every Opex is being multiplied by WI%. Depending on the lease and contract, operating companies might be able to deduct some of the operating costs from the landowner (postproduction deduction). If so, WI must be replaced by NRI.

Total OPEX per month
= [(Gross monthly gas production × WI × total shrinkage factor
× variable lifting cost)] + [(Fixed lifting cost × WI)]
+ [(Gross monthly gas production × WI × total shrinkage factor
× gathering and compression cost)]
+ [(Gross monthly gas production × WI × total shrinkage factor
× processing cost)]
+ [(Gross monthly gas production × WI × total shrinkage factor
× FT cost)] + [(Gross monthly NGL production × WI × NGL OP cost)]
+ [(Gross monthly CND production × WI × CND OP cost)]
+ [(Gross monthly water production × WI × water disposal cost)]

Equation 18.4 Total Opex per month.

Gross monthly gas production = MSCF per month from decline curve
analysis (DCA) or other analyses
WI = Working interest, %
Total shrinkage factor = %
Variable lifting cost = $/MSCF
Fixed lifting cost = $/month/well
G&C cost = $/MSCF, must be grossed up and adjusted for applicable
shrinkages
FT cost = Firm transportation cost, $/MSCF
Processing cost = $/MSCF, must be grossed up and adjusted for appli-
cable shrinkages
Gross monthly NGL production = Monthly volumes of NGL in BBLs
NGL OP cost = NGL operating cost, $/BBL
Gross monthly CND production = Monthly volumes of condensate
in BBLs
CND OP Cost = Condensate operating cost, $/BBL
Gross monthly water production = Monthly volumes of water in BBLs
Water disposal cost = Water operating cost, $/BBL.
Please note that since G&C and processing costs are being multiplied
by total shrinkage, it is very important to gross up the costs based on
applicable shrinkages applied to each category.

Example

Calculate total Opex for the first three months assuming the production volumes located
in Table 18.2 and the following operating costs listed below.

Table 18.2 Gas, CND, and NGL Production Volumes

Time Month	Gross Gas Production MSCF	Gross CND Production BBL	Gross NGL Production BBL
1	350,000	950	19,250
2	330,000	800	18,150
3	300,000	500	16,500

WI = 100%, Inlet BTU = 1240 (Inlet of the plant, BTU factor of 1.240), Outlet BTU = 1100 (Residue gas BTU coming out of the processing plant, BTU factor of 1.1), Compressor burn shrinkage = 1.5%, Liquid shrinkage = 7%, Plant shrinkage = 0.5%, Variable lifting cost = $0.23/MSCF, Fixed lifting cost = $1600/month/well, Gathering and compression cost = $0.30/MMBTU, Processing cost = $0.28/MMBTU, FT cost = $0.25/MMBTU, NGL fractionation and transportation cost = $7/BBL, CND transportation cost = $11/BBL. Assume water disposal cost of $0/BBL since water will be used on an adjacent frac for the first 3 months.

Step 1) Convert all of the units on operating costs to $/MSCF from $/MMBTU and adjust for shrinkage:

G&C cost is provided in $/MMBTU, therefore it must be converted to $/MSCF by multiplying it by the inlet BTU factor of 1.240 and grossed up since the equation discussed multiplies the G&C cost by the total shrinkage.

$$\text{Gathering compression cost} = \frac{0.30 \times 1.240}{(1 - 1.5\%)(1 - 7\%)(1 - 0.5\%)} = \frac{\$0.408}{\text{MSCF}}$$

Processing cost is also provided in $/MMBTU and must be converted to $/MSCF by multiplying it by the inlet plant BTU factor of 1.240 and grossed up for liquid and processing shrinkages.

$$\text{Processing cost} = \frac{0.28 \times 1.240}{(1 - 7\%)(1 - 0.5\%)} = \frac{\$0.375}{\text{MSCF}}$$

FT cost is also provided in $/MMBTU and since the residue gas coming out of the processing plant will be sold, the FT cost provided in $/MMBTU must be multiplied by the outlet BTU factor of 1.1.

$$\text{FT cost} = 0.25 \times 1.1 = \frac{\$0.275}{\text{MSCF}}$$

Step 2) Calculate total shrinkage factor:

$$S_T = (1 - 1.5\%) \times (1 - 7\%) \times (1 - 0.5\%) = 91.1\%$$

Step 3) Calculate total Opex for each month using Eq. (18.4).

$$
\begin{aligned}
&\text{Total OPEX, month 1} \\
&= [(350,000 \times 100\% \times 91.1\% \times 0.23)] \\
&+ [(1600 \times 100\%)] + [(350,000 \times 100\% \times 91.1\% \times 0.408)] \\
&+ [(350,000 \times 100\% \times 91.1\% \times 0.375)] \\
&+ [(350,000 \times 100\% \times 91.1\% \times 0.275)] \\
&+ [(19,250 \times 100\% \times 7)] \\
&+ [(950 \times 100\% \times 11)] = \$557,479
\end{aligned}
$$

Total OPEX, month 2
$= [(330,000 \times 100\% \times 91.1\% \times 0.23)]$
$+ [(1600 \times 100\%)] + [(330,000 \times 100\% \times 91.1 \times 0.408)]$
$+ [(330,000 \times 100\% \times 91.1\% \times 0.375)]$
$+ [(330,000 \times 100\% \times 91.1\% \times 0.275)]$
$+ [(18,150 \times 100\% \times 7)] + [(800 \times 100\% \times 11)] = \$524,661$
Total OPEX, month 3
$= [(300,000 \times 100\% \times 91.1\% \times 0.23)]$
$+ [(1600 \times 100\%)] + [(300,000 \times 100\% \times 91.1 \times 0.408)]$
$+ [(300,000 \times 100\% \times 91.1\% \times 0.375)]$
$+ [(300,000 \times 100\% \times 91.1\% \times 0.275)]$
$+ [(16,500 \times 100\% \times 7)] + [(500 \times 100\% \times 11)] = \$474,610$

SEVERANCE TAX

Severance tax is a production tax imposed on operating companies or anyone with a working or royalty interest in certain states. This tax is essentially applied for the removal of nonrenewable resources such as oil, natural gas, condensate, and so forth. The % of severance tax depends on the state. For example, the severance tax in West Virginia is currently 5%, while some states such as Pennsylvania do not have a severance tax yet (Pennsylvania only pays impact fees); however, there is a possibility of such taxes being imposed to the industry in the future. It is very important to deduct the severance tax from the revenue when performing economic analysis calculation using Eq. (18.5).

Severance tax per month =
(Gross monthly gas production $\times$ adjusted gas pricing $\times$ severance tax
$\times$ NRI $\times$ total shrinkage factor)
$+$ (Gross monthly NGL production $\times$ NGL pricing $\times$ severane tax $\times$ NRI)
$+$ (Gross monthly CND production $\times$ CND pricing $\times$ severance tax $\times$ NRI)

Equation 18.5 Severance tax per month.

Gross monthly gas production = MSCF per month from DCA or other analyses
Adjusted gas pricing = $/MSCF, gas price must be adjusted for BTU of the sold gas
NRI = Net revenue interest, %
Total shrinkage factor = %

Severance tax = %

Gross monthly NGL production = Monthly volumes of NGL in BBLs

NGL pricing = Sold NGL pricing, $/BBL

Gross monthly CND production = Monthly volumes of CND in BBLs

CND pricing = Sold CND pricing, $/BBL.

●●● ───

Example

Calculate severance tax for the first month for a dry gas well using the assumptions listed below.

Gross gas production for month 1 = 250,000 MSCF, Gas pricing = $3.5/MMBTU, Severance tax = 5%, BTU = 1070 (no NGL or CND expected since the gas is dry), Total shrinkage factor = 0.98 (2% shrinkage), WI = 80%, RI = 15%

Step 1) Since pricing is provided in MMBTU, calculate adjusted gas pricing at 1070 BTU (1.07 BTU factor)

$$\text{Adjusted gas pricing} = 3.5 \times 1.070 = \frac{\$3.745}{\text{MSCF}}$$

Step 2) Calculate NRI:

$$\text{NRI\%} = [80\% - (80\% \times 15\%)] \times 100 = 68\%$$

Step 3) Use Eq. (18.5) to calculate severance tax for the first month:

Severance tax for the first month = $(250,000 \times 3.745 \times 5\% \times 68\% \times 0.98) = \$31,196$

───

AD VALOREM TAX

Ad valorem is a Latin phrase meaning *according to value*. This is another form of tax paid when minerals are produced. West Virginia and Texas are examples of states in which ad valorem tax must be paid annually. There are other types of taxes that must be paid in the oil and gas industry in addition to federal income taxes (depending on the state). For example, in Pennsylvania, there is no severance or ad valorem tax (as of the publication date of this book). Instead, there is an impact fee that must be paid. This does not mean that severance or other forms of taxes will not be imposed in the future. As a matter of fact, depending on the person in office for that particular state, such taxes can be added. Ad valorem tax can be calculated using Eq. (18.6).

Advalorem tax per month =
{[(Gross monthly gas production × adjusted gas pricing × NRI
× total shrinkage factor)
+ (Gross monthly NGL production × NGL pricing × NRI)
+ (Gross monthly CND production × CND pricing × NRI)
− Severance tax amount} × Advalorem tax

Equation 18.6 Ad valorem tax.

Gross monthly gas production = MSCF per month from DCA or
other analyses
Adjusted gas pricing = \$/MSCF, gas price must be adjusted for BTU
of the sold gas
NRI = Net revenue interest, %
Total shrinkage factor = %
Ad valorem tax = %
Severance tax amount = \$/month
Gross monthly NGL production = Monthly volumes of NGL in BBLs
NGL pricing = Sold NGL pricing, \$/BBL
Gross monthly CND production = Monthly volumes of CND in BBLs
CND pricing = Sold CND pricing, \$/BBL.

Example

Calculate ad valorem tax for the first month from a dry gas well using the assumptions listed below.

Gross gas production for month $1 = 300{,}000$ MSCF, Severance tax for the first month $= \$35{,}000$, Adjusted gas pricing $= \$2.5$/MSCF, Ad valorem tax $= 2.5\%$, Total shrinkage factor $= 0.98$ (2% shrinkage), NRI $= 42\%$

$$\text{Advalorem tax}_{month\ 1} = [(300{,}000 \times 2.5 \times 42\% \times 0.98) - (35{,}000)] \times 2.5\% = \$6843$$

NET OPEX

Net Opex is referred to the total operating costs including severance and ad valorem taxes and can be simply calculated using Eq. (18.7). Production taxes such as severance and ad valorem taxes are taken into account before federal income tax calculations. Although severance and ad valorem taxes are referred to as taxes, these taxes are deducted as production taxes before federal income tax calculations.

$$Net\ OPEX = Total\ OPEX + severance\ tax\ amount + advalorem\ tax\ amount$$

Equation 18.7 Net Opex.

Net Opex = $/month
Total Opex = $/month
Severance tax amount = $/month
Ad valorem tax amount = $/month

Example

Calculate net Opex for the first 3 months assuming the Opex and production taxes located in Table 18.3.

Net OPEX$_{month\ 1}$ = 501, 564 + 40, 250 + 9, 000 = \$550, 814

Net OPEX$_{month\ 2}$ = 455, 520 + 35, 650 + 8, 560 = \$499, 730

Net OPEX$_{month\ 3}$ = 401, 365 + 30, 000 + 8, 250 = \$439, 615

Table 18.3 Net Opex example

Month	1	2	3
Total Opex ($)	501,564	455,520	401,365
Severance tax ($)	40,250	35,650	30,000
Ad valorem tax ($)	9000	8560	8250

REVENUE

In the oil and gas industry, revenue is the amount of money received from normal business activities, services, and products such as selling hydrocarbon, providing various services to the operating companies, or any other activities that generate money. For example, a big portion of an operating company's revenue comes from selling hydrocarbon. It is very important to avoid confusing revenue with profit because revenue is just the gross money that the company earned and does not take the expenses associated with a project into account. A company's gross revenue can be enormous but the profit might actually be negative because of high amounts of expenses associated with performing that project or any other reasons. For natural gas-producing wells, monthly net gas/NGL/CND production can be calculated using Eqs. (18.8)−(18.10).

Monthly shrunk net gas production
  = Monthly unshrunk gross gas production × total shrinkage factor × NRI

Equation 18.8 Monthly shrunk net gas production.

Monthly shrunk net gas production = MSCF/month
Monthly unshrunk gross gas production = MSCF/month, wellhead volumes
Total shrinkage factor = %
NRI = Net revenue interest, %

Monthly shrunk net NGL production
  = Monthly shrunk gross NGL production × NRI

Equation 18.9 Monthly shrunk net natural gas liquid (NGL) production.

Monthly shrunk net NGL production = BBLs/month, sold volumes
Monthly shrunk gross NGL production = BBLs/month, sold volumes
NRI = Net revenue interest, %

Monthly shrunk net CND production
  = Monthly shrunk gross CND production × NRI

Equation 18.10 Monthly shrunk net CND production.

Monthly shrunk net CND production = BBLs/month, sold volumes
Monthly shrunk gross CND production = BBLs/month, sold volumes
NRI = Net revenue interest, %.

After calculating monthly net gas, NGL, and CND volumes, net revenue can be simply calculated using Eq. (18.11).

Net Revenue = (Monthly shrunk net gas production × adjusted gas pricing)
  + (Monthly shrunk net NGL production × NGL sales pricing)
  + (Monthly shrunk net CND production × CND sales pricing)

Equation 18.11 Net revenue.

Monthly shrunk net gas production = MSCF/month, residue gas
Adjusted gas pricing = $/MSCF
Monthly shrunk net NGL production = BBL/month

NGL sales pricing = $/BBL

Monthly shrunk net CND production = BBL/month

CND sales pricing = $/BBL.

●●●
_____

Example

Calculate net revenue from a retrograde condensate well using Table 18.4 for the first 3 months assuming 80% NRI and total shrinkage factor of 90%.

Monthly shrunk net gas production$_{month\ 1}$ = 450,000 × 80% × 90% = 324,000 MSCF

Monthly shrunk net gas production$_{month\ 2}$ = 435,500 × 80% × 90% = 313,560 MSCF

Monthly shrunk net gas production$_{month\ 3}$ = 395,400 × 80% × 90% = 284,688 MSCF

Monthly shrunk net NGL production$_{month\ 1}$ = 22,500 × 80% = 18,000 BBLs

Monthly shrunk net NGL production$_{month\ 2}$ = 21,775 × 80% = 17,420 BBLs

Monthly shrunk net NGL production$_{month\ 3}$ = 19,770 × 80% = 15,816 BBLs

Monthly shrunk net CND production$_{month\ 1}$ = 950 × 80% = 760 BBLs

Monthly shrunk net CND production$_{month\ 2}$ = 750 × 80% = 600 BBLs

Monthly shrunk net CND production$_{month\ 3}$ = 720 × 80% = 576 BBLs

Net revenue$_{month\ 1}$ = (324,000 × 3.5) + (18,000 × 35) + (760 × 55) = $1,805,800

Net revenue$_{month\ 2}$ = (313,560 × 3.4) + (17,420 × 30) + (600 × 56) = $1,622,304

Net revenue$_{month\ 3}$ = (284,688 × 3.6) + (15,816 × 33) + (576 × 53) = $1,577,333

Table 18.4 Net Revenue Example

Time	Gross Volumes			Sales Pricing		
	Unshrunk Gas	Shrunk NGL	Shrunk CND	Adjusted Gas	NGL	CND
Month	MSCF	BBL	BBL	$/MSCF	$/BBL	$/BBL
1	450,000	22,500	950	3.5	35	55
2	435,500	21,775	750	3.4	30	56
3	395,400	19,770	720	3.6	33	53

NYMEX (NEW YORK MERCANTILE EXCHANGE)

NYMEX stands for New York Mercantile Exchange and is essentially a commodity exchange located in New York. Trading is conducted into two divisions. The first division is the NYMEX division, which is home to the energy (oil and natural gas), platinum, and palladium markets. The second division is called the COMEX (commodity exchange) division where metals such as gold, copper, and silver are traded (Investopedia).

NYMEX is used amongst some operating companies to estimate the future price of natural gas for the purpose of economic analysis evaluation. Many companies have developed their own pricing forecast model based on supply and demand and other various factors. On the other hand, some companies prefer to use flat pricing and perform sensitivity analysis instead of using the NYMEX forecast (strip forecast). If NYMEX is used, NYMEX must to be corrected for the basis. The basis can have substantial impact on a project's economics.

HENRY HUB AND BASIS PRICE

Henry Hub is a natural gas pipeline in Louisiana where onshore and offshore pipelines meet and is the most important natural gas hub in North America. Henry Hub is the pricing point for natural gas futures on NYMEX. The settlement prices at the Henry Hub are used as benchmarks for the entire North American natural gas market. It is very important to understand that when NYMEX is used to estimate the monthly price of natural gas, it is based on the delivery to the Henry Hub. For example, if the price of natural gas at NYMEX is $4/MMBTU in March 2017, this price must be adjusted to represent the price at Henry Hub. The difference between the Henry Hub natural gas price and natural gas price at a specific location is called the *basis differential*. For example, the NYMEX price might be $5/MMBTU; however, a particular pipeline could have a basis differential of $ − 1.5/MMBTU. Therefore, the price at which the gas is sold is $5/MMBTU plus − $1.5/MMBTU, which yields $3.5/MMBTU. If NYMEX is used for the purpose of economic analysis, NYMEX must be adjusted for the basis by taking the NYMEX forecast for each month and adding the basis forecast for each month to NYMEX.

Basis is a function of NYMEX in that a regional basis is that particular region's differential to NYMEX. In a perfectly balanced market (where supply is equal to demand), the basis is the cost of transportation. For example, if the cost of transporting gas from Appalachia to NYMEX/ Henry Hub (Louisiana) was $0.25/MMBTU, in a perfectly balanced market the Appalachian basis would be $0.25/MMBTU. In the Appalachian basin, the basis used to be positive before the development of the Marcellus Shale; however, with the development of the Marcellus Shale and a huge surge in gas supplies, various bases across the basin have become negative. Weather (seasonal variations), geography, natural gas pipeline capacity, product quality, and supply/demand determine the price of natural gas on a particular market (pipeline).

Example

Calculate the actual price of natural gas sold on pipeline ABC for the next 2 years using the provided projected NYMEX forecast and the projected basis forecast using Table 18.5.

As can be seen from the table, pipeline ABC's natural gas price can be calculated by taking the NYMEX forecast per month and adding the basis forecast. In this case, the negative basis prices are due to too much supply and not enough demand. For example, the NYMEX price in October 2017 is listed as $3.90/MMBTU and the basis price is listed as − $0.6/MMBTU. Therefore,

$$\text{Pipeline ABC natural gas price on Oct 2017} = 3.9 + (-0.6) = \frac{\$3.3}{\text{MMBTU}}$$

Table 18.5 NYMEX and Basis Forecast Example

Date	NYMEX ($/MMBTU)	Basis ($/MMBTU)
Jan–17	3.5	− 0.8
Feb–17	3.4	− 0.7
Mar–17	3.51	− 0.6
Apr–17	3.53	− 0.55
May–17	3.6	− 0.58
Jun–17	3.9	− 0.8
Jul–17	3.87	− 0.9
Aug–17	3.88	− 0.95
Sep–17	3.6	− 0.7
Oct–17	3.9	− 0.6
Nov–17	4	− 0.5
Dec–17	4.1	− 0.55

The basis forecast is positive in some places and negative in others depending on the market, and it all boils down to supply and demand at the end. Before the development of unconventional shale plays across the United States, the basis prices used to be positive. However, with the development of unconventional shale plays, the supply has dramatically increased while the demand has not changed in the same proportionality. Therefore, the basis prices have switched from being positive to negative in high-supply markets primarily due to the lack of infrastructure. The basis prices can be easily changed from negative to positive during cold winter months due to too much demand and not much supply. This is one of the main reasons that during the coldest days of winter 2014, the price of natural gas was actually increased to $50/MSCF in Connecticut due to a lack of infrastructure (pipeline) and supply.

CUSHING HUB AND WEST TEXAS INTERMEDIATE (WTI)

Cushing Hub is the largest hub for the distribution of crude oil in the world and is located in Oklahoma. Cushing Hub has always been very important for traders because of its role as the delivery point (just like Henry Hub) in the U.S. benchmark oil futures. This hub, just like Henry Hub, is the pricing point for crude oil futures in West Texas Intermediate (WTI). WTI, also known as Texas light sweet, is used as a benchmark in oil pricing. Light sweet crude oil has a low density of approximately 39.6 API gravity and low sulfur content of about 0.24%. Other essential crude oil benchmarks that serve as a reference price for buyers and sellers of crude oil are Brent crude, Dubai crude, Oman crude, and OPEC Reference Basket. Condensate price is typically a function of oil price and as oil price increases or decreases, condensate price will increase or decrease.

MONT BELVIEU AND OIL PRICE INFORMATION SERVICES (OPIS)

Mont Belvieu is the pricing point for natural gas liquid (NGL) futures. As previously mentioned, the settlement prices at the Henry Hub and Cushing are used as benchmarks for the entire North American natural gas and crude oil market. The settlement prices at the Mont Belvieu are used as benchmarks for the NGL market. Oil Price Information Services (OPIS) has a very similar concept to NYMEX or WTI. OPIS is used as a benchmark for NGLs and is one of the world's biggest sources used for

NGL pricing. OPIS can be used just like NYMEX to estimate the future price of NGL for the purpose of economic analysis calculations. NGL price is also typically a function of oil price.

CAPITAL EXPENDITURE COST (CAPEX)

The next important term in oil and gas economics is capital expenditures (also referred to as Capex). Capital expenditures are the money invested upfront to create future benefits. Capex is not a cost, but it is an investment because companies invest money in projects that are expected to create value for the shareholders. Capital expenditures are as follows:

1. **Acquisition**. Acquisition considered to be a Capex and is referred to as the costs when acquiring the rights to develop and produce oil and natural gas. For example, when a company purchases or leases the right to extract the oil and gas from a property not owned by the company, this will be considered as acquisition Capex. Other examples of acquisition Capex are title search, legal expenses, recording costs, and so forth. The land department is typically responsible and is in charge of dealing with this side of the business. The land department's responsibilities are included but not limited to acquiring/renewing leases, dealing directly and negotiating with landowners, title search, and so on. Before any kind of acquisition is made, reservoir engineers and geologists along with other departments are heavily involved in valuing the asset that is under consideration for acquisition by performing various analyses such as geological potential analysis, type curve analysis, water infrastructure analysis, midstream infrastructure analysis, land analysis, environmental analysis, and finally economic analysis (using the NCF model) for the area. Many acquisition deals occur in a low commodity pricing environment where assets are sold at a discount, which can significantly be cheaper than the intrinsic value of the asset in a regular commodity pricing environment.

2. **Exploration**. Before drilling a well, it is very important to perform some type of seismic to determine the depth of the formation of interest, lithology, formation tops, formation characteristics, directional plan (azimuth, inclination, etc.), and other valuable information. A 2D or 3D seismic is used during the exploration phase to obtain

this information. 3D seismic is more accurate while providing better resolution and more information about a particular prospect. 3D seismic is mostly used and preferred over 2D seismic (when capital is available). Therefore, exploration expenditures are charges related to gathering and analysis of geophysical and seismic data.

3. **Development**. These expenditures are associated with constructing the well sites, building or improving the access roads, drilling/completion, gathering, installation of pipelines, and other expenditures incurred during the developmental phase of the operation. For example, it costs on average about $5–10 million to drill and complete a well in Marcellus Shale depending on the true vertical depth (TVD), lateral length, drilling/completions design, and most importantly market condition. The unconventional shale plays are absolutely promising plays across the United States. However, it is very important to understand that developing unconventional shale plays are very capital intensive. This is due to the fact that not only do these shale formations have to be drilled, but also must be properly hydraulically fractured to produce at an economically feasible rate. Therefore, proper economic evaluation/analysis is an important job that reservoir engineers are responsible for performing.

Net Capex is the net capital expenditure based on the WI% of a well. For example, if an operating company has 40% ownership in a gas well (40% WI), the operating company will only be responsible to pay 40% of the total capital investment on a project. Net Capex can be written as Eq. (18.12).

$$Net\ CAPEX = Gross\ CAPEX \times WI$$

Equation 18.12 Net Capex.

Net Capex = Net capital expenditure, $
Gross Capex = Gross capital expenditure, $
WI = working interest, %.

●●●————————————————————————————

Example

An 8000′ lateral-length well is estimated to be drilled and completed for $7.5 MM. Assuming 40% WI, what is the net Capex?

$$Net\ CAPEX = 7,500,000 \times 0.4 = \$3,000,000$$

OPEX, CAPEX, AND PRICING ESCALATIONS

When performing economic analysis, escalation is a challenging subject in the oil and gas property evaluation. E&P companies typically apply a percentage of escalation on Opex, Capex, and pricing depending on the company's philosophy. When monthly cash flows are used, the Society of Petroleum Evaluation Engineers' best practices recommend that escalation must take place in a "stair-step" fashion on a monthly basis. For example, if prices are assumed to increase at 3% per year, the monthly increase would be based on an effective annual rate of 3% per year with prices increasing every month.

●●●

Example

Perform a stair-step escalation @ $3/MMBTU gas price using 3% effective annual rate for the first 12 months:

Month 1 $= 3 \times (1+3\%)^{\frac{1}{12}} = 3.007$ Month 2 $= 3.007 \times (1+3\%)^{\frac{1}{12}} = 3.015$

Month 3 $= 3.015 \times (1+3\%)^{\frac{1}{12}} = 3.022$ Month 4 $= 3.022 \times (1+3\%)^{\frac{1}{12}} = 3.030$

Month 5 $= 3.030 \times (1+3\%)^{\frac{1}{12}} = 3.037$ Month 6 $= 3.037 \times (1+3\%)^{\frac{1}{12}} = 3.045$

Month 7 $= 3.045 \times (1+3\%)^{\frac{1}{12}} = 3.052$ Month 8 $= 3.052 \times (1+3\%)^{\frac{1}{12}} = 3.060$

Month 9 $= 3.060 \times (1+3\%)^{\frac{1}{12}} = 3.067$ Month 10 $= 3.067 \times (1+3\%)^{\frac{1}{12}} = 3.075$

Month 11 $= 3.075 \times (1+3\%)^{\frac{1}{12}} = 3.082$ Month 12 $= 3.082 \times (1+3\%)^{\frac{1}{12}} = 3.090$

Two to four percent is the typical escalation percentage that is assumed among many operating companies. The percentage escalation is directly related to the inflation rate. Using escalation is an attempt to represent inflationary expectations and should be somewhat in line with the historical long-term trend when nominal cash flows are used. From an economic analysis standpoint and NPV calculation (to be discussed), inflation must be treated consistently. When nominal interest rate is used, nominal cash flows must also be used. On the other hand, when real interest rate is used, real cash flows should be used. Nominal interest rate refers to the actual prevailing interest rate, while real interest rate is adjusted for inflation. For example, the return on a particular investment could be 5%, which is referred to as nominal interest. However, after accounting for inflation of 3%, the real interest is only 2%. Nominal interest rate is written in Eq. (18.13) as follows:

$$\text{Nominal interest rate} = \text{real rate} + \text{inflation}$$

Equation 18.13 Nominal interest rate.

PROFIT (NET CASH FLOW)

Profit or net cash flow is basically revenue minus costs. The most commonly used model in the oil and gas industry to determine profit is the NCF model since this model incorporates the time value of money. Profit in the cash-flow model is also referred to as net cash flow (NCF). As previously mentioned, the NCF model has one unique feature and this unique piece is called time zero. Time zero is the day that the check is written to the contractors to perform a job. Capex is placed in time zero in the NCF model. It is very important that the cash-flow model is used for economic analysis, since it incorporates the time value of money. Profit excluding investment is referred to as operating cash flow and is shown in Eq. (18.14).

$$\text{Profit (excluding investment)} = \text{Net revenue} - \text{net OPEX}$$

Equation 18.14 Profit (excluding investment).

Profit (excluding investment) = Monthly basis, $
Net revenue = Monthly basis, $
Net Opex = Monthly basis, $.

BEFORE FEDERAL INCOME TAX (BTAX) MONTHLY UNDISCOUNTED NET CASH FLOW (NCF)

Before federal income tax (BTAX) monthly cash flows can be calculated by taking the profit in Eq. (18.14) and subtracting net Capex. BTAX monthly undiscounted NCF can be written as Eq. (18.15).

$$\text{BTAX monthly undiscounted net cash flow} = \text{profit} - \text{net CAPEX}$$

Equation 18.15 BTAX monthly undiscounted NCF.

Profit = Monthly basis, $
Net Capex, $.

Net Capex at time zero is equal to net Capex. However, net Capex for subsequent months is zero unless special activities occur after time zero. Examples of such activities are remedial work, refrac, swabbing, artificial lift, etc.

Before starting the most important and beautiful concept in economic analysis, i.e., net present value (net present worth), it is very important to understand discount rate and its significance.

DISCOUNT RATE

Discount rate, also known as interest rate, exchange rate, cost of capital, opportunity cost of capital, cost of money, weighted average cost of capital (WACC), or hurdle rate, is used to discount all of the future cash flows to today's dollar. Discount rate is the basis for all the economic analysis performed in any industry. It is basically the cost of doing business. For example, if a company's cost of capital (discount rate) is 10%, it means the return on a particular project must be greater than 10% or the company will not be creating any value for the shareholders. For the purpose of economic analysis, weighted average cost of capital is typically used to discount all of the future cash flows to the present dollar. Weighted average cost of capital accounts for both time value of money and inflation. Time value of money is not the same as inflation, although they are often confused. Time value of money refers to the fact that a dollar today is worth more than a dollar in the future. It is tied back to the fact that people are impatient about their money. If I were to offer you $1000 today versus $1000 a month from today, the chance that you would like to have your $1000 today is very likely because you are impatient about your money. Therefore, time value of money has to deal with the impatience of people regarding their money. Inflation, on the other hand, refers to the reduction in purchasing power of the money. Ten years or so ago, a $5 bill could have bought much more than a $5 bill today because of the inflation with time. Therefore, the purchasing power of the same $5 bill has decreased due to inflation.

What discount rate should be assumed for a project? This is the task for financial people within a corporation. It is very important to understand the concept of discount rate as it is very significant in the determination of NPV. Every company has a cost of capital. The determination of cost of capital can be tricky and complicated. Cost of capital calculation essentially takes three important factors into account. These three factors are debt, common stock (equity), and preferred stock (equity, if exists, some companies do not have preferred stock). Thus, cost of capital is usually the combination of debt and equity since many companies use a combination of debt and equity to finance their business. If a company only uses debt to finance its projects, cost of capital is referred to as cost of debt. On the other hand, if a company only uses equity to finance its projects, cost of capital is called cost of equity. As previously mentioned, many companies' cost of capital consists of both debt and equity. Cost of capital is sometimes referred to as hurdle rate. Hurdle rate is the minimum discount rate or minimum acceptable rate of return that must be overcome for a company to generate value and return for its investors.

— **Debt**. Companies, just like ordinary people, have to incur debt to finance their projects. Debt can be borrowing money through issuing bonds, loans, and other forms of debts from banks or financial institutions.

— **Preferred stock**. Preferred stock is a type of equity or a class of ownership in a corporation that has a higher claim to a company's assets. The reason this type of equity is referred to as "preferred" is because when a company cannot meet its financial obligations as debts become due (insolvency), preferred stockholders get their money before common stockholders. This means that there is a lower risk associated with preferred stocks in addition to lower rates of return (less potential to appreciate in price) in comparison to common stocks. Furthermore, when the company has excess cash and decides to reward the stockholders by distributing cash in the form of dividends, preferred stockholders are paid before common stockholders. The dividends paid to the preferred stockholders are different and typically more than the dividends paid to common stockholders. Preferred shareholders typically do not have the voting rights, however, under some circumstances these rights can return to shareholders who have not received their dividend.

— **Common stock**. Common stock is also a type of equity or ownership in a corporation in which investors invest their money in a risky stock market. Common stock is riskier than preferred stock due to the fact that those investors will be last in line to get their money back in the event of insolvency. The reward of common stocks would be a higher return compared to preferred stocks and bonds in the long run.

Common stockholders have the power of voting in the election of the board of directors and corporate policy.

To summarize, debt and equity (common and preferred stocks) are used in the computation of cost of capital. Since cost of capital consists of cost of debt and equity, both must be combined into an equation referred to as weighted average cost of capital (WACC), shown in Eq. (18.16).

$$\text{Weighted average cost of capital}$$
$$= \text{WACC} = W_d R_d (1 - T) + W_p R_p + W_c R_c$$

Equation 18.16 Weighted average cost of capital (WACC).

W_d = Weight of debt (% of the company that is debt)
R_d = Cost of debt, %
W_p = Weight of preferred stock (% of the company that is preferred stock)
R_p = Cost of preferred stock, %
W_c = Weight of common stock (% of company that is common stock)
R_c = Cost of common stock, %
T = Corporate tax rate, %.

WEIGHT OF DEBT AND EQUITY

The weight of debt and equity of any company can be obtained using the debt-to-equity ratio, which is publicly available on various financial websites. Weight of debt refers to the percentage of the company that is financed by debt. On the other hand, weight of equity refers to the percentage of the company that is financed by equity. The combination of the two makes up the company's capital structure. It is important to have a balance between the amount of debt and equity that a company has. Low–debt companies are typically safer to invest in. Although debt is tax deductible (beneficial from a tax perspective), having too much debt can cause a company to go through so much capital when the stock of that particular company decreases. For example, imagine a house that you are interested in purchasing is worth $250,000. You were able to obtain a loan from your local bank for $200,000. You placed $50,000 as a down-payment on the house (20% downpayment). Therefore, 80% of the house is financed by debt and only 20% is equity. If the price of the house decreases by 25%, not only have you lost all of the equity that was placed, but also the house is under the market value by an additional 5%. When the debt level is high and the stock of a particular corporation decreases, the company goes through so much capital and they can become history overnight. This concept was very clear when studying some of the financial institutions during the 2008 crash. Many of those institutions had a very high leverage ratio.

Example

Company X currently has a debt-to-equity ratio of 0.65. Calculate the weight of debt and equity of this company.

$$\text{Weight of debt} = \frac{0.65}{0.65 + 1} = 39.4\% \text{ weight of debt}$$

$$\text{Weight of equity} = 1 - 39.4\% = 60.6\% \text{ weight of equity}$$

COST OF DEBT

Cost of debt can be calculated by taking the weighted average of the percentage interest rate paid on debt. For example, if Matt has a house for $400,000 @ 4%, a car for $30,000 @ 2.5%, and a small boat for $25,000 @ 1.5%, his cost of debt will be the weighted average of the percentage interest rates above, which is 3.76%. When tax season arrives, any interest that Matt paid on the house, car, and boat are tax deductible. Companies obtain their cost of debt by taking the weighted average of all their debts.

Example

A bond is issued at $950. After 2 years, your company will pay back $1000 to the investors. What is the cost of debt for this particular bond?

Par value = face value = $950
Maturity value = $1000

$$PV = \frac{FV}{(1+i)^t} \gg 950 = \frac{1000}{(1+i)^2} \gg i = 2.59\%$$

COST OF EQUITY

Cost of equity is the percentage return that shareholders expect from a company. For example, if you decide to invest in a company, you will have some demands from that company in exchange for obtaining the risk of ownership. Let's assume that an investor's required rate of return (demand) in order to be convinced to invest in a company's stock is 20%. This 20% required rate of return is referred to as *cost of equity* for the company. It is important to remember that 20% is not the profit earned by the investors but it is just the demand from the investors for putting their money in a high-risk and volatile stock market. The main reason the cost of equity is considered a cost from a company's perspective is because if the company fails to deliver this kind of return, shareholders will simply sell their stocks (shares) causing the stock price to go down. There are multiple ways to calculate cost of equity but one of the most commonly used models is the capital asset pricing model (CAPM). The general idea about the CAPM model is that investors need to be compensated in two ways:

1. Time value of money $>>$ risk free rate
2. Risk $>>$ the amount of compensation for taking additional risk
   Cost of equity using CAPM is written in Eq. (18.17).

$$K_e = R_f + \beta(R_m - R_f)$$

Equation 18.17 Capital asset pricing model (CAPM).

$K_e =$ Cost of equity, %
$R_f =$ Risk-free rate, %
$\beta =$ Beta of security
$R_m =$ Expected market return, %
$R_m - R_f =$ Risk premium, %.

Risk-free rate or R_f is the theoretical rate of return obtained from an investment that has no risk associated with it. For example, if a theoretical government asks you to invest in a particular bond and receive 2% in return without any risks, the risk-free rate is 2%. In reality, the risk-free rate does not exist because even the safest investment will have a very small amount of risk. Typically the interest rate of U.S. Treasury bills is used as the risk-free rate among many companies. Governmental treasury bills are called risk-free because the U.S. Government has never defaulted on its debt.

Beta or β, also known as the **beta coefficient**, measures the volatility of a company's share price against the whole market. A beta of 1 means the company moves in line with the market. If beta is more than 1, the security's price will be more volatile than the market. Finally, if beta is less than 1, the security's price will be less volatile than the market. Having a high beta means more risks while offering a possibility of a higher rate of return. Oil and gas companies typically have a beta greater than 1. An example with a beta less than 1 is the U.S. Treasury bill since the price does not change much over time. If market return is 10%, a stock whose beta is 1.5 would return 15% because it would go up 1.5 times as high as the market. Beta considers systematic risk and not idiosyncratic risk. *Systematic risk* refers to the overall market risk. However, *idiosyncratic risk* refers to the risk of change in the price of a security due to the special circumstances of a particular security. Idiosyncratic risk can be eliminated through diversification but systematic risk cannot be eliminated.

Risk premium or $R_m - R_f$ is the return for which investors expect to be compensated for having taken the extra risk in investing in the volatile stock market. It is basically the difference between the risk-free rate and market rate. Risk premium accounts for inflation rate and this is the primary reason escalation is very important to use on Opex, Capex, pricing, etc. when cost of capital is used in economic analysis.

Example

A firm's risk free rate (R_f) is 6% and the market risk premium ($R_m - R_f$) is 7%. Assuming a beta of 1.5, what is the cost of equity using the CAPM model?

$$K_e = R_f + \beta(R_m - R_f) = 6\% + 1.5 \times 7\% = 16.5\%.$$

Example

A company wants to raise money. The company will sell $15 million shares of common stock with the expected return of 15%. In addition, the company will issue $10 million of debt with the cost of debt of 12%. Assuming a corporate tax rate of 35%, calculate WACC.

$$\text{Total value of the company} = \$15\ \text{MM} + \$10\ \text{MM} = \$25\ \text{MM}$$

$$\text{Weight of equity} = \frac{15}{25} = 0.6\ \text{or}\ 60\%$$

$$\text{Weight of debt} = \frac{10}{25} = 0.4 \text{ or } 40\%$$

Using Eq. (18.16):

$$\text{WACC} = 40\% \times 12\% \times (1 - 35\%) + (60\% \times 15\%) = 12.12\%$$

CAPITAL BUDGETING

Capital budgeting is an important part of determining whether to invest in a project such as a drilling/completion program, buying machinery, replacing equipment, and so forth. Capital budgeting defines a firm's strategic direction and planning. Capital budgeting typically involves large capital expenditure (Capex), and making wrong decisions can have serious consequences. Without analyzing capital budgeting parameters and presenting the results, the management committee of any public or private company will not approve projects. This is a very simple concept. The management committee of any company would like to see return as a result of investing money into a project to maximize shareholders' value. Important capital budgeting criteria are NPV, internal rate of return (IRR), modified internal rate of return (MIRR), return on investment (ROI), payback, discounted payback, and profitability index (PI). All of these criteria are very important to understand and comprehend in detail for successful capital budgeting decision making. The capital budgeting decision-making process must be evaluated in great detail before investing big lump sums of money on a project to determine whether the project is worth the investment or not. Now that the concept of capital budgeting is clear, let's discuss the most important capital budgeting criteria involved in the decision-making process.

NET PRESENT VALUE (NPV)

Net present value (NPV) also known as net present worth (NPW) is one way of analyzing the profitability of an investment. NPV is basically the value of specific stream of future cash flows presented in today's dollar. NPV is an essential calculation in petroleum economics due to considering time value of money and inflation. Companies are keen to know what an actual project is worth in today's dollar rather than, say, the dollar of 10 years from now. As a simple example, oil and gas operating companies project the future production rates for each well using various techniques such as decline curve, type curve, reservoir simulation, rate transient analysis, material balance, and so forth. Those future production rates that will yield future cash flows must be discounted (using cost of capital) to present value. It really does not make any logical sense for a company to announce that their profit cash flows for doing a project would be $10 million, $8 million, $12 million, and $11 million in subsequently 2, 3, 4, and 5 years. This is because these cash flows are worth less in today's dollar when discounted back (due to time value of money). Instead, it would make much more sense to use the NPV formula to calculate the present value of all of the future cash flows. A simple insight one can use to think about NPV is that cash flows at different dates are like different currencies. What is the summation of 200 U.S. dollars and 200 euros? Not discounting the future cash flows that are occurring at different points in time to today's dollar is like saying that the answer to the posed question is 400. Therefore, just like currencies that must be converted before being able to perform the summation, cash flows at different points in time must be discounted back to today's dollar. NPV calculation assumes that positive cash flows from a project are reinvested at the cost of capital. NPV can be calculated using Eq. (18.18).

$$NPV = \sum_{t=0}^{n} \frac{CF_t}{(1+i)^t}$$

Equation 18.18 Net present value (NPV).

$i =$ Discount rate, %
$CF_t =$ Cash flow @ time t, $
$t =$ Period of time, yearly

$\sum_{t=0}^{n} =$ Summation sign from time 0 (investment) to n.

The reason the term "net" is used in the term *net present value* is that the initial investment is subtracted and taken into account when calculating NPV. NPV is the summation of the present value of all of the future cash flows and initial investment (Capex). Discount rate in NPV calculation is an essential factor that considers the time value of money and inflation. When discount rate increases, the NPV decreases. The discount rate used among some operating companies regardless of cost of capital is typically 10%. Many E&P companies' cost of capital is anywhere from 8% to 12%. Therefore, the industry standard discount rate used in many economic analysis calculations is 10% for simplicity.

The rules of thumb for NPV projects are as follows:
1. Accept independent projects if the NPV is positive.
2. Reject any project that has a negative NPV.
3. Pick the highest positive NPV in mutually exclusive projects that would add the most value.
4. NPV must be considered along with other capital budgeting criteria to make educational decisions.

Although the rule of thumb says to accept any project with positive NPV, would you accept a project that has a NPV of $20,000 after investing $2 billion dollars in that project? Absolutely not, because although the NPV is positive, the project could be so risky that it might end up costing the company more than creating any value for the shareholders. As a result, it is extremely important to comprehend the magnitude of the investment along with other capital budgeting tools before making such decisions (Table 18.6).

Example

Find the NPV for the cash flows (profits) located in Table 18.6 using a 10% discount rate.

$$NPV = \sum_{t=0}^{n} \frac{CF_t}{(1+i)^t} = -100 + \frac{20}{(1+0.1)^1} + \frac{30}{(1+0.1)^2} + \frac{40}{(1+0.1)^3} + \frac{80}{(1+0.1)^4}$$

$$+ \frac{60}{(1+0.1)^5} = \$64.92$$

Table 18.6 NPV Example

Year	Profit ($MM)
0 (investment)	−$100.00
1	$20.00
2	$30.00
3	$40.00
4	$80.00
5	$60.00

Table 18.7 Net Present Value Summary

Year	Profit ($MM)	Present Value ($MM)
0 (investment)	− $100.00	− $100.00
1	$20.00	$18.18
2	$30.00	$24.79
3	$40.00	$30.05
4	$80.00	$54.64
5	$60.00	$37.26
Summation (NPV at 10%)		$64.92

The present value for each year is summarized in Table 18.7.

Example

Imagine you win the million-dollar lottery! Don't get too excited. You will actually get paid $50,000 per year for the next 20 years. If the discount rate is a constant 8% and the first payment will be in year 1, how much have you actually won in present dollars? What are you going to do, take a lump sum or yearly payments for the next 20 years?

This is a classic NPV example. First of all, it is very important to take a lump sum instead of yearly payments for the next 20 years due to the fact that this money can be invested in various projects and make higher returns. Secondly, based on the time value of money concept, people are impatient about their money and would love to have their money as soon as possible instead of yearly payments for the next 20 years. In addition, taxes will have to be paid on the calculated $490,907 present value of the lottery.

$$NPV = \frac{50,000}{(1+0.08)^1} + \frac{50,000}{(1+0.08)^2} + \frac{50,000}{(1+0.08)^3} + \frac{50,000}{(1+0.08)^4} + \frac{50,000}{(1+0.08)^5} + \frac{50,000}{(1+0.08)^6}$$
$$+ \frac{50,000}{(1+0.08)^7} + \frac{50,000}{(1+0.08)^8} + \frac{50,000}{(1+0.08)^9} + \frac{50,000}{(1+0.08)^{10}} + \frac{50,000}{(1+0.08)^{11}}$$
$$+ \frac{50,000}{(1+0.08)^{12}} + \frac{50,000}{(1+0.08)^{13}} + \frac{50,000}{(1+0.08)^{14}} + \frac{50,000}{(1+0.08)^{15}} + \frac{50,000}{(1+0.08)^{16}}$$
$$+ \frac{50,000}{(1+0.08)^{17}} + \frac{50,000}{(1+0.08)^{18}} + \frac{50,000}{(1+0.08)^{19}} + \frac{50,000}{(1+0.08)^{20}}$$
$$= \$490,907$$

Table 18.8 shows the summary for discounted cash flows for the next 20 years at an 8% interest rate. As can be seen from Table 18.8, the present value of $50,000 at year 20 is only $10,727. This example clearly illustrates the main reason why E&P companies would like to make as much money as possible during the first 5−10 years of the life of a well in order to create the most value for the shareholders. Typically during a well's life (which varies from well to well based on type curve and economic assumptions made), up to 70–80% of the value is associated with the first 8 years of production with less than 50% of the EUR produced. The next 42 years (if the reserve life is assumed to be 50 years) delivers approximately 20–30% of present value and over 50% of EUR. A sensitivity analysis on various parameters of this exercise can be made to understand the impact of value creation and EUR throughout the life of a well. As previously mentioned, these percentages (value and EUR) will be different based on type curve and economic parameter assumptions.

Table 18.8 Present Value Example Summary

Year	CF	PV	Year	CF	PV
1	$50,000	$46,296	11	$50,000	$21,444
2	$50,000	$42,867	12	$50,000	$19,856
3	$50,000	$39,692	13	$50,000	$18,385
4	$50,000	$36,751	14	$50,000	$17,023
5	$50,000	$34,029	15	$50,000	$15,762
6	$50,000	$31,508	16	$50,000	$14,595
7	$50,000	$29,175	17	$50,000	$13,513
8	$50,000	$27,013	18	$50,000	$12,512
9	$50,000	$25,012	19	$50,000	$11,586
10	$50,000	$23,160	20	$50,000	$10,727

CF, cash flow; PV, present value

Advantages of NPV:
- NPV accounts for time value of money.
- Cash flows over the economic life of the project are taken into account.
- NPV provides a sense of scale about the value that will be created for the shareholders.
- NPVs can be added. If there are 100 projects with an NPV of $1000 for each project, the total NPV can be easily summed up to be $100,000.
- NPV assumes that all the future cash flows are reinvested at the cost of capital. The same cost of capital does not have to be used for the entire life of the project and different discount rates can be assumed.

Disadvantage of NPV:
NPV does not give any indication on the size of the original investment. For instance, the NPV of a $10 million investment could be $1 million, and the NPV of a $1 billion investment could also be $1 million.

BTAX AND ATAX MONTHLY DISCOUNTED NET CASH FLOW (NCF)

Before tax monthly undiscounted NCF is written in Eq. (18.15). The NPV equation can be used on a monthly basis to calculate before and after tax present value of all the future cash flows as shown in Eq. (18.19). The only difference when discounting monthly cash flows is to divide time in the NPV equation by 12. Calculation of after federal income tax (ATAX) monthly undiscounted NCF is discussed in the tax model.

BTAX or ATAX Monthly Discounted NCF

$$= \frac{\text{BTAX or ATAX Monthly Undiscounted NCF}}{(1 + \text{WACC})^{\frac{\text{Time}}{12}}}$$

Equation 18.19 BTAX or ATAX monthly discounted NCF.

BTAX or ATAX monthly undiscounted NCF = $
WACC = Weighted average cost of capital, $
Time = month

1. BTAX NPV is the summation of all BTAX monthly discounted cash flows.
2. ATAX NPV is the summation of ATAX monthly discounted cash flows.

INTERNAL RATE OF RETURN (IRR)

Internal rate of return (IRR) is known as discounted cash-flow rate of return (DCFROR) or simply rate of return (ROR). Internal rate of return is the discount rate when the NPV of particular cash flows is exactly zero. The higher the IRR, the more growth potential a project has. IRR is an important decision metric on any project. IRR is frequently used for project evaluation and profitability of a project. The formula for calculating IRR is basically the same formula as NPV except that the NPV is replaced by zero and the discount rate is replaced by IRR as shown in Eq. (18.20). As opposed to NPV, IRR assumes that positive cash flows of a project are reinvested at IRR instead of cost of capital. This is one of the **disadvantages** of using the IRR method since it **defectively** assumes that positive cash flows are reinvested at the IRR.

When the NPV of a particular project is exactly zero, the IRR will yield cost of capital of a project. For example, if the cost of capital of a particular publicly traded company is 9.3% and the NPV of a particular project yields zero, IRR will be 9.3% for that particular project. This means the present value of all the cash inflows is just enough to cover the cost of capital. When NPV is zero, no value will be created for the shareholders. IRR must be higher than the cost of capital of a project to create any value for the shareholders. When IRR is less than the cost of capital, no value will be created for the shareholders.

$$0 = \sum_{t=0}^{n} \frac{CF_t}{(1+IRR)^t}$$

Equation 18.20 Internal rate of return (IRR).

IRR rule of thumb:

The rationale behind IRR in an independent project is:

1. If IRR is greater than WACC (IRR > WACC), the project's rate of return will exceed its costs and as a result the project should be accepted.
2. If IRR is less than WACC (IRR < WACC), the project's rate of return will not exceed its costs and as a result the project should be rejected.

For example, if a company's cost of capital (WACC) is 12% and IRR for a particular project is calculated to be 11%, the project must be declined because it would cost more to finance the project (through debt and equity) than the actual return of the project. On the other hand, if a company's cost of capital is 12% and the IRR for a specific project is 20%, the project is approved. A lot of companies have a minimum acceptable IRR before investing in a project. This minimum acceptable IRR for one particular company could be 15% while for others could be 20% or 25% depending on many factors, especially market conditions.

In mutually exclusive projects, the project with higher IRR must be picked. For example, if IRR on project A is 15% and project B is 20%, project B must be selected.

Example

Calculate the IRR for the cash flows listed in Table 18.9.

IRR can be calculated using Eq. (18.20):

$$0 = -500 + \frac{-100}{(1+IRR)^1} + \frac{20}{(1+IRR)^2} + \frac{300}{(1+IRR)^3} + \frac{400}{(1+IRR)^4} + \frac{500}{(1+IRR)^5}$$

As can be determined when manually computing IRR, IRR can be solved either using trial and error or linear interpolation methods. Financial calculators or Excel are recommended to perform this calculation. In this example, if various discount rates are inputted into the above equation when the IRR is 19.89% in the denominator of each term, the equation is equal to 0. This means the IRR for this particular project is approximately 20%.

Table 18.9 IRR Example

Year	Profit ($MM)
0 (investment)	− $500.00
1	− $100.00
2	$20.00
3	$300.00
4	$400.00
5	$500.00
IRR	**19.89%**

Internal rate of return calculation:

As previously discussed, when manually computing IRR, IRR can be calculated using trial and error, which is tedious and time consuming, or linear interpolation. Many commercial economic software packages use linear interpolation, in which the software finds the discount rate when the sign of NPV changes from positive to negative and linearly interpolates between the two discount rates. One of the flaws with this type of calculation is that two users who define different series of discount rates will see different calculated IRR. There are other mathematical methods (not discussed in this book) such as the root finding method, which can be used to perform such calculations. In the above example, let's calculate NPV at different discount rates of 10%, 15%, 18%, and 25%. Afterward, linear interpolation can be used to calculate the discount rate when NPV is zero.

As can be seen from Table 18.10, NPV goes from 37.08 MM @ 18% discount rate to -85.92 MM @ 25% discount rate. After performing linear interpolation to find the discount rate when NPV is 0, IRR is found to be 20.11%, which is close 19.89%. This difference may be expanded at

Table 18.10 NPV at Various Discount Rates Example

Discount Rate	10%	15%	18%	25%
Time 0	− 500.00	− 500.00	− 500.00	− 500.00
Discounted CF, Year 1	− 90.91	− 86.96	− 84.75	− 80.00
Discounted CF, Year 2	16.53	15.12	14.36	12.80
Discounted CF, Year 3	225.39	197.25	182.59	153.60
Discounted CF, Year 4	273.21	228.70	206.32	163.84
Discounted CF, Year 5	310.46	248.59	218.55	163.84
Summation (NPV)	234.68	102.71	37.08	− 85.92

CF, cash flow; NPV, net present value.

higher IRRs and widely spaced discount rates. Therefore, users with various series of discount rates will see different calculated IRRs.

Example

Calculate the internal rate of return using Table 18.11 given the NPV for each discount rate.

IRR is the discount rate at which NPV is equal to zero. In this example, NPV @ 15% discount rate is $20 MM and NPV @ 20% discount rate is $ − 6 MM. Therefore, NPV is equal to zero when the discount rate is in between 15% and 20%. Linear interpolation can be used to find the discount rate when NPV is 0 given the predefined series of discount rates.

$$Y = Y_a + (Y_b - Y_a) \times \frac{X - X_a}{X_b - X_a} = 15 + (20 - 15) \times \frac{0 - 20,000,000}{-6,000,000 - 20,000,000} = 18.85\%$$

From this example, the discount rate when NPV is 0 is equal to 18.85%.

Table 18.11 IRR Example

Discount Rate (%)	0	5	10	15	20	25	30	35
NPV ($MM)	200	150	100	20	−6	−11	−16	−21

NPV Profile:

NPV profile is a graphical representation of project's NPV against various discount rates. Discount rates and NPV are subsequently plotted on the x- and y-axis.

Example

Draw the NPV profile for projects A and B and determine which project is better assuming a cost of capital of 5%.

The first task in this problem is to draw the NPV profile by plotting discount rate (x-axis) versus NPVs for projects A and B (y-axis). IRR is the point at which NPV curves cross the x-axis as shown in Fig. 18.2. There is a point referred to as the crossover point (rate) in Fig. 18.2. Crossover point is the discount rate at which the NPV for both projects is equal (Table 18.12).

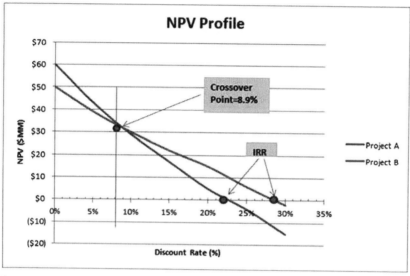

Figure 18.2 Crossover point illustration.

Table 18.12 Net Present Value (NPV) Profile

Rate (%)	NPV (A)	NPV (B)
0	$60	$50
5	$43	$39
10	$29	$30
15	$17	$22
20	$5	$15
25	($4)	$6
30	($15)	($2)

There are three stages in the following NPV profile. The first stage occurs before the crossover point, and in this phase, the NPV of project A is more than the NPV of project B. In this stage, there is a conflict between IRR and NPV since the NPV of project A is more than B, while the IRR of project B is more than A. The company's cost of capital in this example is given to be 5%. When cost of capital is less than cross-over point (rate), a conflict exists. When a conflict exists and the cost of capital is less than the crossover point, the NPV method must be used for decision making. Therefore, project A is superior to project B in this example since the cost of capital is given to be 5%. When the cost of capital is low, delaying cash flows is not penalized as much compared to at a higher cost of capital. When the cost of capital is high (more than the crossover point) delaying cash flows will be penalized.

At the crossover point (second stage), NPV of both projects is equal. Finally, during the third stage, NPV for project B is more than NPV for project A. Please note that if the cost of capital in this problem was given to be 10% instead of 5% (cost of capital > crossover rate), both NPV and IRR methods would have led to the same project selection. It is important to note that it is the difference in timing of cash flows that is causing the crossover between the two projects. The project with faster payback provides more cash flows in the early years for reinvestment. If the interest rate is high, it is vital to get the money back faster because it can be reinvested while if the interest rate is low, there is not such a hurry to get the money back faster.

Advantages of IRR:
- IRR accounts for time value of money.
- Cash flows over the economic life of the project are taken into account.

Disadvantage of IRR:
- IRR does not provide a **sense of scale** about the value created for the shareholders.
- IRRs cannot be added. If there are four projects with IRRs of 15%, 18%, 22%, and 12% the total IRR will not be 67%. Instead, cash flows of all the projects must be combined and IRR can be determined from the combined cash flows.
- IRR assumes that all the future cash flows are reinvested at IRR.
- IRR just like NPV does not give any indication of the size of the original investment.
- IRR cannot be calculated when:
- cash flows are all negative or positive;
- total undiscounted revenues are less than the original investment;
- cumulative cash flow stream changes sign more than once by going positive to negative.

NPV VERSUS IRR

NPV basically measures the dollar benefit (added value) of the project to the shareholders but it does not provide information on the safety margin or the amount of capital at risk. For example, if NPV of a project is calculated to be $2 million, it does not indicate the kind of safety margin that the project has. In contrast, IRR measures the annual rate of return and provides safety margin information. All in all, for mutually exclusive projects and ranking purposes, NPV is always superior to IRR. Unfortunately, in the oil and gas industry, IRR is quite often used for making critical decisions. It is recommended to calculate and understand IRR methodology for each project. However, the ultimate decision whether to perform a project should be determined using NPV calculation.

MODIFIED INTERNAL RATE OF RETURN (MIRR)

Modified internal rate of return (MIRR) is basically an improved version of IRR and is another tool used in capital budgeting. It is very important to understand the difference between IRR and MIRR. As previously mentioned, IRR defectively assumes that positive cash flows from a particular project are reinvested at IRR. In contrast to IRR, MIRR assumes that cash flows from a project are reinvested at cost of capital or a particular reinvestment rate. In addition to this improvement, MIRR only yields one solution. Consequently MIRR can be defined as the discount rate that causes the present value of a project's terminal value to equal the present value of cost. The MIRR concept is fairly complicated and will only make more sense with examples. This is one of the main reasons that IRR is used more frequently in the real world, i.e., since MIRR is not completely understood by a lot of managers. MIRR can be calculated using Eq. (18.21).

$$MIRR = \sqrt{\frac{\text{Future value (positive cash flows @ reinvestment rate)}}{-\text{Present value (negative cash flows @ cost of capital or finance rate)}}} - 1$$

Equation 18.21 Modified internal rate of return (MIRR).

Example

The cash flows for projects A and B are summarized in Table 18.13. Calculate MIRR assuming a cost of capital of 10% and reinvestment rate of 12%.

The first step is to calculate the present value of negative cash flows at cost of capital for both projects:

$$\text{Project A present value} = \frac{-600}{(1+0.1)^0} = -600$$

$$\text{Project B present value} = \frac{-350}{(1+0.1)^0} = -350$$

Next, future values of positive cash flows at reinvestment rate must be calculated for both projects:

Project A future value
$$= 100 \times (1+12\%)^4 + 250 \times (1+12\%)^3 + 320 \times (1+12\%)^2 + 385 \times (1+12\%)^1$$
$$+ 400 \times (1+12\%)^0 = \$1741.19$$

Project B future value
$$= 200 \times (1+12\%)^4 + 225 \times (1+12\%)^3 + 250 \times (1+12\%)^2 + 350 \times (1+12\%)^1$$
$$+ 450 \times (1+12\%)^0 = \$1786.41$$

Using the MIRR equation:

$$\text{Project A MIRR} = \sqrt{\frac{1741.19}{-(-600)}} - 1 = 0.2375 \text{ or } 23.75\%$$

$$\text{Project B MIRR} = \sqrt{\frac{1786.41}{-(-350)}} - 1 = 0.3854 \text{ or } 38.54\%$$

In this example, note that the first-year cash inflow is assumed to be reinvested in 4 years (5-1), the second-year cash inflow is assumed to be reinvested in 3 years (5-2), the third-year cash inflow is assumed to be reinvested in 2 years (5-3), the fourth-year cash inflow is assumed to be reinvested in 1 year (5-4), and finally the fifth-year cash inflow is received at the end of the fifth year and is not available for reinvestment since it accords with the end of the project's life.

Table 18.13 MIRR Example

Year	Project A ($MM)	Project B ($MM)
0	($600)	($350)
1	$100	$200
2	$250	$225
3	$320	$250
4	$385	$350
5	$400	$450

PAYBACK METHOD

Payback method is another capital budgeting method to determine the quick profitability of an original investment. Payout period is the period of time in which a particular project is expected to recover its initial investment. For example, if $7 MM was initially invested on a particular project for drilling and completing a well and it took 3.5 years to earn $7 MM of profit back, the payback period for this project would be 3.5 years. The payback period when cash inflows per period are even can be calculated using Eq. (18.22).

$$\text{Payback period} = \frac{\text{Initial investment}}{\text{Cash inflow per period}}$$

Equation 18.22 Payback period.

Example

Calculate the payback period given the following undiscounted cash flows, assuming uneven cash inflows and using Table 18.14.

Years 1, 2, and 3 are added up to be 90. The transition from year 3 to year 4 is when the cash inflows exceed the original investment of $100 million. Placing this in an algebraic equation,

$$\text{Payback} = 3 + \frac{10}{45} = 3.22 \text{ years}$$

Table 18.14 Payback Period Example

Year	Cash Flow ($MM)
0	($100)
1	$20
2	$30
3	$40
4	$45
5	$70

Therefore, it takes 3.22 years to pay back the original investment.

Strengths of payback method:
- easy to calculate and understand
- provides an intuition of project risk and liquidity.

Weaknesses of payback method:

As can be easily determined, payback method ignores the time value of money.

This method also ignores the cash flows occurring after the payback period.

Discounted payback method:

Discounted payback method, just like the payback method, is another method used in capital budgeting to determine the profitability of an investment. The difference between payback period and discounted payback period is that in discounted payback period, time value of money is taken into account when calculating the number of years it takes to break even from an initial investment. When calculating discounted payback period, discounted cash flows are used instead of undiscounted cash flows.

Example

Calculate the discounted payback period using the undiscounted cash flows as shown in Table 18.15 by assuming a 10% cost of capital.

The first step is to use the PV equation to discount the future cash flows for each period at 10% discount rate (shown in Table 18.16). Afterward, discounted payback period can be easily calculated using algebra.

$$\text{Discounted payback period} = 3 + \frac{90 - 73.02}{34.15} = 3.5 \text{ years}$$

This method, just like the payback method, ignores the cash flows after the discounted payback period; however, it takes time value of money into account.

Table 18.15 Discounted Payback Period Example Problem

Year	Cash Flow ($MM)
0	($90)
1	$20
2	$30
3	$40
4	$50
5	$60

Table 18.16 Discounted Payback Period Example Answer

Year	Cash Flow ($MM)	PV Equation	PV of Cash Flow ($MM)
0	($90)	$-90/1.1^0$	($90)
1	$20	$20/1.1^1$	$18.18
2	$30	$30/1.1^2$	$24.79
3	$40	$40/1.1^3$	$30.05
4	$50	$50/1.1^4$	$34.15
5	$60	$60/1.1^5$	$37.26

PROFITABILITY INDEX (PI)

Profitability index (PI) is another tool used in capital budgeting to measure the profitability of a project. As previously discussed, NPV yields the total dollar figure of a project (absolute measure), but profitability is a relative measure given by a ratio; the higher the PI the higher the ranking. Profitability index essentially tells us how much money will be gained for every dollar invested. For example, PI of 1.4 of a project tells us that for every dollar invested in the project, an expected return of $1.4 is anticipated. PI is well known among financial managers as representing the bang-per-buck measure. Profitability index can be calculated using Eq. (18.23).

$$\text{Profitability index (PI)} = \frac{\text{PV of future cash flows excluding investment}}{\text{Initial investment}}$$

Equation 18.23 Profitability index.

Profitability index rule of thumb:
- Accept projects that have PI more than 1 (PI > 1)
- Reject projects that have PI of less than 1 (PI < 1).

●●●

Example

Calculate the profitability index assuming 10% discount rate and $200 million investment using Table 18.17.

Present value for each cash flow is summarized in Table 18.18 and PI can be calculated as follows:

$$PI = \frac{\text{PV of future cash flows}}{\text{Initial investment}} = \frac{245.97}{200} = 1.23$$

Table 18.17 Profitability Index Example Problem

Year	Cash Flow ($MM)
1	$20
2	$30
3	$55
4	$60
5	$70
6	$55
7	$90

Table 18.18 Profitability Index Example Answer

Year	Cash Flow ($MM)	PV Equation	PV ($MM)
1	$20	20/1.1^1	$18.18
2	$30	30/1.1^2	$24.79
3	$55	55/1.1^3	$41.32
4	$60	60/1.1^4	$40.98
5	$70	70/1.1^5	$43.46
6	$55	55/1.1^6	$31.05
7	$90	90/1.1^7	$46.18
Summation			$245.97

TAX MODEL (ATAX CALCULATION)

The tax model is used for after-tax calculation in oil and gas property evaluation. This model takes into account depreciation, taxable income, corporation tax rate, and discounting. The discounting equations for the tax model are the same as the NCF model. The primary difference between the two models is that depreciation, taxable income, and corporate tax rate are all taken into account in the tax model.

Depreciation

Tangible and intangible capital expenditure must be specified for after-tax calculations. Typically 10–20% is considered tangible with the remaining percentage being intangible Capex. The Internal Revenue Service (IRS) defines depreciation rates using the accelerated recovery method depreciation for 7 years as shown in Table 18.19. Majority of petroleum investments have a 5 or 7 year guideline life when using the accelerated recovery method depreciation, however, other tables such as ACR 3-Year and ACR 10-Year can also be used. Most petroleum engineers primarily

Table 18.19 ACR2 7-Year
Depreciation Rate

Year	ACR2 7-Year (%)
1	14.29
2	24.49
3	17.49
4	12.49
5	8.93
6	8.92
7	8.93
8	4.46

use ACR2 7-Year or ACR2 5-Year depreciation tables to perform after tax economic analysis. Monthly depreciation must be calculated for each year using the defined IRS depreciation rate and appropriate tangible Capex. It is important to classify the items that are considered tangible versus items that are intangible in an attempt to accurately account for depreciation of tangible capital using accelerated recovery method depreciation either over 5 or 7 year (depending on the company). Monthly depreciation can be calculated using Eq. (18.24) which assumes depreciation rate for each year is paid equally on a monthly basis. Depreciation occurrence will be different depending on the time of the year that a well is turned in line but for the simplicity of the monthly depreciation calculation, yearly depreciation rate is divided by 12.

$$
\text{Monthly Depreciation} = \frac{\text{Yearly depreciation rate}}{12} \times \text{tangible investment} \times \text{WI}
$$

Equation 18.24 Monthly depreciation calculation for tax model.

Yearly depreciation rate = IRS-defined accelerated recovery method, 7 years (will vary depending on the company), %
Tangible investment = Typically 10−20% of the total drilling and completions Capex, $
WI = Working interest, %.

Example

Calculate depreciation rate for the first month assuming a total drilling and completions Capex of $7 MM with 15% tangible and 65% WI using ACR2 7-year depreciation table.

$$
\text{Depreciation}_{\text{month 1}} = \frac{14.29\%}{12} \times (7,000,000 \times 15\%) \times 65\% = \$8127
$$

Taxable Income

Once depreciation is calculated, taxable income can be calculated using Eqs. (18.25) and (18.26).

$$\text{Taxable income @ investment date} = -(\text{Intangible investment} \times WI)$$

Equation 18.25 Taxable income @ investment date.

Intangible investment = Typically 80% to 90% of the total drilling and completions Capex, $

WI = Working interest, %

Please note that in Eq. (18.25), profit excluding investment at investment date is 0 and that is why the term (intangible investment★WI) is being multiplied by −1. In addition, depreciation is 0 at investment date. Equation 18.25 assumes that the entire intangible capital is written off when investment is made. Typically the intangible capital is written off in the first year but for the simplicity of calculating monthly taxable income, intangible capital is written off at investment date in this equation.

$$\text{Taxable income after invstment}$$
$$= \text{Profit excluding investment} - \text{depreciation}$$

Equation 18.26 Taxable income after investment.

Taxable income after investment = Monthly taxable income, $

Profit excluding investment = Monthly basis, $.

Example

An 8000′ lateral length well's total Capex is $8.250 MM. Assuming 88% intangible Capex and 100% WI, calculate taxable income at and after initial investment date using Table 18.20 for the first year assuming ACR2,7 YRS depreciation schedule:

Depreciation rate for the first year from ACR 2, 7 YRS schedule is 14.29%.

$$\text{Depreciation}_{\text{month 1 through 12}} = \frac{14.29\%}{12} \times (8,250,000 \times 12\%) \times 100\% = \$11,789$$

$$\text{Taxable income @ investment date} = -(8,250,000 \times 88\% \times 100\%) = \$ -7,260,000$$

$$\text{Taxable income}_{\text{month 1}} = 253,794 - 11,789 = \$242,005$$

$$\text{Taxable income}_{\text{month 2}} = 207,166 - 11,789 = \$195,377$$

$$\text{Taxable income}_{\text{month 3}} = 178,231 - 11,789 = \$166,442$$

The remaining taxable incomes for this example are summarized in Table 18.21.

Table 18.20 Taxable Income Example Problem

Month	Profit Excluding Investment
0 (investment date)	$0
1	$253,794
2	$207,166
3	$178,231
4	$158,156
5	$143,241
6	$131,633
7	$122,288
8	$114,571
9	$108,068
10	$102,499
11	$97,665
12	$93,421

Table 18.21 Taxable Income Example Answer

	Question		Answer	
Month	Profit Excluding Investment	Depreciation	Taxable Income
0 (investment date)	$0	$0	−$7,260,000
1	$253,794	$11,789	$242,005
2	$207,166	$11,789	$195,377
3	$178,231	$11,789	$166,442
4	$158,156	$11,789	$146,367
5	$143,241	$11,789	$131,452
6	$131,633	$11,789	$119,844
7	$122,288	$11,789	$110,499
8	$114,571	$11,789	$102,782
9	$108,068	$11,789	$96,279
10	$102,499	$11,789	$90,710
11	$97,665	$11,789	$85,876
12	$93,421	$11,789	$81,632

Corporation Tax

Corporations, just like individuals and small business owners, have specific tax brackets. Therefore, when performing ATAX calculation corporation tax must also be taken into account. Corporation tax can be calculated using Eq. (18.27).

$$\text{Corporation tax} = \text{Taxable income} \times \text{corporation tax rate}$$

Equation 18.27 Corporation tax.

Taxable income $= \$$
Corporation tax rate $= \%$.

ATAX Monthly Undiscounted NCF

Now that depreciation, taxable income, and corporation tax have all been discussed, the last step before discounting the ATAX future cash flows is to calculate ATAX monthly undiscounted NCF. ATAX monthly undiscounted NCF can be calculated using Eq. (18.28).

$$\text{ATAX Monthly Undiscounted NCF} = \text{BTAX Monthly Undis. NCF}$$

$$- \text{Corporation Tax}$$

Equation 18.28 ATAX monthly undiscounted NCF.

BTAX monthly undiscounted NCF $= \$$
Corporation tax $= \$$.

Example

Assuming a corporation tax rate of 35%, calculate monthly corporation tax and ATAX monthly undiscounted NCF for the first year using the assumptions listed in Table 18.22.

Step 1) Calculate corporation tax for each month (sample calculation below):

$$\text{Corporation tax}_{time\ 0} = -3,897,395 \times 35\% = \$ -1,364,088$$

$$\text{Corporation tax}_{month\ 1} = 248,058 \times 35\% = \$86,820$$

$$\text{Corporation tax}_{month\ 2} = 201,429 \times 35\% = \$70,500$$

Table 18.22 ATAX Monthly Undiscounted NCF Example Problem

Month	BTAX Monthly Undiscounted NCF	Taxable Income	Corporation Tax Rate
0	$- \$4,379,096$	$- \$3,897,395$	35%
1	$253,794	$248,058	35%
2	$207,166	$201,429	35%
3	$178,231	$172,495	35%
4	$158,156	$152,420	35%
5	$143,241	$137,505	35%
6	$131,633	$125,897	35%
7	$122,288	$116,552	35%
8	$114,571	$108,835	35%
9	$108,068	$102,332	35%
10	$102,499	$96,763	35%
11	$97,665	$91,928	35%
12	$93,421	$87,685	35%

Step 2) Calculate ATAX monthly undiscounted NCF (sample calculation below):

ATAX monthly undiscounted $NCF_{time\ 0} = -4,379,096 - (-1,364,088) = \$ -3,015,007$

ATAX monthly undiscounted $NCF_{month\ 1} = 253,794 - 86,820 = \$166,974$

ATAX monthly undiscounted $NCF_{month\ 2} = 207,166 - 70,500 = \$136,665$

Table 18.23 summarizes the results for this example.

Table 18.23 ATAX Monthly Undiscounted NCF Example Answer

	Question		Answer		
Month	BTAX Monthly Undiscounted NCF	Taxable Income	Corporation Tax Rate (%)	Corporation Tax	ATAX Monthly Undiscounted NCF
0	− 4,379,096	− 3,897,395	35	− 1,364,088	− 3,015,007
1	253,794	248,058	35	86,820	166,974
2	207,166	201,429	35	70,500	136,665
3	178,231	172,495	35	60,373	117,858
4	158,156	152,420	35	53,347	104,809
5	143,241	137,505	35	48,127	95,114
6	131,633	125,897	35	44,064	87,569
7	122,288	116,552	35	40,793	81,495
8	114,571	108,835	35	38,092	76,479
9	108,068	102,332	35	35,816	72,252
10	102,499	96,763	35	33,867	68,632
11	97,665	91,928	35	32,175	65,490
12	93,421	87,685	35	30,690	62,732

●●●

Example

A type curve is generated from 200 producing dry gas wells from a field with similar reservoir properties. You are to run economic analysis and figure out whether the management should proceed with drilling and completing the well or not. The type curve generated is for an 8000′ lateral length well with an IP of 14,500 MSCF/D, annual secant effective decline of 58%, and b value of 1.5. Assuming the following parameters, calculate NPV and IRR for the life of the well (assume 50-year life).

Terminal decline = 5%, WI = 100%, RI = 20%, BTU factor = 1.06 (1060 BTU/SCF), Shrinkage factor = 0.985

Fixed variable and gathering cost = \$426/month/well escalated at 3% to the life of the well, Variable lifting cost = \$0.14/MSCF escalated at 3% to the life of the well,

Variable gathering and compression cost = $0.35/MMBTU escalated at 3% to the life of the well, Firm transportation = $0.30/MMBTU escalated at 3% to the life of the well, Gas price = assume $3/MMBTU escalated at 3% to the life of the well, Severance tax = 5%, Ad valorem tax = 2.5%, Tangible investment = $1,500,000, Intangible investment = $6,500,000 (assume the entire intangible capital is written off when investment is made), Apply total investment 3 months before start date (TIL date), Discount all of the future cash flows using mid-point discounting to the date (time) the investment is made, Weighted average cost of capital = 8.8%, Corporation tax rate = 40%

The calculations shown below are for the first 6 months only and the remaining time is recommended to be performed using an Excel spreadsheet to compare the final NPV and IRR reported in this problem. This problem should provide step-by-step guidance on how to perform economic analysis on a new well based on the assumptions listed above. Some of the assumptions used in this example (e.g., ATAX calculation method, discounting method, etc.) can greatly vary from company to company.

Step 1) Calculate monthly nominal secant hyperbolic:

$$D_i = \left[\frac{1}{12b}\right] \times \left[(1-D_{eis})^{-b} - 1\right] = \left[\frac{1}{12 \times 1.5}\right] \times \left[(1-58\%)^{-1.5} - 1\right] = 14.85\%$$

Step 2) Calculate hyperbolic cumulative rate for each month starting with month 1:

$$N_p = \left\{ \left[\frac{IP}{(1-b) \times \text{Monthly Nominal Hyp}}\right] \right.$$

$$\times \left. \left[1-(1+b \times \text{Monthly Nominal Hyp} \times \text{time}\right]^{1-\frac{1}{b}}\right\} \times \frac{365}{12}$$

$$N_{p,\text{month 1}} = \left\{ \left[\frac{14,500}{(1-1.5) \times 14.85\%}\right] \times \left[1-(1+1.5 \times 14.85\% \times 1)\right]^{1-\frac{1}{1.5}} \right\}$$

$$\times \frac{365}{12} = 411,820 \frac{MSCF}{1 \text{ month}}$$

$$N_{p,\text{month 2}} = \left\{ \left[\frac{14,500}{(1-1.5) \times 14.85\%}\right] \times \left[1-(1+1.5 \times 14.85\% \times 2)\right]^{1-\frac{1}{1.5}} \right\}$$

$$\times \frac{365}{12} = 776,199 \frac{MSCF}{2 \text{ month}}$$

$$N_{p,month\ 3} = \left\{ \left[\frac{14,500}{(1-1.5) \times 14.85\%} \right] \times \left[1 - (1+1.5 \times 14.85\% \times 3)^{1-\frac{1}{1.5}} \right] \right\}$$

$$\times \frac{365}{12} = 1,104,815 \frac{MSCF}{3\ month}$$

$$N_{p,month\ 4} = \left\{ \left[\frac{14,500}{(1-1.5) \times 14.85\%} \right] \times \left[1 - (1+1.5 \times 14.85\% \times 4)^{1-\frac{1}{1.5}} \right] \right\}$$

$$\times \frac{365}{12} = 1,405,331 \frac{MSCF}{4\ month}$$

$$N_{p,month\ 5} = \left\{ \left[\frac{14,500}{(1-1.5) \times 14.85\%} \right] \times \left[1 - (1+1.5 \times 14.85\% \times 5)^{1-\frac{1}{1.5}} \right] \right\}$$

$$\times \frac{365}{12} = 1,683,079 \frac{MSCF}{5\ month}$$

$$N_{p,month\ 6} = \left\{ \left[\frac{14,500}{(1-1.5) \times 14.85\%} \right] \times \left[1 - (1+1.5 \times 14.85\% \times 6)^{1-\frac{1}{1.5}} \right] \right\}$$

$$\times \frac{365}{12} = 1,941,935 \frac{MSCF}{6\ month}.$$

Step 3) Calculate monthly rate by subtracting cumulative volumes from the previous month:

$$q_{hyperbolic,month\ 1} = 411,820 \frac{MSCF}{M}$$

$$q_{hyperbolic,month\ 2} = 776,199 - 411,820 = 364,379 \frac{MSCF}{M}$$

$$q_{hyperbolic,month\ 3} = 1,104,815 - 776,199 = 328,616 \frac{MSCF}{M}$$

$$q_{hyperbolic,month\ 4} = 1,405,331 - 1,104,815 = 300,516 \frac{MSCF}{M}$$

$$q_{hyperbolic,month\ 5} = 1,683,079 - 1,405,331 = 277,748 \frac{MSCF}{M}$$

$$q_{hyperbolic,month\ 6} = 1,941,935 - 1,683,079 = 258,856 \frac{MSCF}{M}.$$

Step 4) Calculate monthly nominal decline for each month:

$$D_{month\ 1} = \frac{\text{Monthly nominal hyperbolic}}{1 + b \times \text{monthly nominal hyperbolic} \times \text{time}}$$

$$= \frac{14.85\%}{1 + 1.5 \times 14.85\% \times 1} = 12.1\%$$

$$D_{month\ 2} = \frac{14.85\%}{1 + 1.5 \times 14.85\% \times 2} = 10.3\%$$

$$D_{month\ 3} = \frac{14.85\%}{1 + 1.5 \times 14.85\% \times 3} = 8.9\%$$

$$D_{month\ 4} = \frac{14.85\%}{1 + 1.5 \times 14.85\% \times 4} = 7.9\%$$

$$D_{month\ 5} = \frac{14.85\%}{1 + 1.5 \times 14.85\% \times 5} = 7.0\%$$

$$D_{month\ 6} = \frac{14.85\%}{1 + 1.5 \times 14.85\% \times 6} = 6.4\%.$$

Step 5) Calculate annual effective decline for each month:

$$D_{e,month\ 1} = 1 - (1 + 12 \times b \times D)^{-\frac{1}{b}} = 1 - (1 + 12 \times 1.5 \times 12.1\%)^{-\frac{1}{1.5}} = 53.8\%$$

$$D_{e,month\ 2} = 1 - (1 + 12 \times 1.5 \times 10.3\%)^{-\frac{1}{1.5}} = 50.2\%$$

$$D_{e,month\ 3} = 1 - (1 + 12 \times 1.5 \times 8.9\%)^{-\frac{1}{1.5}} = 47.1\%$$

$$D_{e,month\ 4} = 1 - (1 + 12 \times 1.5 \times 7.9\%)^{-\frac{1}{1.5}} = 44.4\%$$

$$D_{e,month\ 5} = 1 - (1 + 12 \times 1.5 \times 7.0\%)^{-\frac{1}{1.5}} = 42.0\%$$

$$D_{e,month\ 6} = 1 - (1 + 12 \times 1.5 \times 6.4\%)^{-\frac{1}{1.5}} = 39.9\%$$

After calculating the annual effective decline for the remaining life of the well, it appears that at month 145, the annual effective decline reaches 5% terminal decline. The hyperbolic decline equation must be switched to an exponential decline equation for the life of the well starting with month 145.

Step 6) Calculate monthly nominal exponential decline using the equation below:

$$D = -\ln\left[(1 - D_e)^{\frac{1}{12}}\right] = -\ln\left[(1 - 5\%)^{\frac{1}{12}}\right] = 0.427\%$$

Step 7) Calculate exponential decline rate for each month after reaching 5% terminal decline using the equation below:

$$q_{exponential} = \left(IP \times e^{-D \times t}\right) \times \left(\frac{365}{12}\right)$$

The rate at which hyperbolic decline is switched to exponential is 1404 MSCF/D or 42,698 MSCF/M. The first month right after the switch time (month 146) is called month 1 in this particular example, followed by the remaining months for the life of the well.

$$q_{exponential,month\ 1} = \left(1404 \times e^{-0.427\% \times 1}\right) \times \left(\frac{365}{12}\right) = 42,516 \frac{MSCF}{M}$$

$$q_{exponential,month\ 2} = \left(1404 \times e^{-0.427\% \times 2}\right) \times \left(\frac{365}{12}\right) = 42,334 \frac{MSCF}{M}$$

$$q_{exponential,month\ 3} = \left(1404 \times e^{-0.427\% \times 3}\right) \times \left(\frac{365}{12}\right) = 42,154 \frac{MSCF}{M}$$

$$q_{exponential,month\ 4} = \left(1404 \times e^{-0.427\% \times 4}\right) \times \left(\frac{365}{12}\right) = 41,974 \frac{MSCF}{M}$$

$$q_{exponential,month\ 5} = \left(1404 \times e^{-0.427\% \times 5}\right) \times \left(\frac{365}{12}\right) = 41,795 \frac{MSCF}{M}$$

$$q_{exponential,month\ 6} = \left(1404 \times e^{-0.427\% \times 6}\right) \times \left(\frac{365}{12}\right) = 41,617 \frac{MSCF}{M}.$$

Step 8) Calculate net gas production for each month:

Net gas production = Gross gas production × shrinkage factor × NRI%

$$\text{Net gas production}_{month\ 1} = 411,820 \times 0.985 \times 80\% = 324,514 \frac{MSCF}{M}$$

$$\text{Net gas production}_{month\ 2} = 364,379 \times 0.985 \times 80\% = 287,131 \frac{MSCF}{M}$$

$$\text{Net gas production}_{month\ 3} = 328,616 \times 0.985 \times 80\% = 258,949 \frac{MSCF}{M}$$

$$\text{Net gas production}_{month\ 4} = 277,748 \times 0.985 \times 80\% = 236,806 \frac{MSCF}{M}$$

$$\text{Net gas production}_{month\ 5} = 258,856 \times 0.985 \times 80\% = 218,865 \frac{MSCF}{M}$$

$$\text{Net gas production}_{\text{month 6}} = 242,883 \times 0.985 \times 80\% = 203,979 \frac{\text{MSCF}}{\text{M}}.$$

Step 9) Calculate gas pricing incorporating an escalation of 3% using a stair-step escalation:

$$\text{Gas price}_{\text{month 1}} = \frac{\$3}{\text{MMBTU}}$$

$$\text{Gas price}_{\text{month 2}} = 3 \times (1+3\%)^{\frac{1}{12}} = \frac{\$3.007}{\text{MMBTU}}$$

$$\text{Gas price}_{\text{month 3}} = 3.007 \times (1+3\%)^{\frac{1}{12}} = \frac{\$3.015}{\text{MMBTU}}$$

$$\text{Gas price}_{\text{month 4}} = 3.015 \times (1+3\%)^{\frac{1}{12}} = \frac{\$3.022}{\text{MMBTU}}$$

$$\text{Gas price}_{\text{month 5}} = 3.022 \times (1+3\%)^{\frac{1}{12}} = \frac{\$3.030}{\text{MMBTU}}$$

$$\text{Gas price}_{\text{month 6}} = 3.030 \times (1+3\%)^{\frac{1}{12}} = \frac{\$3.037}{\text{MMBTU}}.$$

Step 10) Calculate adjusted gas pricing by accounting for 1060 BTU gas:

$$\text{Adjusted gas price} = \text{Gas price} \times \text{BTU factor}$$

$$\text{Adjusted gas price}_{\text{month 1}} = 3 \times 1.06 = \frac{\$3.18}{\text{MSCF}}$$

$$\text{Adjusted gas price}_{\text{month 2}} = 3.007 \times 1.06 = \frac{\$3.188}{\text{MSCF}}$$

$$\text{Adjusted gas price}_{\text{month 3}} = 3.015 \times 1.06 = \frac{\$3.196}{\text{MSCF}}$$

$$\text{Adjusted gas price}_{\text{month 4}} = 3.022 \times 1.06 = \frac{\$3.204}{\text{MSCF}}$$

$$\text{Adjusted gas price}_{\text{month 5}} = 3.030 \times 1.06 = \frac{\$3.211}{\text{MSCF}}$$

$$\text{Adjusted gas price}_{\text{month 6}} = 3.037 \times 1.06 = \frac{\$3.219}{\text{MSCF}}.$$

Step 11) Calculate net revenue for each month:

Net revenue = (Monthly shrunk net gas production $\times$ adjusted gas pricing)

$$\text{Net revenue}_{\text{month 1}} = 324,514 \times 3.18 = \$1,031,955$$

$$\text{Net revenue}_{\text{month 2}} = 287,131 \times 3.188 = \$915,328$$

$$\text{Net revenue}_{\text{month 3}} = 258,949 \times 3.196 = \$827,526$$

$$\text{Net revenue}_{\text{month 4}} = 236,806 \times 3.204 = \$758,630$$

$$\text{Net revenue}_{\text{month 5}} = 218,865 \times 3.211 = \$702,883$$

$$\text{Net revenue}_{\text{month 6}} = 203,979 \times 3.219 = \$656,691.$$

Step 12) Calculate severance tax for each month:

Severance tax per month
= (Gross monthly gas production $\times$ adjusted gas pricing $\times$ severance tax $\times$ NRI $\times$ total shrinkage factor) OR Net revenue $\times$ severance tax

$$\text{Severance tax}_{\text{month 1}} = 1,031,955 \times 5\% = \$51,598$$

$$\text{Severance tax}_{\text{month 2}} = 915,328 \times 5\% = \$45,766$$

$$\text{Severance tax}_{\text{month 3}} = 827,526 \times 5\% = \$41,376$$

$$\text{Severance tax}_{\text{month 4}} = 758,630 \times 5\% = \$37,931$$

$$\text{Severance tax}_{\text{month 5}} = 702,883 \times 5\% = \$35,144$$

$$\text{Severance tax}_{\text{month 6}} = 656,691 \times 5\% = \$32,835.$$

Step 13) Calculate ad valorem tax for each month:

Advalorem tax per month
= {[(Gross monthly gas production $\times$ adjusted gas pricing $\times$ NRI $\times$ total shrinkage factor)] − Severance tax amount}
$\times$ Advalorem tax OR (Net revenue − severance tax) $\times$ Ad valorem tax

$$\text{Ad valorem tax}_{\text{month 1}} = (1,031,955 - 51,598) \times 2.5\% = \$24,509$$

$$\text{Ad valorem tax}_{\text{month 2}} = (915,328 - 45,766) \times 2.5\% = \$21,739$$

$$\text{Ad valorem tax}_{\text{month 3}} = (827,526 - 41,376) \times 2.5\% = \$19,654$$

$$\text{Ad valorem tax}_{\text{month 4}} = (758,630 - 37,931) \times 2.5\% = \$18,017$$

$$\text{Ad valorem tax}_{\text{month 5}} = (702,883 - 35,144) \times 2.5\% = \$16,693$$

$$\text{Ad valorem tax}_{\text{month 6}} = (656,691 - 32,835) \times 2.5\% = \$15,596$$

Step 14) First perform escalation on fixed, variable, and FT costs:

$$\text{Fixed cost escalatin}_{\text{month 1}} = \$426$$

$$\text{Fixed cost escalatin}_{\text{month 2}} = 426 \times (1+3\%)^{\frac{1}{12}} = \$427.1$$

$$\text{Fixed cost escalatin}_{\text{month 3}} = 427.1 \times (1+3\%)^{\frac{1}{12}} = \$428.1$$

$$\text{Fixed cost escalatin}_{\text{month 4}} = 428.1 \times (1+3\%)^{\frac{1}{12}} = \$429.2$$

$$\text{Fixed cost escalatin}_{\text{month 5}} = 429.2 \times (1+3\%)^{\frac{1}{12}} = \$430.2$$

$$\text{Fixed cost escalatin}_{\text{month 6}} = 430.2 \times (1+3\%)^{\frac{1}{12}} = \$431.3$$

$$\text{Total variable}\left(\frac{\$}{\text{MSCF}}\right)$$

$$= \text{variable lifting}\left(\frac{\$}{\text{MSCF}}\right) + \text{variable gathering}\left(\frac{\$}{\text{MSCF}}\right) + \text{FT}\left(\frac{\$}{\text{MSCF}}\right)$$

$$\text{Total variable cost per MSCF} = \frac{\$0.14}{\text{MSCF}} + \left(\frac{\$0.35}{\text{MMBTU}} \times 1.06\right)$$

$$+ \left(\frac{\$0.3}{\text{MMBTU}} \times 1.06\right) = \frac{\$0.829}{\text{MSCF}}$$

$$\text{Variable cost escalatin}_{\text{month 1}} = \frac{\$0.829}{\text{MSCF}}$$

$$\text{Variable cost escalatin}_{\text{month 2}} = 0.829 \times (1+3\%)^{\frac{1}{12}} = \frac{\$0.831}{\text{MSCF}}$$

$$\text{Variable cost escalatin}_{\text{month 3}} = 0.831 \times (1+3\%)^{\frac{1}{12}} = \frac{\$0.833}{\text{MSCF}}$$

$$\text{Variable cost escalatin}_{\text{month 4}} = 0.833 \times (1+3\%)^{\frac{1}{12}} = \frac{\$0.835}{\text{MSCF}}$$

$$\text{Variable cost escalatin}_{\text{month 5}} = 0.835 \times (1+3\%)^{\frac{1}{12}} = \frac{\$0.837}{\text{MSCF}}$$

$$\text{Variable cost escalatin}_{\text{month 6}} = 0.837 \times (1+3\%)^{\frac{1}{12}} = \frac{\$0.839}{\text{MSCF}}.$$

Step 15) Calculate total Opex for each month:

Total OPEX per month
= [(Gross monthly gas production × WI × total shrinkage factor
× variable lifting cost)][(Fixed lifting cost × WI)]
+ [(Gross monthly gas production × WI × total shrinkage factor
× gathering and compression cost)]
+ [(Gross monthly gas production × WI
× total shrinkage factor × FT cost)]

$$\text{Total OPEX}_{\text{month 1}} = (411,820 \times 100\% \times 0.985 \times 0.829) + (426 \times 100\%)$$
$$= \$336,704$$

$$\text{Total OPEX}_{\text{month 2}} = (364,379 \times 100\% \times 0.985 \times 0.831) + (427.1 \times 100\%)$$
$$= \$298,700$$

$$\text{Total OPEX}_{\text{month 3}} = (328,616 \times 100\% \times 0.985 \times 0.833) + (428.1 \times 100\%)$$
$$= \$270,090$$

$$\text{Total OPEX}_{\text{month 4}} = (300,516 \times 100\% \times 0.985 \times 0.835) + (429.2 \times 100\%)$$
$$= \$247,640$$

$$\text{Total OPEX}_{\text{month 5}} = (277,748 \times 100\% \times 0.985 \times 0.837) + (430.2 \times 100\%)$$
$$= \$229,475$$

$$\text{Total OPEX}_{\text{month 6}} = (258,856 \times 100\% \times 0.985 \times 0.839) + (431.3 \times 100\%)$$
$$= \$214,424.$$

Step 16) Calculate net Opex for each month:

$$\text{Net OPEX} = \text{Total OPEX} + \text{severance tax amount} + \text{advalorem tax amount}$$

$$\text{Net OPEX}_{\text{month 1}} = 336,704 + 51,598 + 24,509 = \$412,811$$
$$\text{Net OPEX}_{\text{month 2}} = 298,700 + 45,766 + 21,739 = \$366,206$$
$$\text{Net OPEX}_{\text{month 3}} = 270,090 + 41,376 + 19,654 = \$331,120$$
$$\text{Net OPEX}_{\text{month 4}} = 247,640 + 37,931 + 18,017 = \$303,589$$
$$\text{Net OPEX}_{\text{month 5}} = 229,475 + 35,144 + 16,693 = \$281,312$$
$$\text{Net OPEX}_{\text{month 6}} = 214,424 + 32,835 + 15,596 = \$262,855.$$

Step 17) Calculate operating cash flow or profit excluding investment:

Profit (excluding investment) = Net revenue − net OPEX

$$\text{Profit}_{\text{month 1}} = 1,031,955 - 412,811 = \$619,145$$

$$\text{Profit}_{\text{month 2}} = 915,328 - 366,206 = \$549,122$$

$$\text{Profit}_{\text{month 3}} = 827,526 - 331,120 = \$496,406$$

$$\text{Profit}_{\text{month 4}} = 758,630 - 303,589 = \$455,041$$

$$\text{Profit}_{\text{month 5}} = 702,883 - 281,312 = \$421,570$$

$$\text{Profit}_{\text{month 6}} = 656,691 - 262,855 = \$393,836.$$

Step 18) Calculate net Capex (since WI is 100%, net Capex is equal to gross Capex):

$$\text{Net CAPEX} = \text{Gross CAPEX} \times \text{WI}$$

$$\text{Net CAPEX} = (1,500,000 + 6,500,000) \times 100\% = \$8,000,000$$

Apply $8,000,000 total net investment 3 months prior to start date (the date where the production begins).

Step 19) Calculate BTAX monthly undiscounted NCF:

BTAX monthly undiscounted net cash flow = profit − net CAPEX

Net Capex at time zero is equal to net Capex. However, net Capex for subsequent months is zero.

$$\text{Investment date (time 0)} = 0 - 8,000,000 = -\$8,000,000$$

$$\text{BTAX undiscounted NCF}_{\text{month 2}} = \$0$$

$$\text{BTAX undiscounted NCF}_{\text{month 3}} = \$0$$

$$\text{BTAX undiscounted NCF}_{\text{month 4 from investment date}} = 619,145 - 0 = \$619,145$$

$$\text{BTAX undiscounted NCF}_{\text{month 5 from investment date}} = 549,122 - 0 = \$549,122$$

$$\text{BTAX undiscounted NCF}_{\text{month 6 from investment date}} = 496,406 - 0 = \$496,406$$

$$\text{BTAX undiscounted NCF}_{\text{month 7 from investment date}} = 455,041 - 0 = \$455,041$$

$$\text{BTAX undiscounted NCF}_{\text{month 8 from investment date}} = 421,570 - 0 = \$421,570$$

$$\text{BTAX undiscounted NCF}_{\text{month 9 from investment date}} = 393,836 - 0 = \$393,836.$$

Step 20) Calculate BTAX monthly discounted (midpoint) NCF. To perform midpoint discounting, subtract 0.5 from each month as shown in the equation below:

$$\text{BTAX Monthly Discounted NCF}$$

$$= \frac{\text{BTAX Monthly Undiscounted NCF}}{(1+\text{WACC})^{\frac{\text{Time}-0.5}{12}}}$$

$$\text{BTAX Discounted NCF}_{\text{investment date}} = -\$8,000,000$$

$$\text{BTAX Discounted NCF}_{\text{month 2}} = \$0$$

$$\text{BTAX Discounted NCF}_{\text{month 3}} = \$0$$

$$\text{BTAX Discounted NCF}_{\text{month 4 from investment date}} = \frac{619,145}{(1+8.8\%)^{\frac{4-0.5}{12}}} = \$604,100$$

$$\text{BTAX Discounted NCF}_{\text{month 5 from investment date}} = \frac{549,122}{(1+8.8\%)^{\frac{5-0.5}{12}}} = \$532,026$$

$$\text{BTAX Discounted NCF}_{\text{month 6 from investment date}} = \frac{496,406}{(1+8.8\%)^{\frac{6-0.5}{12}}} = \$477,583$$

$$\text{BTAX Discounted NCF}_{\text{month 7 from investment date}} = \frac{455,041}{(1+8.8\%)^{\frac{7-0.5}{12}}} = \$434,720$$

$$\text{BTAX Discounted NCF}_{\text{month 8 from investment date}} = \frac{421,570}{(1+8.8\%)^{\frac{8-0.5}{12}}} = \$399,923$$

$$\text{BTAX Discounted NCF}_{\text{month 9 from investment date}} = \frac{393,836}{(1+8.8\%)^{\frac{9-0.5}{12}}} = \$370,997.$$

Step 21) Calculate depreciation for each month starting with production date:

$$\text{Monthly Depreciation} = \frac{\text{Yearly depreciation rate}}{12} \times \text{tangible investment} \times \text{WI}$$

$$\text{Depreciation}_{\text{month 1}} = \frac{14.29\%}{12} \times 1,500,000 \times 100\% = \$17,863$$

Depreciation for the next 11 months will be the same using IRS-defined accelerated recovery method.

Step 22) Calculate taxable income starting with when the investment is made:

$$\text{Taxable income @ investment date} = -(\text{Intangible investment} \times \text{WI})$$

$$= -(6,500,000 \times 100\%) = -\$6,500,000$$

$$\text{Taxable income after invstment} = \text{Profit excluding investment} - \text{depreciation}$$

$$\text{Taxable income}_{month\ 2} = \$0$$

$$\text{Taxable income}_{month\ 3} = \$0$$

$$\text{Taxable income}_{month\ 4} = 619,145 - 17,863 = \$601,282$$

$$\text{Taxable income}_{month\ 5} = 549,122 - 17,863 = \$531,260$$

$$\text{Taxable income}_{month\ 6} = 496,406 - 17,863 = \$478,544$$

$$\text{Taxable income}_{month\ 7} = 455,041 - 17,863 = \$437,179$$

$$\text{Taxable income}_{month\ 8} = 421,570 - 17,863 = \$403,708$$

$$\text{Taxable income}_{month\ 9} = 393,836 - 17,863 = \$375,974.$$

Step 23) Calculate corporation tax for each month starting with the investment date:

$$\text{Corporation tax} = \text{Taxable income} \times \text{corporation tax rate}$$

$$\text{Corporation tax}_{investment\ date} = -6,500,000 \times 40\% = -\$2,600,00$$

$$\text{Corporation tax}_{month\ 2} = \$0$$

$$\text{Corporation tax}_{month\ 3} = \$0$$

$$\text{Corporation tax}_{month\ 4} = 601,282 \times 40\% = \$240,513$$

$$\text{Corporation tax}_{month\ 5} = 531,260 \times 40\% = \$212,504$$

$$\text{Corporation tax}_{month\ 6} = 478,544 \times 40\% = \$191,417$$

$$\text{Corporation tax}_{month\ 7} = 437,179 \times 40\% = \$174,871$$

$$\text{Corporation tax}_{month\ 8} = 403,708 \times 40\% = \$161,483$$

$$\text{Corporation tax}_{month\ 9} = 375,974 \times 40\% = \$150,389.$$

Step 24) Calculate ATAX undiscounted NCF for each month starting with the investment date:

ATAX Monthly Undiscounted NCF

$$= \text{BTAX Monthly Undis. NCF} - \text{Corporation Tax}$$

ATAX Monthly Undiscounted $\text{NCF}_{\text{Invetsment date}}$

$$= -8,000,000 - (-2,600,000) = -\$5,400,000$$

ATAX Monthly Undiscounted $\text{NCF}_{\text{months 2}} = \$0$

ATAX Monthly Undiscounted $\text{NCF}_{\text{months 3}} = \$0$

ATAX Monthly Undiscounted $\text{NCF}_{\text{months 4}} = 619,145 - 240,513$
$$= \$378,632$$

ATAX Monthly Undiscounted $\text{NCF}_{\text{months 5}} = 549,122 - 212,504$
$$= \$336,618$$

ATAX Monthly Undiscounted $\text{NCF}_{\text{months 6}} = 496,406 - 191,417$
$$= \$304,989$$

ATAX Monthly Undiscounted $\text{NCF}_{\text{months 7}} = 455,041 - 174,871$
$$= \$280,170$$

ATAX Monthly Undiscounted $\text{NCF}_{\text{months 8}} = 421,570 - 161,483$
$$= \$260,087$$

ATAX Monthly Undiscounted $\text{NCF}_{\text{months 9}} = 393,836 - 150,389$
$$= \$243,447.$$

Step 25) Calculate ATAX monthly discounted NCF for each month starting with investment date:

$$\text{ATAX Monthly Discounted NCF} = \frac{\text{ATAX Monthly Undiscounted NCF}}{(1 + \text{WACC})^{\frac{\text{Time} - 0.5}{12}}}$$

ATAX Discounted $\text{NCF}_{\text{investment date}} = -\$5,400,000$

ATAX Discounted $\text{NCF}_{\text{month 2}} = \$0$

ATAX Discounted $\text{NCF}_{\text{month 3}} = \$0$

$$\text{ATAX Discounted NCF}_{\text{month 4 from investment date}} = \frac{378,632}{(1+8.8\%)^{\frac{4-0.5}{12}}} = \$369,431$$

$$\text{ATAX Discounted NCF}_{\text{month 5 from investment date}} = \frac{336,618}{(1+8.8\%)^{\frac{5-0.5}{12}}} = \$326,138$$

$$\text{ATAX Discounted NCF}_{\text{month 6 from investment date}} = \frac{304,989}{(1+8.8\%)^{\frac{6-0.5}{12}}} = \$293,424$$

$$\text{ATAX Discounted NCF}_{\text{month 7 from investment date}} = \frac{280,170}{(1+8.8\%)^{\frac{7-0.5}{12}}} = \$267,658$$

$$\text{ATAX Discounted NCF}_{\text{month 8 from investment date}} = \frac{260,087}{(1+8.8\%)^{\frac{8-0.5}{12}}} = \$246,732$$

$$\text{ATAX Discounted NCF}_{\text{month 9 from investment date}} = \frac{243,447}{(1+8.8\%)^{\frac{9-0.5}{12}}} = \$229,329$$

BTAX NPV is the summation of all BTAX monthly discounted cash flows for 50 years. ATAX NPV is the summation of ATAX monthly discounted cash flows for 50 years. The summaries of both BTAX and ATAX NPVs are listed in Table 18.24.

Table 18.24 ATAX and BTAX NPV Profile Example
NPV Profile (Mid-discounting)

Discount Rate (%)	BTAX NPV	ATAX NPV
0	$42,158,086	$25,294,851
5	$17,876,005	$10,637,205
8.8	**$11,250,715**	**$6,608,750**
10	$9,910,182	$5,789,584
15	$6,163,010	$3,487,676
20	$3,962,247	$2,124,255
25	$2,488,747	$1,205,150
30	$1,418,424	$534,014
40	($56,802)	($395,914)
50	($1,043,607)	($1,020,815)
60	($1,760,296)	($1,475,760)
70	($2,309,467)	($1,824,769)
80	($2,746,487)	($2,102,619)
90	($3,104,192)	($2,330,036)
100	($3,403,442)	($2,520,232)

BTAX and ATAX IRR can now be easily calculated by interpolating between 30% and 40% discount rate, in which the BTAX and ATAX NPV switch sign from positive to negative.

$$Y = Y_a + (Y_b - Y_a) \times \frac{X - X_a}{X_b - X_a}$$

$$\text{BTAX IRR} = 30\% + (40\% - 30\%) \times \frac{(0 - 1,418,424)}{(-56,802 - 1,418,424)} = 39.61\%$$

$$\text{ATAX IRR} = 30\% + (40\% - 30\%) \times \frac{(0 - 534,014)}{(-395,914 - 534,014)} = 35.74\%.$$

EXAMPLE PROBLEMS

1. What type of frac fluid system is used in formations with high Young's modulus and low Poisson's ratio? Why?
2. What type of frac fluid system is used in formations with low Young's modulus and high Poisson's ratio? Why?
3. What type of frac fluid system is ideal for underpressured and depleted reservoirs? Why?
4. Why is foam fracturing ideal for water-sensitive formations with high percentage of clay?
5. What is the most common foam quality (FQ) used in coalbed methane (CBM) wells?
6. Why is a foam frac fluid system considered to have a better fluid efficiency and lower fluid-loss feature? Give an example.
7. You are treating a frac stage in the field using slick water fluid system. The stage treats very well at lower sand concentrations of up to 1.5 ppg. However, as soon as a higher sand concentration of 2 ppg reaches the perforations, surface-treating pressure starts rising without any relief. What would you do to solve the problem and be able to place the remaining sand into the formation? Explain in detail.
8. Is a slick water frac fluid system considered to have a laminar flow pattern or turbulent? Why?
9. What kind of flow pattern is expected from a cross-linked fluid system? Why?
10. What is the main essence of slick water hydraulic frac?
11. What is the purpose of the acidization stage in slick water frac? What kind of reaction will typically be observed in the surface-treating pressure when acid hits the perforations?
12. What is the reaction of surface-treating pressure if the formation is known to be very limy (limestone)?
13. How much acid and water volume will be needed given 28% hydrochloric acid in order to obtain 1500 gallons of 15% acid?
14. Is it possible to pump 6 ppg sand concentrations with slick water frac? Why?
15. You are a completions engineer in the field responsible for treating a frac stage located in the Barnett Shale in Texas. Calculate the flush volume if 5½", 20 lb/ft, P-110 (ID = 4.778") production casing is used. The bottom perforation MD of the stage is 15,640'.

16. Why do many companies overflush (by 10–40 barrels) on top of the calculated volume per stage?

17. What is the main purpose of proppant in hydraulic frac jobs?

18. The closure pressure from a shale formation is calculated to be 5432 psi from the DFIT (G function analysis). What type of proppant would you design for your job?

19. What is the importance of specific gravity of proppant in hydraulic fracturing? What happens to proppant in the fractures as SG increases?

20. What type of sand is famous for encapsulating fines?

21. Part (a) You have been notified via the production engineer that a lot of sand (frac sand and not formation sand) is being produced with the flowback water in a particular area. As a completions engineer, what type of proppant will you use to prevent this problem on an upcoming pad?
    Part (b) What if not enough capital (Capex) is available to pump this type of proppant? What other approach will you take to prevent excessive proppant flowback during production? Explain why.

22. Name the three main categories of proppant from lowest conductivity to highest conductivity.

23. What are the main applications of 100 mesh sand?

24. What could be the reasons why so many E&P companies have seen great production success with pumping large percentages of 100 mesh in some areas?

25. What kind of sand size is highly recommended in highly naturally fractured formations? Why?

26. Based on Stokes' law, 40/70 mesh creates more surface area compared to 30/50. Why?

27. What type of analysis is used on a daily basis in the field to ensure proper sand size and is reported on each frac ticket?

28. Define conductivity. What is dimensionless frac conductivity?

29. What happens to conductivity as closure pressure increases? Explain why.

30. Why is it highly recommended to test bigger sand sizes in liquid-rich areas (high BTU)?

31. Explain in detail the characteristics that ISO conductivity test does not account for.

32. Define proppant stress. When is the most sensitive time during flowback in relation to proppant crushing?

33. Part (a) Calculate proppant stress during the following two periods:
  Period (1)
      Closure pressure = 5450 psi
      Net pressure = 500 psi
      P_{wf} (flowing bottom-hole pressure) = 3000 psi
  Period (2)
      Closure pressure = 5450 psi
      Net pressure = 500 psi
      P_{wf} (flowing bottom-hole pressure) = 2500 psi.
  Part (b) What is the most recommended practice when lowering flowing bottom-hole pressure?

34. Calculate dimensionless frac conductivity with the following reservoir and completions properties (assume the conductivity is reduced to 1 lb/ft^2 and 80% reduction of conductivity due to time degradation):
  Frac conductivity = 800 md-ft @ 2 lb/ft^2
  $K = 0.003$ md
  $X_f = 500$ ft

35. What is the main purpose of friction reducer during hydraulic fracturing?

36. You are treating a live frac stage at 84 bpm and 8500 psi surface-treating pressure. Suddenly, due to FR pump malfunctioning, FR is completely lost. What happens to surface-treating pressure? What would you do? Explain the steps.

37. Part (a) Calculate the estimated pipe friction pressure (assuming no FR) using the following parameters:
  Total MD (measured depth of the well) = 14,000'
  Fluid density = 8.5 ppg
  Designed rate (flow rate) = 85 bpm
  ID of the production casing = 4.778"
  Fluid viscosity = 1.1 cp.
  Part (b) Can you pump this job without the use of FR? Why?
  Part (c) Recalculate your pipe friction pressure assuming 20,000' of pipe instead of 14,000'.
  Part (d) What is the biggest challenge with long lateral-length wells in any formation from a completions perspective?
  Part (e) What will happen to the pipe friction pressure if the decision is made to use a smaller size production casing (with smaller ID)?

38. Calculate the expected surface-treating pressure for each of the following measured depths (MD) and assuming the following parameters:

TVD (ft)	MD (ft)	Pipe Friction Pressure, Lab Test	Estimated Surface-Treating Pressure
12,500	20,000	7080	?
12,500	19,500	6931	?
12,500	19,000	6782	?
12,500	18,500	6633	?
12,500	18,000	6484	?
12,500	17,500	6335	?
12,500	17,000	6186	?
12,500	16,500	6037	?
12,500	16,000	5888	?
12,500	15,500	5739	?
12,500	15,000	5590	?
12,500	14,500	5441	?
12,500	14,000	5292	?
12,500	13,500	5143	?
12,500	13,000	4994	?
12,500	12,500	4845	?

ISIP $= 9258$ psi (from DFIT test), TVD $= 12,500'$, Water density $= 8.65$ ppg, Designed rate $= 90$ bpm, Casing ID $= 5$ $1/2"$, 23 #/ft (ID $= 4.670"$), $D_p = 0.42"$, $N = 40$ perforations (holes)/stage, $C_d = 0.8$, ΔPnet $= 200$ psi

39. The total amounts of water and sand have both been optimized for a Barnett Shale slick water frac schedule using production data and rate transient analysis. Calculate the remaining schedule and show step-by-step work in the frac design schedule table shown below.

40. Part (a) Calculate slurry density and hydrostatic pressure of 2.75 ppg sand stage using sintered bauxite (SG $= 3.0$) mixed with produced water (density $= 8.85$ ppg). The TVD of this particular frac stage 13,001'.

Part (b) Calculate the surface-treating pressure increase when sand is cut and the well is being flushed.

90 bpm Frac Design Schedule

Stage Name	Pump Rate (bpm)	Fluid Name	Stage Fluid Clean Vol (BBLs)	Stage Fluid Slurry Vol (BBLs)	% of Total Clean Vol (BBLs)	Prop Conc. (ppg)	Stage Proppant (lbs)	% of Total Prop %	Cumulative Prop (lbs)	Stage Time (min)
Pump ball	15	Slickwater	24			0				
5% HCl acid	90	Acid	60			0				
Pad	90	Slickwater	410			0				
100 mesh	90	Slickwater	450			0.25				
100 mesh	90	Slickwater	500			0.5				
100 mesh	90	Slickwater	475			0.75				
100 mesh	90	Slickwater	450			1				
100 mesh	90	Slickwater	450			1.25				
100 mesh	90	Slickwater	450			1.5				
40/70 mesh	90	Slickwater	400			0.25				
40/70 mesh	90	Slickwater	375			0.5				
40/70 mesh	90	Slickwater	450			0.75				
40/70 mesh	90	Slickwater	475			1				
40/70 mesh	90	Slickwater	485			1.25				
40/70 mesh	90	Slickwater	490			1.5				
40/70 mesh	90	Slickwater	495			1.75				
40/70 mesh	90	Slickwater	480			2				
40/70 mesh	90	Slickwater	470			2.25				
40/70 mesh	90	Slickwater	453			2.5				
40/70 mesh	90	Slickwater	450			3				
Flush	90	Slickwater	300			0				

Total clean volume   BBLs

Sand/water ratio   %

Pad percentage   OR   %

OR

100 mesh (lbs)	OR
40/70 mesh (lbs)	OR
Total (lbs)	

Total stage time (min)

0.0

Stage length (ft.)	200
Water/ft.	0	BBL/ft.
Sand/ft.	0	lb/ft.

41. From the G-function plot below, answer the following questions:

   a. What is the bottom-hole (BH) closure pressure?

   b. How did you determine BH closure pressure? Explain the steps and show on the plot.

   c. What is the minimum horizontal stress from this plot? Explain the steps and show on the plot.

   d. What is the pressure-dependent leak-off (PDL) pressure (if any) from this plot? Explain the steps and show on the plot.

   e. Define anisotropy. What is the anisotropy in this figure?

   f. Knowing your closure and PDL pressures, do you expect to get a complex fracture network system or biwing fracture system from this plot? Why? Explain in detail.

   g. Do you expect to see a higher breakdown pressure from frac jobs or lower? Why?

   h. Do you see PDL? Explain how you determined that there is PDL?

   i. What strategies will you take to design the frac job based on this figure? What type and size of proppant will you design just from a completions perspective? (Discard the economic perspective.) Explain why.

   j. What is the fluid efficiency from this plot?

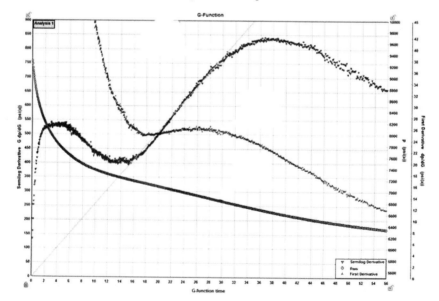

42. There are three unique leak-off regimes that can be noted on the G-function plot. Explain each leak-off regime in detail based on your understanding of the concept.

43. What are the first and the second derivatives in DFIT analysis used for?

44. Explain in detail the type of gauge that is used to record pressure fall-off (surface-pressure fall-off) after the injection test. How do you calculate bottom-hole pressure from surface pressure gauge?

45. Instantaneous shut-in pressure (ISIP) from DFIT is determined to be 4155 psi. If 8.9 ppg fluid density was used to pump the DFIT at 6900' TVD, calculate the bottom-hole ISIP.

46. Why is it not recommended to pump large fluid volume during DFIT jobs in unconventional shale reservoirs with low or very low permeability?

47. In before-closure analysis (BCA), what kind of flow regimes do $\frac{1}{2}$ and $\frac{1}{4}$ slope represent?

48. What does -1 slope of the second derivative of the log–log plot represent?

49. What if pseudoradial flow is not observed from a log–log plot. What other technique can be used to get an approximate measure of pore pressure?

50. Estimate the effective permeability and fluid efficiency from the G-function plot provided with the following properties. Once the frac job starts, will the frac fluid be effective in creating hydraulic fractures based on the calculated fluid efficiency? Why?

    $\mu = 1$ cp, ISIP $= 4225$ psi, TVD $= 7340'$, Fluid density $= 8.5$ ppg, $P_c = 6100$ psi, $C_t = 0.0000254$ 1/psi, $G_c = 3.7$, $E = 5$ MMpsi, $r_p = 1$ (since PDL exists), Porosity $= 20\%$

51. Your boss asks you to determine whether 200' stage spacing is economically better than 300' stage spacing. As a completions engineer, you are trying to determine whether the additional production gain from 200' stage spacing offsets the additional Capex. You are given two sets of cash flows by the reservoir engineer. Calculate NPV and IRR given the tables below and determine whether 200' stage spacing is economically better or 300'. Explain why. Cash flows below are yearly. Assume 12% weighted average cost of capital (WACC).

Year	200′ Stage Spacing CF (M$)	300′ Stage Spacing CF (M$)
0 (Capex)	−10,000	−9000
1	5000	4500
2	4200	4150
3	3400	3500
4	2500	2500
5	2000	2000

52. What makes the determination of various fracture parameters very difficult in unconventional shale reservoirs?
53. How many quantiplex pumps would you design for your job if the calculated surface-treating pressure is around 10,500 psi at the designed rate of 80 bpm assuming each pump has 2250 HHP?

REFERENCES

Abe, H., Mura, T., Keer, L., 1976. Growth-rate of a penny-shaped crack in hydraulic fracturing of rocks. J. Geophys. Res. 81 (29), 5335—5340, 1976.

Adachi, J., Detournay, E., 2008. Plane strain propagation of a hydraulic fracture in a permeable rock. Eng. Fract. Mech. 75, 4666—4694, 2008.

Adesida, A., Akkutlu, I.Y., Resasco, D.E., Rai, C.S., 2011. Characterization of Barnett Shale Pore Size Distribution using DFT Analysis and Monte Carlo Simulations. SPE 147397.

Akkutlu, I.Y., Fathi, E., 2012. Multiscale gas transport in shales with local Kerogen heterogeneities. SPE J. 17 (4).

Ambrose, R.J., Hartman, R.C., Diaz-Campos, M., Akkutlu, I.Y., Sondergeld, C.H., 2010. New pore-scale considerations in Shale gas in-place calculations. SPE-131772.

Ambrose, R.J., Hartman, R.C., Diaz-Campos, M., Akkutlu, I.Y., 2012. Shale gas in-place calculations. Part I—New pore-scale considerations. SPE J. 17 (1), 219—229.

Anderson, M.A., 1988. Predicting reservoir condition,pv compressibility from hydrostatic stress laboratory. Data society of petroleum Engns. Journal (SPE). 3, 1078—1082.

Anderson, J.R., Pratt, K.C., 1985. Introduction to Characterization and Testing of Catalysts. Academic Press, Sydney.

Aria, A., Gharib, M., 2011. Carbon nanotube arrays with tunable wettability and their applications. NSTI-Nanotech. 1.

Arnold, R., Anderson, R., 1908. Preliminary report on Coalinga oil district. U.S. Geol. Survey Bull. 357, 79.

Arps, J.J., 1944. Analysis of Decline Curves—Petroleum Engineering. A.I.M.E. Houston Meeting.

Aziz, K., Settari, A., 1979. Petroleum Reservoir Simulation. Applied Science Publishers Ltd., London.

Bailey, S., 2009. Closure and Compressibility Corrections to Capillary Pressure Data in Shales. October 19, Colorado School of Mines.

Bao, J.Q., Fathi, E., Ameri, S., 2014. A coupled method for the numreical simulation of hydraulic fracturing with a condensation technique. Eng. Fract. Mech. 131, 269—281 (2014).

Bao, J.Q., Fathi, E., Ameri, S., 2015. Uniform investigation of hydraulic fracturing propagation regimes in the plane strain model. Int. J. Numer. Anal. Meth. Geomech. 39, 507—523, 2015.

Bao, J.Q., Fathi, E., Ameri, S., 2016. A unified finite element method for the simulation of hydraulic fracturing with and without fluid lag. Eng. Fract. Mech. 162, 164—178.

Barree, B., 2013. Overview of Current DFIT Analysis Methodology.

Barree, R.D., Baree, V.L., Craig, D., 2007. Holistic fracture diagnostics. In: Rocky Mountain Oil & Gas Technology Symposium, April 16—18, Denver, Colorado. SPE-107877.

Belyadi, F., 2014. Impact of gas desorption on production behavior of shale gas. PhD Dissertation submitted to West Virginia University.

Belyadi, H., Yuyi, S., Junca-Laplace, J., 2015. Production Analysis using Rate Transient Analysis. SPE Eastern Regional Meeting. SPE-177293.

Binder, R.C., 1973. Fluid Mechanics. Prentice-Hall, Inc, Englewood Cliffs, NJ.

Besler, M., Steele, J., Egan, T., Wagner, J., 2007. Improving Well Productivity and Profitability in the Bakken-A Summary of Our Experiences Drilling, Stimulating, and Operating Horizontal Wells, SPE Annual Technical Conference and Exhibition held in Anaheim, SPE-110679.

Belyadi, H., Fathi, E., Belyadi, F., 2016. Managed Pressure Drawdown in Utica/Point Pleasant with Case Studies. SPE Eastern Regional Meeting. SPE-184054.

Belyadi, H., Yuyi, J., Ahmad, M., Wyatt, J., 2016. Deep Dry Utica Well Spacing Analysis with Case Study. SPE Eastern Regional Meeting. SPE-184045.

Blanton, T.L., 1982. An experimental study of interaction between hydraulically induce and pre-existing fractures. SPE 10847. SPE/DOE Unconventional Gas Recovery Symposium, Pittsburgh, PA.

Brace, W.W., 1968. Permeability of granite under high pressure. J. Geophys. Res. 73, 2225.

Britt, L.K., 2011. Keys to successful multi-fractured horizontal wells in tight and unconventional reservoirs. NSI Fracturing & Britt Rock Mechanics Laboratory presentation.

Bui, K., Akkutlu, I.Y., 2015. Nanopore wall effect on surface tension of methane. J. Mol. Phys.

Bunger, A.P., Detournay, E., Garagash, D.I., 2005. Toughness-dominated hydraulic fracture with leak-off. Int. J. Fract. 134, 175−190, 2005.

Cheng, Y., 2010. Impacts of the number of perforation clusters and cluster spacing on production performance of horizontal shale gas wells. In: 138843-MS SPE Conference Paper.

Cheng, Y., 2012. Mechanical interaction of multiple fractures-exploring impacts of the selection of the spacing number of perforation clusters on horizontal shale-gas wells. SPE J. 17 (4), 992−1001.

Cinco-Ley, H., Samaniego, V.F., 1981. Transient pressure analysis for fractured wells. J. Petrol. Technol. 33 (09).

Cramer, D.D., 1987. The application of limited-entry techniques in massive hydraulic fracturing treatments. In: SPE Production Operations Symposium, March 8−10, Oklahoma City, OK.

Cronquist, C., 2001. Estimation and Classification of Reserves of Crude Oil, Natural Gas and Condensates. SPE Book Series, Houston, TX, pp. 157−160.

Culter Jr., W.W., 1924. Estimation of underground oil reseves by well production curves. U.S. Bur. Mines Bull. 228.

Culter, W.W., Johnson, H.R., 1940. Estimating Recoverable Oil of Curtailed Wells. Oil Weekly.

Curtis, M.E., Ambrose, R.J., Sondergeld, C.H., Rai, C.S., 2010. Structural Characterization of Gas Shales on the Micro- and Nano-Scales. CUSG/SPE 137693.

Curtis, M.E., Sondergeld, C.H., Rai, C.S., 2013. Investigation of the microstructure of shales in the oil window. In: Unconventional Resources Technology Conference, August 12−14, Denver, CO.

Dahi, A., Olson, J., 2011. Numerical modeling of multistranded-hydraulic-fracture propagation: accounting for the interaction between induced and natural fractures. SPE J. 16 (03).

Daneshy, A.A., 1974. Hydraulic fracture propagation in the Presence of Planes of Weakness. In: SPE 4852, SPE-European Spring Meeting, Amsterdam.

Dontsov, E.V., Peirce, A.P., 2015. An enhanced pseudo-3D model for hydraulic fracturing accounting for viscous height growth, non-local elasticity, and lateral toughness. Eng. Fract. Mech. 42, 116−139.

Drake, L.C., Ritter, H.L., 1945. Macropore-size distributions in some typical porous substances. Ind. Eng. Chem. Anal. Ed. 17 (12), 787−791, 1945.

Dubinin, M.M., 1960. The Potential Theory of Adsorption of Gases and Vapors for Adsorbents with Energetically Nonuniform Surfaces. Chem. Rev. 60 (2), 235−241.

Dubinin, M.M., 1966. Chemistry and Physics of Carbon. Marcel Dekker, New York.

Duong, A.N., 2011. Rate-decline analysis for fracture-dominated shale reservoirs. SPE Reserv. Eval. Eng. 14 (03).

Ellsworth, W.L., 2013. Injection-induced earthquakes. Science. 341 (6142).

Economides, M., Martin, T., 2007. Modern Fracturing. ET Publishing, Houston,TX.

Ely, J., 2012. Proppant Agents. Waynesburg, PA.

Fathi, E., Akkutlu, I.Y., 2009. Nonlinear sorption kinetics and surface diffusion effects on gas transport in low permeability formations. In: SPE Annual Technical Conference held in New Orleans, LA, October 4–7.

Fathi, E., Akkutlu, I.Y., 2011. Gas transport in shales with local Kerogen heterogeneities. In: SPE Annual Technical Conference and Exhibition (ATCE) 2011 in Denver, CO. SPE-146422-PP.

Fathi, E., Akkutlu, I.Y., 2013. Lattice Boltzmann method for simulation of shale gas transport in Kerogen. SPE J. 18 (1).

Fathi, E., Akkutlu, I.Y., 2014. Multi-component gas transport and adsorption effects during CO_2 injection and enhanced shale gas recovery. Int. J. Coal Geol. 123 (2014), 52–61.

Fathi, E., Tinni, A., Akkutlu, I.Y., 2012. Correction to Klinkenberg slip theory for gas dynamics in nano-capillaries. Int. J. Coal Geol. 103, 51–59.

Feng, F., Akkutlu, I.Y., 2015. Flow of hydrocarbons in nanocapillary: a non-equilibrium molecular dynamics study. In: SPE Asia Pacific Unconventional Resources Conference and Exhibition, November 9–11, Brisbane, Australia.

Finsterle, S., Persoff, P., 1997. Determining permeability of tight rock samples using inverse modeling. Water Resour. Res. 33 (8).

Fisher, M.K., Heinze, J.R., Harris, C.D., Davidson, B.M., Wright, C.A., Dunn, K.P., 2004. Optimizing horizontal completion technologies in the Barnett shale using microsismic fracture mapping. In: Annual Technical Conference and Exhibition, Houston, TX. SPE 90051.

Gadde, P., Yajun, L., Jay, N., Roger, B., Sharma, M., 2004. Modeling proppant settling in water-fracs. In: Proc. SPE-89875-MS, SPE Annu. Tech. Conf. Exhib, September 26–29, Houston, TX.

Gan, H., Nandie, S.P., Walker Jr., P.L., 1972. Nature of porosity in American coals. Fuel. 51, 272–277.

Gao, Q., Cheng, Y., Fathi, E., Ameri, S., 2015. Analysis of stress-field variations expected on subsurface faults and discontinuities in the vicinity of hydraulic fracturing. SPE-168761 SPE Reserv. Eval. Eng. J.

Garagash, D., 2006. Propagation of a plane-strain hydraulic fracture with a fluid lag: Early-time solution. Int. J. Solid. Struct. 43 (43), 5811–5835, 2006.

Garagash, D., 2007. Plane-strain propagation of a fluid-driven fracture during injection and shut-in: asymptotics of large toughness. Eng. Fract. Mech. 74, 456–481.

Gijtenbeek, K., Shaoul, J., Pater H., (2012). Overdisplacing Propped Fracture Treatments-Good Practice or Asking for Trouble?. SPE EAGE Annual Conference and Exhibition. SPE-154397.

Goodway, B., Perez, M., Varsek, J., Abaco, C., 2010. Seismic petrophysics and isotropic–anisotropic AVO methods for unconventional gas exploration. Lead. Edge. 29 (12), 1500–1508.

Gudmundsson, J.S., Hveding, F., Børrehaug, A., 1995. Transport or natural gas as frozen hydrate. In: The Fifth International Offshore and Polar Engineering Conference, June 11–16, The Hague, the Netherlands.

Haimson, B., Fairhurst, C., 1967. Initiation and extension of hydraulic fractures in rocks. Soc. Petrol. Eng. J. 7, 310, 1967.

Hartman, R.C., Ambrose, R.J., Akkutlu, I.Y., Clarkson, C.R., 2011. Shale gas in place calculations. Part II: Multicomponent gas adsorption effects. SPE 144097, SPE Unconventional Gas Conference, Woodland, TX.

Hayashi, K., Haimson, B.C., 1991. Characteristics of shut-in curves in hydraulic fracturing stress measurements and determination of in situ minimum compressive stress. J. Geophys. Res. 96 (B11), 18311–18321, 1991.

Heidbach, O., 2008. Helmholtz Centre Potsdam GFZ German Research Centre for Geosciences.

Homfray, I.F., Physik, Z., 1910. Chemistry. 74, 129.

Ilk, D., Rushing, J.A., Perego, A.D., Blasingame, T.A., 2008. Exponential vs. hyperbolic decline in tight gas sands—Understanding the origin and implications for reserve estimates using Arps' decline curves. In: SPE Annual Technical Conference and Exhibition held in Denver, CO, September 21–24, 2008.

Jones, S., 1997. A technique for faster pulse-decay permeability measurements in tight rocks. SPEFE.19–25.

Kang, S.M., Fathi, E., Ambrose, R.J., Akkutlu, I.Y., Sigal, R.F., 2010. CO_2 applications. Carbon dioxide storage capacity of organic-rich shales. SPE J. 16 (4), 842–855.

Kim, B.H., Kum, G.H., Seo, Y.G., 2003. Adsorption of Methane and Ethane into Single-Walled Carbon Nanotubes and Slit-Shaped Carbonaceous pores. Korean J. Chem. Eng. 20, 104–109.

Kim, Y.I., Amadei, B., Pan, E., 1999. Modeling the effect of water, excavation sequence and rock reinforcement with discontinuous deformation analysis. Int. J. Rock. Mech. Min. 36, 949–970, 1999.

King, G., 2010. 30 Years of gas shale fracturing: what have we learnt? SPE 133456. In: SPE Annual Technical Conference and Exhibition, September 19–22, Florence, Italy.

Kong, B., Fathi, E., Ameri, S., 2015. Coupled 3-D numerical simulation of proppant distribution and hydraulic fracturing performance optimization in Marcellus shale reservoirs, Int. J. Coal Geol., 147-148, pp. 35–45 (2015).

Krumbein, W.C., Sloss, L.L., 1963. Stratigraphy and Sedimentation. W.H. Freeman, San Francisco.

Lamont, N., Jessen, F., 1963. The Effects of Existing Fractures in Rocks on the Extension of Hydraulic Fractures. JPT, pp. 203–209, February.

Langmuir, I., 1916. The constitution and fundamental properties of solids and liquids. Part I. Solids. Am. Chem. Soc. 2221–2295.

Larkey, C.S., 1925. Mathematical determinatio of production decline curves. Trans. AIME. 71, 1315.

Legarth, B., Huenges, E., Zimmermann, G., 2005. Hydraulic fracturing in sedimentary geothermal reservoir: results and implications. Int. J. Rock Mech. Mining Sci. 42 (7–8), 1028–1041, 2005.

Levasseur, S., Charlier, R., Frieg, B., Collin, F., 2010. Hydro-mechanical modelling of the excavation damaged zone around an underground excavation at Mont Terri Rock Laboratory. Int. J. Rock Mech. Mining Sci. 47 (3), 414–425, 2010.

Luffel, D.H., 1993. Matrix permeability measurement of gas productive shales. Paper SPE 26633, SPE Annual Technical Conference and Exhibition. SPE, Houston, TX.

Mallet, J.L., 2002. Geomodeling. Oxford University Press, Inc, New York, NY.

Mavor, M.J., Owen, L.B., Pratt, T.J., 1990. Measurement and evaluation of coal sorption isotherm data. SPE-20728, Paper presented during the Annual Technical Conference and Exhibition of the SPE held in New Orleans, LA, September 23–26.

McClure, B., 2010. Investors Need A Good WACC. Investopedia Newsletter.

McNeil, R., Jeje, O., Renaud, A., 2009. Application of the power law loss-ratio method of decline analysis. In: Canadian International Petroleum Conference, June 16–18, Calgary, Alberta.

Miller, M., 2010. Gas shale evaluation techniques-things to think about. In: OGS Workshop presented July 28, 2010, Norman, OK.

Mobbs, A.T., Hammond, P.S., 2001. Computer simulations of proppant transport in a hydraulic fracture. SPE Prod. Facil. 16 (2).

Morrill, J., Miskimins, J., 2012. Optimizing hydraulic fracture spacing in unconventional shales. In: Paper SPE 152595 presented at Hydraulic Fracturing Technology Conference, Woodlands, February 6–8.

Murdoch, L.C., 2002. Mechanical analysis of ideaized shallow hydraulic fracture. J. Geotechnic. Geoenviron. Eng. 128 (6), 289–313, 2002.

Mutalik, P.N., Gibson, B., 2008. Case history of sequential and simultaneous fracturing of the Barnett shale in Parker county. Presented at the 2008 SPE Annual Technical Conference and Exhibition held in Denver, September 21–24.

Myers, A.L., Prausnitz, J.M., 1965. Thermodynamics of mixed-gas adsorption. AICHE J. 11 (1), 121–127.

Newman, G.H., 1973. Pore-volume compressibility of consolidated, friable and unconsolidated reservoir rock under hydrostatic loading. J. Petrol. Technol. 25 (2), SPE-3835.

Ning, X., 1992. The measurement of matrix and fracture properties in naturally fractured low permeability cores using a pressure pulse method. Ph.D. Thesis. Texas A&M University.

Olson, J., Dahi, A., 2009. Modeling simultaneous growth of multiple hydraulic fractures and their interaction with natural fractures. In: Paper SPE 119739 presented at Hydraulic Fracturing Technology Conference, Woodlands, January 19–21.

Ousina, E., Sondergeld, C., Rai, C., 2011. An NMR study on shale wettability. In: Canadian Unconventional Resources Conference Calgary, Alberta, November.

Ozkan, E., Brown, M., Raghavan, R., Kazemi, H., 2009. Comparison of fractured horizontal-well performance in conventional and unconventional reservoirs. In: Paper SPE 121290 presented at the SPE Western Regional Meeting, San Jose, March 24–26.

Passey, Q.R., Bohacs, K.M., Esch, W.L., Klimentidis, R., Sinha, S., 2010. From oil-prone source rock to gas-producing shale reservoir-geologic and petrophysical characterization of unconventional shale-gas reservoirs. In: SPE 131350 presented at the CPS/SPE International Oil & Gas Conference and Exhibition, Beijing, China, June 8–10.

Phani, B.G., Liu, Y., Norman, J., Bonnecaze, R., Sharma, M., 2004. Modeling proppant settling in Water-Fracs. In: 89875-MS SPE Conference Paper, 2004.

Pirson, S.J., 1935. Production decline curve of oil well may be extrapolated by loss-ratio. Oil Gas J. September 15.

Potluri, N., Zhu, D., Hill, A.D., 2005. Effect of Natural Fractres on Hydraulic Fracture Propagation. SPE 94568, SPE European Formation Damage Conference, The Netherlands, 25-27 May 2005.

Rafiee, M., Soliman, M.Y., Pirayesh, E., 2012. Hydraulic fracturing design and optimization: a modification to zipper frac. SPE 159786.

Rahmani Didar, B., Akkutlu, I.Y., 2013. Pore-size dependence of fluid phase behavior and properties in organic-rich shale reservoirs. SPE-164099, paper prepared for presentation at the SPE Int. Symposium on Oilfield Chemistry held in Woodlands, TX, USA, April 8–10.

Rickman, V.R., Mullen, M.J., Petre, J.E., Erik, J., Grieser, W.V., Vincent, W., Kundert, D., 2008. A practical use of shale petrophysics for stimulation design optimization: All Shale Plays Are Not Clones of the Barnett Shale. In: SPE Annual Technical Conference and Exhibition, September 21–24, Denver, CO.

Rodvelt, G., Ahmad, M., Blake A., (2015). Refracturing Early Marcellus Producers Accesses Additional Gas. SPE Eastern Regional Meeting. SPE-177295.

Roussel, R.N.P., Sharma, M., 2011. Strategies to minimize frac spacing and stimulate. Natural Fractures in Horizontal Completions. SPE 146104.

Ruthven, D.M., 1984. Principles of Adsorption and Adsorption Processes. John Wiley & Sons, Inc, New York.

Rylander, E., Singer, P., Jiang, T., Lewis, R., McLin, R., 2013. NMR T2 Distributions in the Eagle Ford Shale: Reflections on Pore Size. SPE 164554 presented at the Unconventional Resources Conference, Woodlands, TX, April 10−12, 2013.

Santos, J.M., Akkutlu, I.Y., 2012. Laboratory measurement of sorption isotherm under confining stress with pore volume effects. SPE 162595, Calgary, Canada.

Saunders, J.T., Tsai, B.M.C., Yang, R.T., 1985. Adsorption of gases on coals and heattreated coals at elevated temperature and pressure: 2. Adsorption from hydrogen−methane mixtures. Fuel. 64 (5), 621−626.

Schlebaum, W., Scharaa, G., Vanriemsdijk, W.H., 1999. Influence of nonlinear sorption kinetics on the slow desorbing organic contaminant fraction in soil. Environ. Sci. Technol. 33, 1413−1417.

Seshadri, J., Mattar, L., 2010. Comparison of power law and modified hyperbolic decline methods. Paper SPE 137320 presented at the Canadian Unconventional Resources & International Petroleum Conference, Calgary, Alberta, Canada, October 19−21.

Shen, Y., 2014. A variational inequality formulation to incorporate the fluid lag in fluid-driven fracture propagation. Comp. Methods Appl. Mech. Eng. 272 (HF-69), 17−33, 2014.

Shi, J.Q., Durucan, S., 2003. A bidisperse pore diffusion model for methane displacement desorption in coal by CO_2 injection. Fuel. 82, 1219−1229.

Siebrits, E., Peirce, A.P., 2002. An efficient multi-layer planar 3D fracture growth algorithm using a fixed mesh approach. Int. J. Num. Meth. Eng. 53, 691−717, 2002.

Sing, K.S., 1985. Reporting physisorption data for gas/solid systems with special reference to the determination of surface area and porosity. Pure Appl. Chem. 2201−2018.

Singh, S.K., Sinha, A., Deo, G., Singh, J.K., 2009. Vapor − liquid phase coexistence, critical properties, and surface tension of confined alkanes. J. Phys. Chem. C. 113 (17), 7170−7180, 2009.

Singha, P., Van Swaaijb, W.P.M., (Wim) Brilmanb, D.W.F., 2013. Energy efficient solvents for CO_2 absorption from flue gas: vapor liquid equilibrium and pilot plant study. Energy Procedia. 37, 2021−2046.

Soliman, M.Y., East, L., Adams, D., 2004. Geomechanics aspects of multiple fracturing of horizontal and vertical wells. In: SPE International Thermal Operations and Heavy Oil Symposium and Western Regional Meeting, March 16−18, Bakersfield, CA.

Soltanzadeh, H., Hawkes, C.D., 2009. Induced poroelastic and thermoelastic stress changes within reservoirs during fluid injection and production. Porous Media: Heat and Mass Transfer, Transport and Mechanics (Chapter 2). Nova Science Publishers, Inc, Hauppauge, NY.

Srinivasan, R., Auvil, S.R., Schork, J.M., 1995. Mass transfer in carbon molecular sieves—an interpretation of Langmuir kinetics. Chem. Eng. J. 57, 137−144.

Stevenson, M.D., Pinczewski, W.V., Somers, M.L., Bagio, S.E., 1991. Adsorption/desorption of multi-component gas mixtures at in-seam conditions. SPE23026, SPE Asia-Specific Conference, Perth, Western Australia, November 4−7.

Tadmor, R., 2004. Line energy and the relation between advancing, receding, and young contact angles. Langmuir. 20 (18), 7659−7664, 2004.

Taghichian, A., 2013. On the geomechanical optimization of hydraulic fracturing in unconventional shales. A thesis submitted to University of Oklahoma.

Tinni, A., Fathi, E., Agrawal, R., Sondergeld, C., Akkutlu, I.Y., Rai, C., 2012. Shale permeability measurements on plugs and crushed samples. SPE-162235 Selected for presentation at the SPE Canadian Resources Conference held in Calgary, Alberta, Canada, 30 October−1 November.

Todd Hoffman, B., 2012. Comparison of various gases for enhanced recovery from shale oil reservoirs. 154329-MS SPE C SPE Improved Oil Recovery Symposium, April 14−18, Tulsa, OK.

Valkó, P.P., 2009. Assigning value to stimulation in the barnett shale: a simultaneous analysis of 7000 plus production histories and well completion records. Paper SPE 119369 presented at the SPE Hydraulic Fracturing Technology Conference, The Woodlands, TX, January 19−21.

Virk, P.S., 1975. Drag reduction fundamentals. AIChE J. 21 (4), 625−656.

Waples Douglas, W., 1985. Geochemistry in Petroleum Exploration. Brown and Ruth laboratories, Inc, Denver, CO.

Warpinski, N.R., Teufel, L.W., 1987. Influence of geologic discontinuities on hydraulic fracture propagation. J. Petrol. Technol. 39 (2), 209−220.

Warren, J.E., Root, P.J., 1963. The behavior of naturally fractured reservoirs. SPE 426-PA. SPE J. 3 (3), 245−255. Available from: <http://dx.doi.org/10.2118/426-PA>.

Washburn, E.W., 1921. Note on the method of determining the distribution of pore sizes in a porous material. Proc. Natl. Acad. Sci. U.S.A. 7, 115−116.

Waters, G., Dean, B., Downie, R., Kerrihard, K., Austbo, L., McPherson, B., 2009. Simultaneous Hydraulic Fracturing of Adjacent Horizontal Wells in the Woodford Shale. SPE 119635.

Xu, W., Tran, T.T., Srivastava, R.M., Journel, A.G., 1992. Integrating seismic data in reservoir modeling: the collocated cokriging alternative. Soc. Petrol. Eng.paper no. 24,742.

Yamada, S.A., 1890. A review of a pulse technique for permeability measurements. SPE J., 357−358.

Yamamoto, K., Shimamoto, T., Maezumi, S., 1999. Development of a true 3D hydraulic fracturing simulator. SPE Asia Pacific Oil. 1−10.

Yang, J.T., 1987. Gas Separation by Adsorption Processes. Butterworth Publishers.

Yang, Y., Aplin, A.C., 1998. Influence of lithology and compaction on the pore size distribution and modelled permeability of some mudstones from the Norwegian margin. Mar. Pet. Geol. 15, 163−175.

Yee, D., Seidle, J.P., Hanson, W.B., 1993. Gas sorption on coal measurements and gas content. In: Law, B.E., Rice, D.D. (Eds.), Hydrocarbons from Coal, AAPG Studies in Geology, pp. 203−218 (Chapter 9).

Yuyi, J., Belyadi, H., Blake, A., Wyatt, J., Dalton, J., Vangilder, C., Roth, B., (2016). Dry Utica Proppant and Frac Fluid Design Optimization. SPE Eastern Regional Meeting. SPE-184078.

Zamirian, M., Aminian, K., Ameri, S., Fathi, E., 2014. New steady-state technique for measuring shale core plug permeability. SPE-171613-MS, SPE/CSUR Unconventional Resources Conference, September−2 October Calgary, Canada, 2014.

Zamirian, M., Aminian, K., Fathi, E., Ameri, S., 2014. A fast and robust technique for accurate measurement of the organic-rich shales characteristics under steady-state conditions. SPE 171018, SPE Eastern Regional Meeting, October 21−23, 2014, Charleston, WV.

Zimmerman, R.W., 1991. Compressibility of Sandstones. Elsevier Science Publishing Company Inc, New York, NY.

INDEX

Note: Page numbers followed by "*f*" and "*t*" refer to figures and tables, respectively.

A

Above-ground storage tanks (ASTs), 270–271, 272*f*
Absolute volume factor (AVF), 143–144
Acid solubility, 82
Acidic buffer, 118
Acidization stage, 64–65
Ad valorem tax, 339–341
Adjustable choke, 195
Adjusted gas price, 330–331
Adsorbed gas density, 16, 29–31, 33–35, 41
Adsorbed layer correction, 31
Adsorption isotherm, 18–19, 18*f*
Adsorption layer thickness, 30–31
Adsorption process, 16
After-closure analysis (ACA), 244–249
Amott wettability index, 47–48
Anisotropy, 237
Arps decline curve equations, 314–323
   exponential decline equations, 314
   hyperbolic decline equations, 315–323
Artificial neural network (ANNs), 253
ATAX monthly undiscounted NCF, 362, 377–392

B

Back pressure regulator (BPR), 199, 201*f*
Bakken Shale, 38–39, 51–56, 169, 176, 202–203, 213–214
Barton meter, 198–199
Basic buffer, 118
Basis differential, 344
Bean choke, 195
Before federal income tax (BTAX)
   ATAX monthly discounted net cash flow, 362
Before federal income tax (BTAX) monthly undiscounted net cash flows (NCF), 350–351

Before-closure analysis (BCA), 227–244
   fluid leak-off regimes on G-function plot, 235–244
      height recession leak-off, 239–240
      pressure-dependent leak-off (PDL), 236–239
      tip extension, 240–244
   G-function analysis, 235–244
   log–log plot, 229–234
   square root plot, 228–229
Bernoulli's principle, for incompressible flow, 272–273, 273*f*
Beta coefficient, 356
Biocide, 114–115
Biot's constant, 217–218
Bi-wing fracture system, 56, 56*f*
Blender, 275–276
Blender sand concentration, 156
Blender tub, 282, 283*f*
Blowing sand, 276–277
Boost pump, 275–276, 284–285
Bottom-hole frac pressure (BHFP), 125–126
Bottom-hole pressure (BHP), 92–93, 228, 287
   linear flow-time function versus, 247–248
   radial flow-time function versus, 248–249
Bottom-hole static temperature (BHST), 154–155
Bottom-hole treating pressure (BHTP), 125–126, 133–134, 154–155
Boundary-dominated flow, 309
Bounded versus unbounded, 188–190, 189*f*
Boyle's law, 20–21
Boyle's law porosimeter, 26–27, 26*f*
Brady sand, 73–74
British thermal units (BTUs), 1–4, 330
Brittleness and fracability ratios, 212–214

Brown sand, 73–74
Buffer, 118
Bulk density, 20–21, 82

C

Candy canes, 303–304
Capillary pressure, 41–47
Capital asset pricing model (CAPM), 355
Capital budgeting, 357
Capital expenditure cost (CAPEX),
    347–348
Carter's leak-off model, 260
Carter's model, 260
Cash inflow, 325–326
Cash outflow, 325–326
Casing capacity, 70
Casing selection, 98
Cementing operation, 101
Centralized impoundment, 269
Centralized tanks, 271
Ceramic proppant, 76, 77f, 188
    intermediate-strength, 76
    lightweight ceramic proppant (LWC), 76
Check valves, 293–294
Chemical (chem.) chart, 288–289
Chemical coordination, 295–296
Chemical injection ports, 282–284
Chemical sorption, 16
Chemical totes, 282–284, 283f
Choke manifold, 195, 195f
Clay content of shale, 11
Clean rate (no proppant), 157–158
Clean rate (with proppant), 158
Clean rate, 144–145
Clean side, 275–276
Clean volume, 144–147
Closure pressure, 62, 133–136, 229
Cluster, 176
Cluster spacing, 178–179, 185
Clustering, 83
Coal, 8
Coalbed methane, 1–4, 8, 10–11
Coalification, 8
Cohesive zone models (CZM), 256
Cold explosion, 5–6
COMEX (commodity exchange),
    343–344

Common stock, 352–353
Completions and flowback design
    evaluation, 183
    bounded versus unbounded, 188–190
    cluster spacing, 185
    entry-hole diameter (EHD), 186
    flowback design, 192–194
    flowback equipment, 194–204
        choke manifold, 195, 195f
        flare stack, 202
        high-stage separator, 196–202
        low-stage separator, 202
        oil tanks, 203–204, 203f
        sand trap, 195–196, 197f
    flowback equipment spacing guidelines,
        204
    landing zone, 184
    perforation phasing, 186, 186f
    perforations, number of, 186
    proppant size and type, 187–188
    sand and water per foot, 186–187
    stage spacing, 185
    tubing analysis, 204–206
    up dip versus down dip, 190, 190f
    water quality, 192
    well spacing, 190–192
Composite bridge (frac) plug, 170–173,
    172f
    simultaneous frac, 173
    stack fracing, 171
    zipper fracing, 171–172
Compressed natural gas (CNG), 4–5
Condensate price, 346
Conductivity, 87
    dimensionless fracture conductivity, 88
    finite vs infinite conductivity, 95–96
    fracture conductivity, 87
    fracture conductivity testing, 89
Conductor casing, 98–99
Constant-rate decline, 309
Contact time, 64–66
Conventional enhanced oil recovery
    technique, 7
Conventional plug and perf method,
    169–171, 175
Core data, 207–208
Core plug pulse-decay permeameter, 24f

Corporation tax, 376—377
Correction factor for adsorbed layer
    thickness, 31
Cost of capital, 351—353
Cost of debt, 351—354
Cost of equity, 351—353, 355—357
Counts per second (CPS), 11—12
Crack-tip open displacement (CTOD),
    256
Crack-tip plasticity (CTP) method, 254
Critical drawdown pressure, 193—194
Cross-linked fluid system, 55, 119, 119f
Cross-linked gel fluid system, 55—56
Crush resistance, 81—82
Cubic sugar models, 10—11
Curable resin-coated sand (CRCS), 74—75
Cushing Hub, 346
Cyclic stress, 92—93

D

Darcy friction factor, 113
Darcy's equation, 23
Darcy's law, 22—23, 24f, 27
Data mining, 253
Debt, 352
Debt-to-equity ratio, 353—354
Decision trees (DTs), 253
Decline curve analysis (DCA), 183—184,
    305—306
    Arps decline curve equations, 314—323
        exponential decline equations, 314
        hyperbolic decline equations,
            315—323
    Duong decline, 313
    effective decline, 306—307
    hyperbolic exponent, 307
    instantaneous production (IP), 306
    late transient period, 308
    nominal decline, 306
    power law exponential decline model
        (PLE), 311—312
    primary types of DCA, 309—314
        exponential decline, 309
        harmonic decline, 311
        hyperbolic decline, 309—310
        modified hyperbolic decline curve,
            311

pseudosteady state, 309
shape of the decline curve, 307—308
stretched exponential, 312—313
unsteady state period, 308
Delayed cross-linked, 118
Density-logging tool, 216
Densometer, 69—70, 71f, 284
Department of Environment Protection
    (DEP), 99—100
Depreciation, 373—374
Desorption, 16
Desorption isotherm, 18—19
Diagnostic fracture injection test (DFIT),
    123—124, 188, 216—217, 225
    after-closure analysis (ACA), 244—249
    before-closure analysis (BCA),
        227—244
        fluid leak-off regimes on G-function
            plot, 235—244
        G-function analysis, 235—244
        height recession leak-off, 239—240
        log—log plot, 229—234
        pressure-dependent leak-off (PDL),
            236—239
        square root plot, 228—229
    data recording and reporting, 226—227
    Horner plot, 245—246
    linear flow-time function versus
        bottom-hole pressure, 247—248
    radial flow-time function versus BHP,
        248—249
Dimensionless fracture conductivity, 88
Dirty (slurry) versus clean frac fluid,
    144—145
Dirty rate, 144—145
Dirty side, 276
Dirty volume, 147
Discharge pump, 275—276
Discharge side, 276
Discontinuous displacement (DD) method,
    251, 264, 266—267
Discount rate, 351—353
Discounted payback method, 371
Double-cell Boyle's law porosimeter,
    26—27, 26f
Down dip versus up dip, 190, 190f
Drill-out, 192—193

Dual-porosity single-permeability models, 37
Dump valve, 198−199
Duong decline, 313
Dynamic contact angle measurement, 44*f*
Dynamic Young's modulus calculation, 208

E

Economic analysis, 177−178, 185, 193−194, 305
Economic evaluation, 325
  ad valorem tax, 339−340
  before federal income tax (BTAX) monthly cash flows, 350−351
  British thermal unit (BTU) content, 330
  BTAX and ATAX monthly discounted net cash flow, 362
  capital budgeting, 357
  capital expenditure (CAPEX), 347−348
  cost of debt, 354
  cost of equity, 355−357
  Cushing Hub, 346
  discount rate, 351−353
  Henry Hub and basis price, 344−346
  internal rate of return (IRR), 362−367
    NPV versus, 368
  modified internal rate of return (MIRR), 368−369
  Mont Belvieu, 346−347
  net cash-flow model (NCF), 325−326
  net Opex, 340−341
  net present value (NPV), 357−361
    versus IRR, 368
  net revenue interest (NRI), 329−330
  NYMEX (New York Mercantile Exchange), 343−344
  Oil Price Information Services (OPIS), 346−347
  Operating expense (Opex), 332−335
  OPEX, CAPEX, and pricing escalations, 349
  payback method, 370−372
  profit, 350
  profitability index (PI), 372−373
  revenue, 341−343
  royalty, 326−327
  severance tax, 338−339
  shrinkage factor, 331−332
  tax model, 373−392
    ATAX monthly undiscounted NCF, 377−392
    corporation tax, 376−377
    depreciation, 373−374
    taxable income, 375−376
  total Opex per month, 335−338
  weight of debt and equity, 353−354
  West Texas Intermediate (WTI), 346
  working interest, 327−329
Effective decline, 306−307
Effective pore volume, 20−21
Energy Information Administration (EIA), 6−7
Enhanced oil recovery technique, 7
Entry-hole diameter (EHD), 128, 186
Environmental Protection Agency (EPA), 101
Equilibrium contact angle, 43−44
ESD (emergency shut-down valve), 301−302, 302*f*
Estimated ultimate recovery (EUR), 311
Exploration and production (E&P) companies, 107, 190−192, 269
Exponential decline, 309, 310*f*
Exponential decline curve, 305−306
Exponential decline equations, 314
Extended finite element method (XFEM), 265−266
Extended sweep, 68

F

Fanning friction factor, 111−113
Fault reactivation, 102−104
Field stratigraphic and geological structure modeling, 252−253
Fines migration, 93−94
Finite vs infinite conductivity, 95−96
Firm transportation (FT) cost, 334
Fixed lifting cost, 328
Flare, 203*f*
Flare stack, 202
Flow cross, 301−302, 302*f*
Flowback, 193
Flowback after screening out, 296−299
Flowback design, 192−194

Flowback equipment, 194—204
  choke manifold, 195, 195*f*
  flare stack, 202
  high-stage separator, 196—202
  low-stage separator, 202
  oil tanks (upright tanks), 203—204, 203*f*
  sand trap, 195—196, 197*f*
Flowback equipment spacing guidelines, 204
Fluid efficiency, 67
Fluid flow in hydraulic fractures, 259—260
Fluid lag, 260
Fluid-loss capability, in foam-fracturing fluid system, 59
Fluid—rock molecular collision, 15
Flush stage, 69—71
Flush volume, 69—70
Foam frac schedule and calculations, 154—155
Foam quality, 60—61
Foam stability, 61
Foam volume, 155
Foam-fracturing fluid system, 58—60
  fluid-loss capability in, 59
  proppant transport in, 59—60
Focused ion beam scanning electron microscopy (FIB/SEM), 13—15
Formation microimager (FMI) log, 221—222
Formation modulus calculation, 208
Formation water, 52
40/70 mesh, 78—80, 82, 187—188
Frac ball, 174*f*, 175
Frac equipment setup, 291*f*
Frac fluid selection, 72
Frac job, 51, 78, 82, 269, 295, 299—300, 303—304
Frac manifold, 285—286, 286*f*
Frac stage spacing, 176
Frac van, 281—282, 286—289
Frac wellhead, 299—304, 304*f*
  flow cross, 301—302
  frac head, 303—304
  hydraulic valve, 300—301
  lower master valve, 299—300
  manual valve, 302—303
  tubing head, 299

Fracability ratios, 212—214
Frac-packing process, 50—51
Fracture closure, 228—229, 237
Fracture conductivity, 87, 94—95
  dimensionless, 88
Fracture conductivity testing, 89, 89*f*
Fracture extension pressure, 133, 135*f*
Fracture gradient (FG), 125
Fracture orientation, 221
Fracture pressure analysis and perforation design, 121
  bottom-hole treating pressure (BHTP), 125—126
  closure pressure, 133—136
  fracture extension pressure, 133, 135*f*
  fracture gradient (FG), 125
  hydrostatic pressure, 122
  hydrostatic pressure gradient, 122—123
  instantaneous shut-in pressure (ISIP), 123—125, 123*f*, 124*f*
  near-wellbore friction pressure (NWBFP), 133
  net pressure, 137—139
  number of holes and limited entry technique, 131—132
  open perforations, 128—129
  perforation design, 130—131
  perforation diameter and penetration, 132
  perforation efficiency, 129
  perforation erosion, 132
  perforation friction pressure, 127—128
  pipe friction pressure, 127
  pressure, 121—122
  surface-treating pressure (STP), 139—141
  total friction pressure, 126—127
Fracture toughness, 211—212
Fracture treatment design, 143
  absolute volume factor (AVF), 143—144
  blender sand concentration, 156
  clean rate (no proppant), 157—158
  clean rate (with proppant), 158
  dirty (slurry) versus clean frac fluid, 144—145
  foam frac schedule and calculations, 154—155

Fracture treatment design (*Continued*)
  foam volume, 155
  nitrogen rate (with and without
      proppant), 159−166
  nitrogen volume, 155−156
  sand per foot, 148
  sand-to-water ratio (SWR), 149
  slick water frac schedule, 150−154
  slurry (dirty) density, 145−146
  slurry factor (SF), 157
  slurry rate (with proppant), 159
  stage fluid clean volume, 146−147
  stage fluid slurry (dirty) volume, 147
  stage proppant, 147−148
  water per foot, 148−149
Fracture width, 83−84, 87, 87*f*
Fracture-tip extension leak-off, 240−244
  effective permeability estimation from
      G-function plot, 242−244
  identifying and dealing with, 242
Freundlich isotherm, 18
Friction pressure, 111, 126−127
  near-wellbore friction pressure
      (NWBFP), 133
  perforation friction pressure, 127−128
  pipe friction pressure, 127
Friction reducer (FR), 52, 108, 192,
      288−289
  flow loop test, 108−110, 109*f*
  FR breaker, 114

G

Gamma ray log, 11−12, 12*f*
gAPI, 11−12
Gas chromatograph, 1−4, 3*f*
Gas hydrates, 1−6, 8
Gas Research Institute (GRI) technique,
      26−27
Gas resources pyramid, 7*f*
Gas-in-place calculation, 29−33
Gas-transport technique, 5−6
Gathering and compression cost (G&C), 334
Gel breaker, 117−118
Gel damage, 91−92
Gelling agents, 61, 114, 116−118
General and administrative (G&A) cost,
      335

G-function analysis, 235−244
  fluid leak-off regimes, 235−244
    height recession leak-off, 239−240
    pressure-dependent leak-off (PDL),
      236−239
G-function plot, 228
  effective permeability estimation from,
      242−244
  fluid leak-off regimes on, 235−244
    height recession leak-off, 239−240
    pressure-dependent leak-off (PDL),
      236−239
  tip extension, 240−244
Gibbs model, 16
Goat head. *See* Frac wellhead: Frac head
"Good" resources, 7−8
Gravel packing, 50−51
Guar, 116

H

Harmonic decline, 305−306, 311
Height recession leak-off, 239−240, 240*f*
  dealing with, 239−240
  identifying, 239
Helium, 20−21
Henry Hub and basis price, 344−346
Henry's law isotherm, 16
High-pressure, high-temperature (HPHT)
      pulse-decay permeameter, 25*f*
High-pressure mercury injection, 14, 21
High-stage separator, 196−202
High-strength proppant, 76−77
Hooke's law, 255−256
Hopper and blender screws, 278*f*
Horizontal separators, 196−198, 200*f*
Horizontal well multistage completion
      techniques, 169
  cluster spacing, 178−179
  composite bridge (frac) plug, 170−173
    simultaneous frac, 173
    stack fracing, 171
    zipper fracing, 171−172
  conventional plug and perf, 170
  frac stage spacing (plug-to-plug spacing),
      176
  refrac overview, 179−181
  shorter stage length (SSL), 176−178

sliding sleeve, 174
sliding sleeve advantages, 175
sliding sleeve disadvantages, 175—176
    hybrid design, 176
    multi-entry point frac sleeve, 176
    single entry-point frac sleeve, 175
    toe sleeve/valve, 175
Horner plot, 245—246, 246*f*
100 mesh, 78—80, 78*f*, 187—188
Hurdle rate, 351—353
Hybrid fluid system, 56—58
Hydration unit, 273—275
Hydraulic and natural fracture interactions,
    265—266
Hydraulic coupling used in
    multicontinuum approach, 40*f*
Hydraulic ESD, 301—302, 302*f*
Hydraulic fracture stage merging, 266—267
    and stress shadow effects, 266—267
Hydraulic fracturing, 90—91
    and aquifer interaction, 101—102
    and fault reactivation, 102—104
    and low-magnitude earthquakes,
        105—106
Hydraulic fracturing chemical selection
        and design, 107
    biocide, 114—115
    buffer, 118
    cross-linker, 119, 119*f*
    FR breaker, 114
    FR flow loop test, 108—110, 109*f*
    friction reducer, 108
    gel breaker, 117—118
    iron control, 120
    linear gel, 116—117
    pipe friction pressure, 111
    relative roughness of pipe, 112—114
    Reynolds number, 112
    scale inhibitor, 115—116
    surfactant, 120
Hydraulic fracturing fluid systems, 49
    acidization stage, 64—65
    cross-linked gel fluid system, 55—56
    flush stage, 69—71
    foam fracturing, 58—60
    foam quality, 60—61
    foam stability, 61

frac fluid selection, 72
hybrid fluid system, 56—58
pad stage, 66—68
proppant stage, 68—69
slick water fluid system, 51—54
tortuosity, 61—63, 63*f*
typical slick water frac steps, 64
Hydraulic fracturing regimes, 261—263,
    263*f*
Hydraulic horsepower (HHP), 102, 140
Hydraulic valve, 285—286, 300—301, 301*f*
Hydrocarbon-in-place calculation, 29
Hydrostatic pressure, 122
Hydrostatic pressure gradient, 122—123
Hyperbolic decline, 309—311, 310*f*
Hyperbolic decline curve, 305—306
Hyperbolic decline equations, 315—323
Hyperbolic exponent, 307
Hysteresis types, 19*f*

I

Ideal adsorption solution (IAS) theory, 34
Idiosyncratic risk, 356
In-situ stresses, 102—103, 207, 221—222
Inflation, 349, 351
In-ground pit, 270, 272*f*
Instantaneous production (IP), 306
Instantaneous shut-in pressure (ISIP), 62,
    123—125, 123*f*, 124*f*
Interfacial tension (IFT), 41—47
Intermediate casing, 100
Intermediate-strength ceramic proppant,
    76
Internal rate of return (IRR), 362—367
    advantages, 367—368
    calculation, 364
    disadvantage, 367
    NPV versus, 368
Internal Revenue Service (IRS), 373—375
International organization for
        standardization conductivity test, 89
Iron control, 120

J

Joint operating agreement (JOA), 328—329
Jones technique, 25

K

Kane method pad volume, 66—67
Kerogen, 8—10, 13—14
  different types of, 9*t*
Khristianovic—Geertsma de Klerk (KGD)
    model, 257, 258*f*
Kickoff point (KOP), 100—101
Knudsen diffusion, 14—15

L

Landing zone, 184
Langmuir, Irvin, 16—17
Langmuir equation, 17, 31
Langmuir isotherm, 17, 17*f*
Langmuir model, 16—18
Langmuir volume, 35—36, 36*f*
Langmuir/Freundlich isotherm, combined,
    18
Laplace's law, 41—42
Leak-off, 53—54, 67
  height recession leak-off, 239—240
  pressure-dependent leak-off (PDL),
    236—239
Lifting cost, 328
  fixed, 328
  variable, 328
Lightweight ceramic proppant (LWC), 76
Limestone, 11—12
Limited entry, 130—132
Linear base gel system, 117, 117*f*
Linear elastic fracture mechanics (LEFM),
    254, 256
Linear flow-time function versus bottom-
    hole pressure, 247—248
Linear gel, 116—117, 273—275
Liquefied natural gas (LNG), 4—6
Liquefied petroleum gas (LPG), 4—5
Liquid additive (LA) pumps, 273—275
Liquid level controller (LLC), 198—199,
    200*f*
Log (BH ISIP-BHP) versus log (time),
    229—234
Log—log paper, 305—306
Log—log plot, 229—234
Longitudinal fractures, 221—224
  versus transverse fractures, 223*f*
Lower master valve, 299—300

Low-magnitude earthquakes, hydraulic
    fracturing and, 105—106
Low-stage separator, 202
Low-viscosity fluid, 52
Lubrication theory, 258—260

M

Manual valves, 293—294, 302—303, 303*f*
Marcellus Shale, 11—12, 51, 55—57, 85,
    102, 120, 196—198, 344—345
Measured depth (MD), 122
Mechanical pop-off, 199, 201*f*, 290—292,
    292*f*
Mercury injection, 45—46
  high-pressure, 21
Meter, 199
Microemulsion, 120
Minimum horizontal stress, 216—217
Minipad, 68
Missile, 284—285, 285*f*
Modern hydraulic fracturing, 49—50
Modified hyperbolic decline curve, 311
Modified internal rate of return (MIRR),
    368—369
Molecular dynamic simulations, 34, 42—43
Mont Belvieu, 346—347
Monthly depreciation, 373—375
Multicontinuum approach, 40*f*
Multicontinuum modeling of shale
    reservoirs, 38—41, 39*f*
Multiphase flow, 90—91
Multiscale fluid flow and transport in
    organic-rich shale, 37
  interfacial tension (IFT) and capillary
    pressure, 41—47
  multicontinuum modeling of shale
    reservoirs, 38—41, 39*f*
  wettability effects on shale recovery,
    47—48
Multistage hydraulic fracturing, 169,
    253—254
  stress shadowing effect in, 266—267

N

Nanoorganic pore-size distribution, 35—36
Natural fracture, 265—266

Natural gas, 1−4
  different types of, 4−5
Natural gas components, 2t
  BTU of, 3t
Natural gas liquid (NGL), 4−5, 346−347
Natural gas transport, 5−6
Near-wellbore friction pressure (NWBFP), 133
Net bottom-hole pressure (NBHP), 287, 289f
Net Capex, 348
NET cash flow (NCF), 350
Net cash-flow model (NCF), 325−326, 326f
Net Opex, 340−341
Net present value (NPV), 325−326, 357−361
Net present worth (NPW). See Net present value (NPV)
Net pressure, 137−139
  charts, 138−139
Net revenue interest (NRI), 329−330
Newtonian fluid Reynolds number, 112
Nitrogen foam-fracturing fluid system, 58−59
Nitrogen gas measurements, 154−155
Nitrogen rate (with and without proppant), 159−166
Nitrogen volume, 154−156
Nodal analysis, 205f
Nolte method pad volume, 66−67
Nominal decline, 306
Nominal interest rate, 349
Non-Darcy flow, 90
Normal fault environment, 220
Normal stress, 254−255
Nuclear magnetic resonance (NMR), 14−15, 47−48
Number of holes and limited entry technique, 131−132
Numerical simulation, 251
  development of simulators, 253−257
  fluid flow in hydraulic fractures, 259−260
  hydraulic and natural fracture interactions, 265−266
  hydraulic fracture stage merging and stress shadow effects, 266−267
  pseudo−three-dimensional hydraulic fracturing models, 261−263
  solid elastic response, 260−261
  stratigraphic and geological structure modeling, 252−253
  three-dimensional hydraulic fracturing models, 263−265
  two-dimensional hydraulic fracturing models, 257−259
NYMEX (New York Mercantile Exchange), 343−345

O

Oil and natural gas, 1
Oil Price Information Services (OPIS), 346−347
Oil tanks (upright tanks), 203−204, 203f
Open perforations, 128−129
Operating expense (Opex), 332−335
Operations and execution, 269
  blender, 275−276
  chemical coordination, 295−296
  flowback after screening out, tips for, 296−299
    post−screen-out injection test, 298−299
  frac manifold, 285−286
  frac van, 286−289
  frac wellhead, 299−304
    flow cross, 301−302
    frac head, 303−304
    hydraulic valve, 300−301
    lower master valve, 299−300
    manual valve, 302−303
    tubing head, 299
  hydration unit, 273−275
  missile, 284−285
  overpressuring safety devices, 290−294
    check valves and manual valves, 293−294
    pressure transducer, 292−293
  sand coordination, 295
  sand master, 276−277
  stage treatment, 296
  T-belt, 277−284
    chemical injection ports, 282−284
    densometer, 284

Operations and execution (*Continued*)
  sand screws, 279–282
  water coordination, 294–295
  water sources, 270–273
    water storage, 270–273
Optimized hydraulic fracture spacing, 267
Original gas-in-place (OGIP) calculation,
    29–30, 35–36
Original oil- and gas-in-place (OOIP and
    OGIP), 13
Ottawa sand, 73–74
Overpressuring safety devices, 290–294
  check valves and manual valves,
    293–294
  pressure transducer, 292–293

P
Pad stage, 66–68
Payback method, 370–372
Payback period, 370–371, 370*f*
Penny-shaped model, 259
Perforating gun, 178
Perforation design, 130–131
Perforation diameter and penetration, 132
Perforation efficiency, 129
Perforation erosion, 132
Perforation friction pressure, 127–128
Perforation guns, 173*f*
Perforation phasing, 186, 186*f*
Perforations, number of, 186
Perkins and Kern (PKN) model, 257–258,
    258*f*
Permeability, defined, 22–23
Physical sorption, 16
Pipe friction pressure, 111, 127
Pipeline, 271–273
Plug-and-perf technique, 193
Pneumatic ESD, 301–302
  vs. hydraulic ESD, 302*f*
Poiseuille's law, 258–260
Poisson's ratio, 11, 209–211
Polyacrylamide, 108
Polyvinyl chloride (PVC), 294–295
"Poor" resources, 7–8
Pop-off, defined, 196
Pop-offs. *See* Pressure-relief valves (PRVs)
Pore compressibility, defined, 22

Pore compressibility measurements of
    shale, 22
Pore-size distribution measurement of
    shale, 13–16
Porosity, defined, 20–21
Post–screen-out injection test, 298–299
Pounds per revolution (PPR), 279–282
Power law exponential decline model
    (PLE), 311–312
Precision Petrophysical Analysis
    Laboratory, 27
Precured resin-coated sand (PRCS), 74
Preferred stock, 352
Pressure, 51, 121–122
Pressure transducer, 292–293, 293*f*
Pressure-dependent leak-off (PDL),
    236–239
Pressure-dependent leak-off, 236*f*
Pressure-relief valves (PRVs), 290–292
Pressure–volume–temperature (PVT) cell
    measurements, 34–35
Principal stress reversal, 267
Principal stresses, 255
Processing cost, 330
Production casing, 100–101
Profit, 350
Profitability index (PI), 372–373
Property, leasing, 327
Proppant, 276–279
Proppant characteristics, 80–83
Proppant characteristics and application
    design, 73
  100 mesh, 78, 78*f*
  20/40 mesh, 80
  30/50 mesh, 79–80
  40/70 mesh, 78–79
  curable resin-coated sand (CRCS),
    74–75
  cyclic stress, 92–93
  dimensionless fracture conductivity, 88
  fines migration, 93–94
  finite vs infinite conductivity, 95–96
  fracture conductivity, 87
  gel damage, 91–92
  high-strength proppant, 76–77
  intermediate-strength ceramic proppant,
    76

international organization for
    standardization conductivity test, 89
lightweight ceramic proppant (LWC), 76
multiphase flow, 90−91
non-Darcy flow, 90
precured resin-coated sand, 74
proppant characteristics, 80−83
proppant particle-size distributions, 83
proppant size, 78−80
    and type, 187−188
proppant stage, 68−69
proppant transport and distribution in
    hydraulic fracture, 83−86
reduced proppant concentration, 91
sand, 73−74
time degradation, 94−95
Proppant crushing embedment, 93f
Proppant particle-size distributions, 83
Pseudo−three-dimensional hydraulic
    fracturing models, 261−263
Pulse-decay permeability measurement
    technique, 24−25
Pump trips, 290

R

Radial flow-time function versus BHP,
    248−249
Radial fracture geometry, 259, 259f
Rapid phase transition, 5−6
Rate and pipe friction pressure, 111
Rate transient analysis (RTA), 183−184
Recovery factor, 35−36
Reduced proppant concentration, 91
Refrac, 179−181
Relative permeability curve, 90f
Relative roughness of pipe, 112−114, 113t
Reservoir pressure, 12
Reservoirs, unconventional, 6−12
Resin-coated proppant, 76
Resin-coated sand
    curable, 74−75
    precured, 74
Revenue, 341−343
Reverse (thrust) fault environment, 221
Reynolds number, 83−84, 112
Risk mitigation practice, 328−329
Risk premium, 356−357

Risk-free rate, 355
Rock brittleness, 11
Rock mechanical properties, 207
    Biot's constant, 217−218
    brittleness and fracability ratios,
        212−214
    fracture orientation, 221
    fracture toughness, 211−212
    longitudinal fractures, 222−224
    maximum horizontal stress, 218−220
    minimum horizontal stress, 216−217
    Poisson's ratio, 209−211
    stress states, 220−221
    transverse fractures, 221−222
    vertical stress, 214−216
    Young's modulus, 207−209
Round per minute (rpm) calculation, 280f
Roundness, 81
Royalty, 326−327

S

Sand, characteristics, 73−74
Sand and water per foot, 186−187
Sand coordination, 295
Sand light, 281−282
Sand master, 276−277, 277f
    and T-belt, 278f
Sand per foot, 148
Sand screws, 279−282, 279f
Sand trap, 195−196, 197f
Sand-and-water-per-foot design, 186−187
Sandstone, 11−12
Sand-to-water ratio (SWR), 149
Saturation pressures, 18−19
Scale inhibitor, 115−116
Scanning transmission electron microscopy
    imaging (STEM), 15
Scrubber pot, 199
Series coupling, 38−39
Settling velocity, 83−84
Severance tax, 338−339
Shale, 8−10
    permeability measurement techniques,
        22−27
    pore compressibility measurements, 22
    pore-size distribution measurement,
        13−16

Shale (*Continued*)
  porosity measurements, 20−21
  sorption measurement techniques, 16−20
Shale gas reservoirs, 8, 38
Shale initial gas-in-place calculation, 29
  adsorbed gas, density of, 33−35
  recovery factor, 35−36
  total gas-in-place calculation, 29−33
Shale matrix bulk volume, 30f
Shale recovery, wettability effects on, 47−48
Shale reservoirs, 10−11, 22−23, 183−184
  multicontinuum modeling of, 38−41
Shear stress, 254−255
Shell method pad volume, 66−67
Shorter stage length (SSL), 176−178
Shrinkage factor, 331−332
Sieve analysis, 82, 83f
Silt and fine particles, 82
Slick water fluid system, 51−54
Slick water frac, 52, 269
  schedule, 150−154
Sliding sleeve, 169, 174
  advantages, 175
  disadvantages, 175−176
    hybrid design, 176
    multi-entry point frac sleeve, 176
    single entry-point frac sleeve, 175
    toe sleeve/valve, 175
Slurry (dirty) density, 145−146
Slurry factor (SF), 157
Slurry rate (with proppant), 159
Soap, 59
Society of Petroleum Engineers (SPE),
  6−7
Solid elastic response, 260−261
Sonic log, 207−209
Sorption, 16
Specific gravity, 82
Sphericity, 81
Square root plot, 228−229
Stack fracing, 171
Stage fluid clean volume, 146−147
Stage fluid slurry (dirty) volume, 147
Stage proppant, 147−148
Stage spacing, 185
Stage treatment, 296
Standard cubic feet (SCF), 1−4, 154−155

Static Young's modulus conversion, 209f
Stratigraphic and geological structure
  modeling, 252−253
Stress, defined, 254
Stress and strain relationship, 255−256
Stress directions, 221−222
Stress shadow effects, 266−267
Stress states, 220−221
Stretched exponential, 312−313
Strike slip (shear) environment, 220
Suction pump, 275−276
Support vector machines (SVMs), 253
Surface casing, 99−100
Surface-treating pressure (STP), 121,
  139−141
  chart, 287, 288f
Surfactant, 120
Systematic risk, 356

T

Tank batteries, 271, 273f
Tax model, 373−392
  ATAX monthly undiscounted NCF,
    377−392
  corporation tax, 376−377
  depreciation, 373−374
  taxable income, 375−376
Taxable income, 375−376
T-belt, 277−284
  chemical injection ports, 282−284
  densometer, 284
  sand screws, 279−282, 279f
Tensile strength, 256
Terminal decline, 311
Thermal maturity (TM) of shale, 8−10, 21
30/50 mesh, 76, 79−80, 82, 187−188
Three-dimensional (3D) geological
  models, 252−253
3D seismic, 347−348
Three-dimensional hydraulic fracturing
  models, 263−265
Three-phase separator, 196−198
Tight gas sands, 8
Time degradation, 94−95
Time value of money, 351, 357−358
Time zero, 325−326, 350
Tortuosity, 61−63, 63f

Total depth (TD), 100—101
Total dissolved solids (TDS), 192
Total friction pressure, 126—127
Total operating costs, 335—336
Total Opex per month, 335—338
Total organic contents (TOC), 8, 35—36
Totes, 273—275
  chemical, 283f
Transverse fractures, 221—222
  versus longitudinal fractures, 223f
Trucking, 271—272
True vertical depth (TVD), 102, 122
Tubing analysis, 204—206
Tubing head, 299
  with production tubing hung, 300f
20/40 mesh, 80, 87, 187—188
Two-dimensional hydraulic fracturing
    models, 257—259
2D seismic, 347—348
Two-phase separator, 196—198

U

Unconventional reservoir development, 97
  casing selection, 98
  conductor casing, 98—99
  hydraulic fracturing and aquifer
      interaction, 101—102
  hydraulic fracturing and fault
      reactivation, 102—104
  hydraulic fracturing and low-magnitude
      earthquakes, 105—106
  intermediate casing, 100
  production casing, 100—101
  surface casing, 99—100
Unconventional reservoirs, 6—12
Unconventional shale reservoir
    characterization, 13
Up dip versus down dip, 190, 190f
US Bureau of Mines (USBM) wettability
    index measurements, 47
US Energy Information Administration
    (EIA), 6—7
US Geological Survey, 6—7

Utica Shale operation, 202—203
Utica/Point Pleasant, 51

V

Van der Waals equation of state, 31, 33
Vapor destructive unit (VDU), 203—204
Variable lifting cost, 328
Vertical stress, 214—216
Vitrinite reflectance, 8—10

W

Water coordination, 294—295
Water delivery, 271
Water disposal cost, 187, 335
Water flooding, 7
Water per foot, 148—149
Water quality, 192
Water sources, 270—273
Water storage, 270—273
Weak coupling, 263—264
Weight of debt and equity, 353—354
Weighted average cost of capital (WACC),
    353
Well spacing, 190—192
Wellbore integrity, 98
West Texas Intermediate (WTI), 346
Wettability effects on shale recovery,
    47—48
Working interest (WI), 327—329

Y

Yield strength, 256
Young's modulus, 11, 137—138, 207—209,
    255—256
Young—Dupré equation, 43—44
Young—Laplace equation, 44—45

Z

Zipper fracing, 171—172

Edwards Brothers Malloy
Ann Arbor MI. USA
November 15, 2016